About Island Press

Island Press is the only nonprofit organization in the United States whose principal purpose is the publication of books on environmental issues and natural resource management. We provide solutions-oriented information to professionals, public officials, business and community leaders, and concerned citizens who are shaping responses to environmental problems.

In 2003, Island Press celebrates its nineteenth anniversary as the leading provider of timely and practical books that take a multidisciplinary approach to critical environmental concerns. Our growing list of titles reflects our commitment to bringing the best of an expanding body of literature to the environmental community throughout North America and the world.

Support for Island Press is provided by The Nathan Cummings Foundation, Geraldine R. Dodge Foundation, Doris Duke Charitable Foundation, Educational Foundation of America, The Charles Engelhard Foundation, The Ford Foundation, The George Gund Foundation, The Vira I. Heinz Endowment, The William and Flora Hewlett Foundation, Henry Luce Foundation, The John D. and Catherine T. MacArthur Foundation, The Andrew W. Mellon Foundation, The Moriah Fund, The Curtis and Edith Munson Foundation, National Fish and Wildlife Foundation, The New-Land Foundation, Oak Foundation, The Overbrook Foundation, The David and Lucile Packard Foundation, The Pew Charitable Trusts, The Rockefeller Foundation, The Winslow Foundation, and other generous donors.

The opinions expressed in this book are those of the author(s) and do not necessarily reflect the views of these foundations.

Animal Behavior
AND Wildlife Conservation

Animal Behavior
AND Wildlife Conservation

EDITED BY
Marco Festa-Bianchet
AND Marco Apollonio

ISLAND PRESS
Washington • Covelo • London

Copyright © 2003 Island Press

All rights reserved under International and Pan-American Copyright Conventions. No part of this book may be reproduced in any form or by any means without permission in writing from the publisher: Island Press, 1718 Connecticut Ave., NW, Suite 300, Washington, DC 20009

Island Press is a trademark of The Center for Resource Economics.

Library of Congress Cataloging-in-Publication Data
Animal behavior and wildlife conservation / edited by Marco
Festa-Bianchet and Marco Apollonio.
    p. cm.
Includes bibliographical references and index.
    ISBN 1-55963-958-X (hardcover : alk. paper) — ISBN 1-55963-959-8
(pbk. : alk. paper)
    1. Animal behavior. 2. Wildlife conservation. I. Festa-Bianchet,
Marco. II. Apollonio, Marco.
    QL751.A6497 2003
    636.9—dc21

                                                    2003005623

British Cataloguing-in-Publication Data available.

No copyright claim is made in the work of Steven L. Monfort and Thomas J. Roffe, employees of the federal government.

Printed on recycled, acid-free paper

Manufactured in the United States of America
09 08 07 06 05 04 03   10 9 8 7 6 5 4 3 2 1

Contents

Preface

The Ettore Majorana Center for Scientific Culture in Erice, Sicily, is known worldwide as a place where important meetings happen, scientific collaborations begin, and new ideas are generated. The unique character of Erice, a walled town perched high upon a hill overlooking the sea, the efficiency and professionalism of the Center's staff, and the great facilities available (from conference rooms to restaurants) combine to provide a wonderfully stimulating atmosphere.

During a 1998 workshop in Erice on vertebrate mating systems, we two editors, Marco Festa-Bianchet and Marco Apollonio, saw the need for more reflection on how the study of animal behavior could facilitate conservation. When Professor Danilo Mainardi, director of the School of Ethology at the Center, asked us to organize another workshop, we said, in unison, "Behavior and conservation," even though we had not previously discussed the idea.

The invitations we extended were met with widespread enthusiasm—many researchers in animal behavior believe their work has useful applications in wildlife conservation. Most of our invited speakers began their scientific career by looking at fundamental questions in the evolution of animal behavior, and then moved to more "applied" research questions. Often, that switch in emphasis was motivated by the realization that animal populations were disappearing as their habitat was being altered by human activities and while many conservation programs continued to ignore the importance of animal behavior, particularly that of individual differences.

We asked all speakers to first review the theoretical foundations of their subject, then explore its implications for wildlife conservation. We also asked all authors to emphasize both the advantages and the limitations of applying knowledge in animal behavior to conservation.

The workshop on Animal Behavior and Conservation was held in Erice in November 2000. All participants provided a chapter for this book, which we rounded out with contributions from researchers who did not take part in the workshop. Financial support for the workshop in Erice was provided by the Erice Center, the Regional Government of Sicily, and the Italian Ministry of University and Scientific Research.

We are very grateful to our colleagues who provided very constructive reviews of earlier drafts of individual chapters: Erin Bayne, Merav Ben-David, David Coltman, Steeve Côté, Tim Coulson, André Desrochers, John Fryxell, Jean-Michel Gaillard, Brendan Godley, Rich Harris, Keith Hobson, Jeff Hutchings, Petr Komers, Wendy King, Gordon Luikart, Sandro Lovari, Dan Mazerolle, Bruce McLellan, Jan Murie, and Bill Sutherland.

Part I

Why Animal Behavior Is Important for Conservation

1.

General Introduction

Marco Festa-Bianchet and Marco Apollonio

Many of the species with whom we share our planet are going extinct because we overexploit them or destroy their habitat (Ehrlich and Wilson 1991, Caughley 1994). Species extinction and habitat destruction have an immediate impact upon many economic and social activities because various uses of wildlife provide income, enjoyment, or recreation for millions of people (Geist 1994). It is therefore not surprising that interest in the conservation of biodiversity is increasing among the general public as well as among behavioral ecologists who study wild animals and their environment.

Two related disciplines, wildlife conservation and wildlife management, use ethological knowledge to limit the impact of humans on ecosystems. Wildlife conservation is concerned with the preservation of species and their habitat in the face of threats from human development. Wildlife management, including fisheries management, seeks sustainable strategies to exploit wild species while ensuring their persistence and availability for future use. Ideally, these strategies should also not damage components of the ecosystem other than the exploited species. Although the distinction between the two disciplines is often blurred, wildlife management is often oriented toward specific objectives for one or a few species of economic interest. The goals of conservation are broader and include the preservation of genetic diversity so that species will maintain their ability to evolve in response to environmental change. Recently, however, wildlife conservation and management are coalescing into a single discipline. Management is often a component of conservation strategies (for example, limited sport harvest of some high-profile species can be used to generate funds for habitat preservation [Lewis and Alpert 1997]), and the conservation of genetic diversity or interpopulation connectivity is often a goal of wildlife management. For simplicity, we will use the term *management* in this introduction to refer both to situations where wild animals are the subject of some form of exploitative management, and to situations where they are of concern because they are at risk of extinction.

Regardless of how one defines *wildlife management* or *wildlife conservation,* however, practical application of these terms inevitably involves the consideration of both animal and human behavior. This book explores how knowledge of animal behavior can help prevent species extinction and sustainably exploit wildlife populations. It is clear to us, however, that human behavior plays a far greater role than animal behavior in both conservation and management.

The Role of Animal Behavior in Wildlife Management

It is important to define the role that animal behavior can play in wildlife conservation and management. Problems in wildlife management are a subset of the global environmental problems that are of interest to conservation biology. Major ecological problems include the wholesale loss of species through habitat destruction; the pollution of air, soil, and water; the introduction of exotic species (including domestic animals, parasites, and pathogens); and the alteration of global biogeochemical cycles. Knowledge of animal behavior is not the sole key to solving global conservation problems; but then, paradoxically, neither is any branch of ecology or any other science. Indeed, biologists do not make the important decisions that affect species extinction and people's continued ability to benefit from functional ecosystems. Such decisions are the purview of politicians and business leaders, who are primarily interested in political and economic goals and are therefore much more influenced by political and economic processes than by science (Morowitz 1991).

Changes in socioeconomic circumstances are also important. For example, immediately following World War II, agriculture was the main occupation in several southern European countries. People were widely distributed over the countryside. Almost all natural resources were exploited, including lands with low productivity. Following industrialization in the mid-1960s, much of the land that was either hilly or mountainous was abandoned as people sought a more comfortable lifestyle in cities. Space and resources in the abandoned countryside became available for wildlife. Urbanization may thus explain the recent recovery of wildlife in Europe more than any other economic or biological process. In North America, increased affluence, good rural road networks, and ability to work from home are instead leading to suburbanization of wildlife habitat, with negative consequences for biodiversity, especially of large predators.

Everyone can make "minor" decisions with environmental consequences, from not eating seafood caught with methods causing extensive bycatch of nontarget species, to not building a home on critical habitat, to family planning, to voting patterns in democratic societies. Zoologists, including animal behaviorists, clearly play a major role in the conservation of biodiversity by informing decision makers and the general public about the ecological consequences of human activities. Solving the global conservation problems that threaten our quality of life, and in some cases our very lives, will require scientific knowledge, but first and foremost it will require a better system of economic valuation of goods and services. Economic externalities such as pollution, habitat destruction, and the loss of ecological functions (including those that provide clean air, safe drinking water, and a stable climate) must be incorporated in the evaluation of different activities (Chichilnisky and Heal 1998). Perhaps the greatest contribution that ecologists can make to environmental conservation is to convince decision makers at all levels, from heads of state to individual consumers, to think about the long-term consequences of their decisions.

Behavioral ecologists typically study the long-term evolutionary consequences of different animal behaviors. As a result, when examining the consequences of human actions, they usually consider a longer timescale than the few years to the next election, or this year's balance sheet, or the time it takes to win one particular court case. It is essential that they transmit such long-term thinking to other sectors of society.

Students of animal behavior can provide an extremely important approach to wildlife conservation because of their tendency to examine individual differences, to emphasize the role of variability, and to think in terms of trade-offs between different behavioral strategies. Such emphasis on the behavior of individuals and the strategies they adopt to maximize fitness plays an important role when a species' natural behavior can lead to conservation problems in habitats altered by humans. In extreme and rare cases, the best management strategy may be to interfere with a species' natural behavior.

The study of animal behavior is most usefully applied to the conservation and management of populations because it both identifies and provides ways to deal with a key characteristic of animals: they are not all alike. Individual differences in age, sex, size, aggressiveness, learning ability, past experience, heterozygosity, and a myriad of other variables can all affect how an animal reacts to a given situation and may determine the success or failure of a management strategy or a conservation initiative. Conservation of animal populations thus often depends on meeting the challenge of how to incorporate individual differences in wildlife management. The importance of individual differences in wildlife conservation is a central theme of this book.

There is a hierarchy of levels of individual heterogeneity, and all are important to wildlife management and conservation. One may start by considering behavioral differences between similar species. For example, two North American canids, the wolf (*Canis lupus*) and the coyote (*C. latrans*) react in opposite ways to urbanization and intensive agriculture: wolves disappear, coyotes prosper (Tremblay, Crête, and Huot 1998; Mladenoff, Sickley, and Wydeven 1999). One may argue that the coyote's greater behavioral adaptability is the key to its success because it allows coexistence with humans, whereas the wolf's behavior leads to its demise: wolves range over a wide area, hunt in packs, and are intolerant of humans. Within the same species, however, there are often behavioral differences between broad geographical areas: wolves in southern Europe coexist with human population densities that are much greater than densities that wolves tolerate in North America (Promberger and Schroeder 1992). The animals belong to the same species, but their behaviors are very different. Southern European wolves resemble North American coyotes in their ability to survive alongside dense human populations. At a smaller geographical scale, variables such as prey type and level of human exploitation can affect pack size, turnover rates, and social structure, which in turn can determine the level of genetic diversity by varying the opportunities for dispersers to recruit into packs. Indeed, it has been suggested that high levels of shooting and trapping in eastern Canada may artificially increase the rate of hybridization of wolves with coyotes (Wilson et al. 2000). Finally, the sex/age composition of each pack, individual preferences, and previous experience may affect variables such as prey selection or space-use patterns, which may in turn affect vulnerability to human harvest or the probability of conflict with humans because of livestock depredation.

Specialist predators that appear to form a "search image" for a particular type of prey are a very good example of how animal behavior can affect wildlife management on a local scale. Marco Festa-Bianchet has studied the ecology and behavior of bighorn sheep (*Ovis canadensis*) in the Sheep River population since 1981 (Festa-Bianchet et al. 1995). From 1982 to 1995, cougars (*Puma concolor*) were studied in the same area. Most cougars in the

Sheep River drainage had radio collars. From 1982 to 1993, they killed only zero to two sheep a year. From 1993 to 1995, one adult female cougar suddenly switched from hunting deer (*Odocoileus* spp.) and wapiti (*Cervus elaphus canadensis*) to preying upon bighorn sheep, and was almost single-handedly responsible for a 20% decline in the bighorn population (Ross, Jalkotzy, and Festa-Bianchet 1997). A similar phenomenon occurred in another study area, Ram Mountain, from 1997 to 1999: again, following a sudden increase in cougar predation, mortality of adult females doubled, mortality of adult males tripled, and the bighorn population declined by almost 50%, although factors other than cougar predation were likely also involved. Almost no cougar predation was recorded at Ram Mountain from 1972 to 1997, but cougar signs were seen in almost every year. In both cases, the increase in predation was apparently due to an individual cougar's specialist behavior. Predation was not associated with increased availability of bighorn sheep as prey or, apparently, a decline in alternate prey.

Because the behavior of bighorn sheep is very different from that of cervids, a cougar must change hunting technique to prey on sheep. Hunting bighorn sheep requires specialized, learned skills that not all cougars have. Indeed, one male cougar attempted to kill a lamb and was itself killed when he and his victim fell off a cliff. From a management viewpoint, the experience both at Sheep River and at Ram Mountain suggests that a generalized predator-control program would have had little effect without removal of the sheep-killing individual (Ernest et al. 2002). Finally, in both cases cougar predation led to an increase in bighorn mortality despite low population density: because the increased predation was due to individual behavior, it was independent of population density.

At about the same time, some cougars in southwestern Alberta started preying on domestic dogs, possibly as a response to increased residential development on cougar range, which is currently a problem in many areas in western North America. Included among the victims were the hounds used to capture cougars at Sheep River from 1985 to 1994. The normal reaction of a cougar pursued by hounds is to climb a tree. It is likely that tree-climbing by cougars has been selected as an adaptive response to pursuit by packs of wolves. Wolves compete with cougars for the same prey and can kill a cougar if they can catch it. Cougars may react to dogs as they would react to wolves. Once a cougar learns that domestic dogs are easily killed, however, it may change its behavior and fight rather than run. Clearly, dog kills lead to rural residents' intolerance of cougars in general. Faced with a difficult social situation, it would be very valuable for managers to know whether the dog-killing behavior is generalized or limited to a few specialist cougars. It would also be very useful to know how to prevent the development of dog-killing behavior in wild cougars. These examples show how behavior, even the behavior of single individuals, can affect many aspects of wildlife management.

Goals of This Book

Our principal objective in assembling this volume was a simple one: to provide a broad overview of how knowledge of animal behavior can improve our ability to manage wildlife. Most chapters explore how conservation strategies either are or should be affected by animal behavior and how particular aspects of behavior affect the viability and growth of populations. Others explore the limits of animal behavior's contribution to conservation biology. In particular, the book addresses practical aspects of conservation and explores the role of animal behavior in the conservation of various ecosystems. Contributors examine both the importance of general principles and the key role played by specific characteristics of different species. Conservation is not a biological problem, it is a human problem. We do not subscribe to the view that wildlife management must improve natural systems, but rather believe that management actions are required either to remedy environmental damages caused by humans or to lessen the impact of human exploitation on natural systems. Because behavior can affect the reactions of wildlife to different conservation strategies, behavior must be taken into account for both remedial and preventive management. The chapters herein will outline the circumstances in which animal behavior affects conservation biology, and identify which behaviors are particularly important to ensure either the continued survival or the sustainable exploitation of wildlife.

Because conservation biology arises from a need to prevent, or at least lessen, human impact on ecosystems, an exploration of the role of animal behavior in conservation must take into account the diversity of situations that are faced in different areas of the world. Human attitudes, societal orientations, economic diversity, and traditions are all very important aspects of wildlife conservation. Social attitudes also determine what people want to protect or exploit, which wild species have economic or cultural value, and the acceptance of different management strategies. These social and economic factors interplay with animal behavior to affect the consequences of human actions on biodiversity. To partially account for diversity in both biology and culture, we attempted to select contributors interested in different aspects of animal behavior, based in different countries, and with expertise in animal behavior in a variety of geographical and political settings. We were only partially successful, mostly because researchers interested in and able to pursue studies in animal behavior are most often based in Western countries. The contributors bring to bear their own scientific expertise as well as their personal experience. Just as differences in behavior can affect the success of alternative conservation strategies, differences in societal attitudes are often the main reason why a conservation strategy can work in one human setting and fail in another.

Structure of the Book

The book is organized into five parts. In part I, chapter 1 provides a general introduction. In chapter 2, Morris Gosling explores the main reason why animal behavior is important to conservation: because individuals differ, models attempting to predict population dynamics, genetic variability, and the risk of population extinction can be improved by a consideration of individual behavior.

Part II (chapters 3–7) considers how resource-use strategies affect wildlife conservation. Rosie Woodroffe examines how dispersal behavior, particularly of carnivores, can have both positive and negative implications for conservation. Dispersing individuals can in some situations ensure gene flow and sustain a metapopulation structure, but in other cases dispersal movements bring carnivores into conflict with humans. When most habitat has been destroyed, the chances of successful dispersal are so low that emigration becomes essentially a source of mortality.

Paolo Luschi details the example of marine turtles, which migrate over huge distances over very long periods of time, requiring international coordination of protective measures. For exploited populations there is uncertainty over national "ownership" of different stocks, because individuals traverse the territorial waters of several countries.

André Desrochers considers how different bird species behave near edges of different types of habitat to show how this behavior affects the ability of different species to cope with habitat fragmentation brought about by forest harvesting. This is a very important topic in many boreal forests where forestry activities are expanding, often with unknown consequences for biodiversity.

Isabelle Côté examines fisheries management with and without taking into account details of fish mating systems. Norman Owen-Smith underlines the importance of foraging behavior for the reintroduction of extirpated large herbivores to remaining habitat. Both chapters argue that a knowledge of animal behavior is essential for the success of management programs: availability of suitable habitat is not necessarily all that is needed to guarantee the persistence of some animal populations.

Part III (chapters 8–12) examines practical applications of animal behavior in wildlife management. Jean-Michel Gaillard and coauthors provide an eloquent illustration of how individual differences, including differences in age/sex composition of ungulate populations, can improve the ability of models to provide a realistic representation of population demography.

Joel Berger and colleagues look at how behavior of individuals affects their reaction to potential predators. Although large predators have been extirpated over large tracts of their historic range, in a few parts of the world this trend has recently been reversed. Successful reintroduction and habitat restoration programs, together with a changing societal attitude toward large carnivores, has allowed the return of bears, wolves, and other large carnivores to areas from which they had been extirpated. Berger and colleagues argue that the impact of recolonizing carnivore populations on prey species is partly a function of how naive prey individuals react to their first encounters with predators.

Marco Apollonio and coauthors examine several recent southern European experiences in the management of large mammals. They point out that animal behavior is often ignored in reintroduction and harvesting programs. Their chapter exposes the problems caused by not taking into account available knowledge of animal behavior, and proposes ways to incorporate behavior into wildlife management in a European context.

The importance of behavioral ecology for exploitative management of carnivores is illustrated by Jon Swenson, who argues that sexually selected infanticide in bears and other large carnivores is a major management concern in cases where adult males are the preferred target of harvesting programs. The traditional view that male bears are "expendable," given the polygynous mating system of this species, is challenged by suggesting that killing of male bears may increase cub mortality by promoting infanticide committed by surviving males.

Closing this section on the consequences of harvesting programs, Marco Festa-Bianchet suggests that human harvest of wild animals is a major selective force that may shape both the morphology and the reproductive strategies of harvested species. Wildlife managers are interested in the population consequences of sport hunting, but few have considered the possibility that hunting may be a selective pressure. Because hunters select for specific morphology (such as horn or antler size), and because the mortality caused by hunting is very different from natural mortality, hunting could be a very strong agent of evolutionary change.

Part IV (chapters 13–16) explores individual variability in genotype and phenotype. Conservation biologists have long been concerned with heterozygosity and genetic variability because of the negative consequences associated with inbreeding and low genetic variability. Stephen Dobson and Bertram Zinner examine how differences in social structure of mammals can affect the maintenance of genetic variability in wildlife populations, concentrating on how social structure can affect the difference between census size and effective population size.

Claus Wedekind considers the genetic consequences of mate choice and reproductive skew for conservation programs, particularly for small, free-ranging populations and the management of captive-bred species. Nonrandom mating and reproductive skew are the norm in most wild populations, but within the confines of captivity, or when populations have been reduced to a very small size by human activities, these behaviors are not necessarily to be encouraged by wildlife managers.

Two chapters examine techniques to measure and account for individual differences. Brian Steele and John Hogg offer a detailed look at the uses of Generalized Linear Mixed Models, based on repeated observations of marked individuals. Rather than being affected by the statistical problems of pseudoreplication, these models take advantage of individual heterogeneities to better understand natural variation in different types of behavior, but their use is not for the statistically faint-hearted! In the next chapter, Peter Arcese uses a long-term data set on individually marked song sparrows to search for both a definition and a measure of individual quality. The theme of individual differences is pervasive throughout the book and is picked up again in the last two chapters.

The book's final part is a concluding chapter. Marco Festa-Bianchet provides an overview of the possible future contributions to wildlife management of research in animal behavior. He calls for greater cooperation between managers, researchers, and all people interested in the preservation of biodiversity.

2.

Adaptive Behavior and Population Viability

Leonard Morris Gosling

This chapter focuses on areas where an understanding of adaptive patterns of behavior is important, and sometimes essential, for predicting the demographic and genetic processes that determine population viability. Although many aspects of behavior are important for conservation, I will concentrate on adaptive behavior rather than development or mechanism because it is here that the potential benefits are greatest. The explosion of research in the fields of sociobiology and behavioral ecology over the past 30 years has revolutionized our understanding of animal behavior. But this fundamental understanding has not been fully incorporated into conservation biology, the applied science that aims to provide a scientific underpinning for practical conservation (Clemmons and Buchholz 1997, Caro 1998b, Sutherland and Gosling 2000). Although advances in behavioral ecology are of little use to conservation in the face of wanton overhunting or total habitat destruction, such knowledge is important in circumstances where reason prevails and where careful strategic planning is needed. We are living through an extinction crisis of unprecedented dimensions and must deploy all of the tools at our disposal.

Conservation is concerned principally with the viability of populations, communities, and habitats. All these entities are of daunting complexity, and ecological theorists and practitioners have generally looked for ways to describe and understand their key processes in a relatively tractable fashion. Tractability usually equates with simplicity, and the most obvious way to achieve this is to deal with higher-level processes rather than attempt a reductionist, individual-based approach. This has been the dominant trend in ecology to date and it has achieved considerable success. Thus, in population dynamics, most demographers consider birth rates and death rates of average individuals, or broad patterns of gene–environment interactions rather than dynamics based on the differences between individuals and their interactions.

Increasingly, however, it has become impossible to ignore studies of individual variation on population dynamics, particularly those that address functional issues within behavioral ecology. In addition, many of the problems considered by conservationists involve populations reduced to low levels and in these the behavior of individuals is relatively more important than in larger populations. Estimates suggest that effective population sizes must be at least 5000 to maintain adaptive potential in the face of mutation and random genetic drift (Lande 1995). Many populations or subpopulations of conservation concern are well below this level, and it may only be possible to maintain them under management using detailed information about factors that affect their viability. Many of the models now used for population viability analysis incorporate demographic stochasticity to simulate the sort of chance events that affect individuals within the context of average population values for the main population processes. Only recently have we begun to consider the effect of variation in the mating system and patterns of mate choice in such models (Legendre et al. 1999, Durant 2000a).

Newly developed population models use individual decision rules to derive population processes. Because decisions by individuals have been shaped by selection, an understanding of individual behavior and the incorporation of decision rules into population models should enhance our confidence in predictions about population responses to environmental change (Goss-Custard and Sutherland 1997; Pettifor, Norris, and Rowcliffe 2000; Bradbury et al. 2001). These models also have the advantage that some crucial features, such as frequency-dependent occupation of habitat patches, can be modeled using game theoretic approaches. Individual behavior-based models have had their greatest success in predicting for shorebirds the fitness and demographic consequences of human-caused changes in their foraging environment (Stillman et al. 2000, 2001, West et al. 2002). Although the approach is currently restricted to wintering migratory coastal birds, there is no reason in principle why it should not be applied to a wider range of taxa and habitats.

Such models are likely to be more successful in predicting population changes under novel circumstances than statistically based phenomenological models based on empirical data from a limited time period and a limited range of environmental conditions (Goss-Custard and Sutherland 1997). The approach has also been used to explore carrying capacity in migratory animals and as a guide to habitat management (Goss-Custard et al. 2002).

Perhaps the most compelling reason for believing that incorporating individual behavior into population models enhances our ability to predict population viability is that, under the powerful evolutionary force of sexual selection, the genetic fitness of individuals does not equate with population viability. Indeed, the enhanced fitness of individuals may actually depress the viability of a population and make it more vulnerable to extinction. For example, the evolution of costly display traits may generally be at the expense of other components of fitness, and these may affect population viability (Møller 2000). This argument and others about the negative potential of sexual conflict and intrasexual competition are developed in the following text. When we understand the circumstances under which such phenomena occur, it may become possible to intervene to ameliorate their effect, particularly where their action is conditional on ecological variation. Basic studies of sexual selection can also show where current practice in conservation management may be misguided. Mate choice may allow selection for heritable viability traits or selection against deleterious mutations, particularly where these are expressed in display traits. Where potential mates differ in their heritable viability, the current practice of maximizing outbreeding in conservation breeding programs may thus be an inferior management strategy to one based on free choice (Wedekind 2002a).

Lastly, I will discuss reasons for the failure of advances in behavioral ecology to be used in conservation and conservation biology. The primary issues may be sociological: behavioral ecology is an academic discipline practiced in universities, whereas conservation is a practical discipline that proceeds largely by trial and error in the field. I will debate this proposition and explore ways of reaching a working accommodation.

Density-Dependent Behavior

The best-known link between population density and viability owes its origins to W.C. Allee who observed that many animals suffer a decrease in per capita population growth rate when population density reaches a low level. Under these circumstances, the rate of increase can decline to zero or even to negative values. Although this effect, known as the Allee effect, can easily be mimicked using simple deterministic models (Courchamp et al. 1999), its occurrence and consequences can only be predicted where details of the mechanism are known and understood. The *Allee effect* is an umbrella term that has limited predictive value until its mechanisms are unraveled. These mechanisms differ between contexts and may not always involve behavior. Genetic inbreeding and forms of demographic stochasticity that do not act on behavior directly (for example, some variation in primary sex ratios) may exert the key effect (Lande 1998a).

Behavioral mechanisms of the Allee effect include the effects of reduced foraging efficiency due to an increased need for vigilance by individuals in small groups and higher rates of predation on such animals. Other behavioral mechanisms are the loss of cooperating individuals (such as nest helpers) and consequences of the mating system or of sexual conflict. For example, males in polygynous mating systems may have to range more widely at low densities to mate with females and may thus be more likely to encounter sit-and-wait predators. In the case of the coypu (*Myocastor coypus*), an environmental pest that was reduced to low densities in a trapping campaign in England, this effect reduced the proportion of males in the adult population to an extent that significantly reduced conceptions and accelerated the population's decline (Gosling and Baker 1989).

Mate choice may also be a key mechanism of the Allee effect. There is considerable empirical support for the involvement of (1) an inability to find a suitable mate, which reduces the frequency of reproduction and (2) poor reproductive success due to differential parental investment by females that do not find suitable or preferred mates (Møller and Legendre 2001). When females are forced to mate with a nonpreferred male, they reproduce at reduced levels. Examples include a reduction of 58% when female zebra finches (*Poephila guttata*) are mated experimentally with nonpreferred males (Burley 1986) and a 35% reduction for female barn swallows (*Hirundo rustica*) (summarized in Møller and Legendre 2001). Models of populations with either random or choosy mating with respect to phenotype show that the probability of population extinction as a function of initial population size was significantly increased with mate choice. Behavioral mate choice is typically ignored in conservation breeding programs, and the high rate of failure to breed species such as giant pandas (*Ailuropoda melanoleuca*) in captivity may be because females do not have the opportunity to choose their mates. In such cases the chances of finding a preferred mate may already be small because of the low numbers available. Reduced reproductive success among females with restricted choice of mates is such that to achieve a given minimum risk of extinction, initial population size must be more than twice as that in the absence of such effects (Møller and Legendre 2001).

There are many consequences for practical conservation from behaviorally induced forms of the Allee effect (e.g., Stephens and Sutherland 1999). In the case of the coypu already mentioned, increased male mortality led to more rapid decline than expected and earlier eradication of an unwelcome environmental pest (Gosling and Baker 1989). However, where population growth of an introduced species is slower than expected in its early stages, this may lead to the unwarranted assumption that the species will not thrive and thus to missing the best opportunity for eradication when the population is small and subject to Allee constraints. Failure to recognize Allee effects in exploited populations may also drive them to extinction. This process may have been responsible for the failure of many fisheries operating under the principle of maximum sustainable yield (Liermann and Hilborn 1997). Negative density dependence also affects the critical population size required to manage rare or fragmented populations, including the number required for successful introductions. However, simple adjustments of the number of animals may not be possible (for example when the remaining world population is very low) and in any case may not be necessary. Only when the mechanism of the Allee effect is known can predictions be made and specific corrective action be taken. For example, in the reintroduction of the bush-tailed phascogales (*Phascogales tapoatafa*), a carnivorous marsupial, it is important to allow females to establish territories before releasing males, to prevent the males from dispersing into unoccupied areas (Soderquist 1994).

Behavior under Spatial Constraints

All animals are adapted to particular habitats, and their lives are constrained by the spatial limits of these habitats. Typically, populations are also divided between sub-areas of suitable habitat. The dynamics of such metapopulations becomes partially dependent on the behavioral rules that govern joining and leaving habitat patches and the costs and benefits of moving between patches. In general, the benefit:cost ratio declines as patch size declines and the distance between patches increases—factors that become relevant in networks of protected areas. An understanding of the movements of individuals within and between patches and their population consequences is essential for conservation management. For example, some butterflies are restricted to habitat patches that contain essential larval foods. In each generation a proportion of individuals emigrate from the natal patch and the proportion that leaves depends on the perimeter:patch area ratio. When this ratio is too high, the numbers that leave quickly drive the population to extinction (Thomas and Hanski 1997). Unfortunately, conservation measures are rarely planned using this sort of understanding. More usually we obtain practical information post hoc by observing the consequences of conservation measures (such as the population viability consequences of adopting particular areas for protected areas) that have been designed without taking into account the natural movements of animals. Thus we tend to measure the decline of populations when they occur in areas that are too small rather than estimating in advance the area required for populations with defined viability criteria.

Often we simply do not know enough about the behavior and ecology of the animals that we seek to conserve. Getting sufficient quantitative information to predict the behavior of individuals and populations under spatial constraints takes time and effort. Putting such information to practical use may be even more difficult. The best example of a system where all these elements have been achieved is in the use of individual behavior-based models in predicting the dynamics of populations of wading birds. Existing models consider patches of habitat within a single estuary with variable resource densities in which birds compete to maximize food intake. An individual bird's access to food patches of varying quality is determined by the bird's own physiological needs, its own foraging efficiency, its dominance status, and the decisions made by its competitors, which in turn depend on their own dominance rankings (Goss-Custard et al. 1995b, Stillman et al. 2000). This sort of model can predict the fitness and population consequences of habitat or demographic changes that have not previously been observed. For example, these models can predict the effect of removing a part of a feeding ground and thus have important relevance to practical conservation, which is often concerned with the loss to habitat to alternative forms of land use. Such approaches have also been extended to entire populations of migratory geese using different wintering sites across large parts of northwestern Europe (Pettifor, Norris, and Rowcliffe 2000).

A major problem for conventional models of animal populations is that it is difficult to estimate density dependence, especially when populations are relatively constant. In addition, any measures of density dependence will tend to apply only to the range of population densities and environmental conditions over which they were measured. These limitations do not apply to individual behavior-based models because the full range of individual experience can be measured and thus included in a model, even when a population is constant. For example, an individual may experience a wide range of local population densities and a wide range of prey densities due to small-scale habitat variation. By combining data from observations of such individuals, the responses of entire populations to similar variation can be simulated (Goss-Custard and Sutherland 1997; Pettifor, Norris, and Rowcliffe 2000). Further, individuals in the models respond flexibly in relation to environmental change, and the demographic consequences of these changes can be used to generate new density-dependent functions. These allow predictions of future changes of abundance under altered spatial and other environmental constraints, such as different regimes of disturbance (West et al. 2002), that are outside the range of conditions under which the input data were collected (Stillman et al. 2000).

The principles of resource availability and depletion, and frequency-dependent competition based on physiological need, also apply to migratory species. However, in this case, the economic considerations of accumulating nutrient reserves needed to fuel movement between resource patches becomes critical (Piersma and Baker 2000). The costs to individuals of movement between patches are often so large that survival and further movement become critically dependent on adequate resources in a series of patches. Where such staging posts are degraded or destroyed, individuals may starve, become more susceptible to disease, or arrive in such poor condition at the end of migration that they are unable to breed. The conservation of staging posts along migratory routes poses particular problems because often these routes cross several countries, and conservation priorities tend to be agreed upon at a national rather than an international level. There is also inherent instability in a system that depends on a linear series of interconnected events. Even when most staging posts are intact, damage to one essential link could result in population extinction. To date, the number of migratory species that have been lost remains low; but the extinction or near extinction of Eskimo curlews (*Numenius borealis*) over a span of only 30 years, due in part to the destruction of their tallgrass spring staging habitat, is an indication of how vulnerable even very abundant species can be (Piersma and Baker 2000). The slender-billed curlew (*Numenius tenuirostris*) may suffer a similar fate.

The problems of conserving animal populations increase as populations are confined to small protected areas and as species ranges are subdivided and fragmented with intervening human developments that prevent wide-ranging movements or dispersal. Sometimes problems arise when individuals of particular species have very large ranges and are thus likely to wander outside a protected area. Once outside they are at risk from poachers or, more generally, from being in an area with conflicting land-use priorities. A particularly clear analysis of this effect was carried out by Woodroffe and Ginsberg (2000), who show that the probability of extinction of carnivore populations is affected most strongly by an interaction between range size and reserve size. They argue that, although demographic stochasticity may sometimes exert an effect, it is less important than edge effects in understanding and thus trying to prevent carnivore extinctions.

These problems have an additional dimension when combined with sexually selected dispersal. Where one sex disperses more widely than the other, this sex is likely to die at a faster rate, and in small populations this could potentially lead to Allee effects where the numbers of one sex limit population fecundity. In the case of elephants, there is a further social dimension: males are the dispersing sex, but mortality is biased toward large, socially dominant males because these animals have large tusks and are targeted by poachers. These large males are preferred by females (Poole 1989) and their loss reduces population growth rate and the speed at which populations can recover from overexploitation (Dobson and Poole 1998). The loss of animals with particular genetically determined traits may also select against the trait, and ivory poaching has led to an increase in adults without tusks (Jachmann et al. 1995). The selective loss of males could also lead to unforeseen genetic effects such as the loss of sex-linked genetic variation.

Edge effects may also have less obvious, indirect effects on population viability within the protected area. For example, male lions are often killed by sport hunting outside the Serengeti National Park, leading to a higher takeover rate of prides by coalitions of males. Because males taking over prides often kill existing young to ensure that females invest in their offspring, infanticide may affect population growth (Caro and Durant 1995).

The most general lesson from these examples is the importance of understanding animal movements before planning spatially based conservation measures. This has sometimes been achieved at a population or community level. For example, the modern boundaries of Amboseli National Park, Kenya, were based on a long-term study of the area required by the large mammal community for its year-round food and water supply (Western 1994). The area selected for the park included a dry season food and water reserve and part of the wet season dispersal area. In practice, the wet season dispersal area was so large that the entire area could not be protected, but since this is the least productive part of the range, there have been relatively few conflicts of interest with landowners. Most problems have been caused by elephants when they try to feed on crops outside park boundaries (D. Western, 2002, pers. comm.). Where conflicts of interest are perceived as being too great, an alternative is simply to fence a protected area to eliminate the conflict. However, the demographic costs of fencing an area that does not include all of the seasonal resources needed can be high since flexible movements to limiting resources are crucial adaptations for animals that live in strongly seasonal environments. For example, the fencing of Etosha National Park in Namibia in the 1970s (Berry 1997) was undertaken to prevent animals straying onto farms where they would inevitably have been killed. As a result, seasonal movements to food and water were restricted and there has been a significant reduction in population sizes. Burchell's zebra (*Equus burchelli*) declined from 25,000 to 5000 and gemsbok (*Oryx gazella*) from 5000 to 2200

(Berry 1997). This effect occurred despite the fact that, at more than 22,000 km² in area, Etosha National Park is one of the largest protected areas in the world. When animals suffer in such vast areas, it becomes clear that natural movements of large animals will rarely be possible in fenced areas. Under such circumstances, management intervention may succeed in manipulating movements to some extent (for example, by controlled burning or the provision of artificial water supplies), but reduced carrying capacity is usually unavoidable. This is because movements to seasonally important food resources are individual adaptations that maximize both individual fitness and, generally speaking, population viability.

Very large protected areas would help prevent many of these problems, but the example of Etosha shows that even the largest parks impose spatial constraints on behavior that ultimately limit population size. Often large parks are an impossibility, and, in general, the trend will be to reduce fully protected areas as human populations increase. As protected area size is reduced, population sizes decline and the chance of Allee effects and extinction increases. Data on African parks show that species loss can be predicted from protected area size (Newmark 1996) as would be predicted from relationships between species number and the area of natural islands (Soulé, Wilcox, and Holtby 1979; Cowlishaw 1999). Some form of multiple land use outside protected areas that includes the conservation of dispersing animals may be the only solution. Although it may be possible to negotiate such arrangements with appropriate compensation, as in the case of wolves in parts of Europe or North America, it will often be impossible where the costs of damage are very high, as in the case of African elephants, which sometimes kill people as well as damage crops. Efforts to provide a benefit to people for sharing their land with wild animals (such as the CAMPFIRE scheme in southern Africa) are in their infancy and have had mixed results (Gillingham and Lee 1999; Campbell, Sithole, and Frost 2000; Getz et al. 2000).

Nonrandom Mating and the Maintenance of Genetic Variation

Population viability depends partly on the maintenance of genetic variation. Variation increases the chance that some animals will survive when faced with short-term environmental vicissitudes such as droughts or diseases and also allows the possibility of long-term evolvability; for example, in response to long-term climatic change or the appearance of novel pathogens. Thus behaviors that affect levels of genetic variation may be of vital importance for population viability, particularly when populations are simultaneously affected by other negative factors such as reduced size. Here I consider the effects of intrasexual competition and mate choice, factors that are rarely considered by conservationists but that may have profound genetic effects on populations. These effects are not simply of theoretical interest because, when they are understood, their effects, or the negative consequences of losing their effects, can potentially be ameliorated by management intervention.

Intrasexual competition and mate choice are the main driving forces of sexual selection. Both sometimes lead to some individuals gaining a disproportionate share of matings and result in strong directional selection on the characteristics of the successful individuals (Andersson 1994). The most dramatic illustration of these effects occurs in lek breeding where a small proportion of males gain most matings. The characteristics of the males chosen include conspicuous displays (e.g., sage grouse [*Centrocercus urophasianus*], Gibson and Bradbury 1985) and display structures (e.g., the peacock's tail) (Petrie, Halliday, and Sanders 1991). Sometimes females choose particular territories in a lek rather than the characteristics of a male, perhaps because the outcome of competition for such territories is a more reliable indicator of fitness than any phenotypic trait. An example is the lek breeding system of topi (*Damaliscis lunatus*) where females strongly favor central lek territories even when a successful male is temporarily replaced by a smaller male (Gosling and Petrie 1990).

Other mating systems are less dramatic but may also result in mating skew. For example, birds in monogamous relationships often show extra-pair copulations, which boost the reproductive success of particular males (Petrie and Kempenaers 1998); female great reed warblers (*Acrocephalus arundinaceus*) choose for extra-pair copulations males who have more elaborate songs than their own male (Catchpole 1986).

Increased mating skew must result in reduced effective population size and, theoretically, reduced genetic variation. Some empirical support for this has come from estimates of variation in neutral genetic markers; for example in ungulates (Apollonio and Hartl 1993). This effect may become extreme in resource defense mating systems where limiting resources are manipulated in protected areas. For example, male springbok (*Antidorcas marsupialis*) can defend highly successful territories next to water holes, resulting in an extreme mating skew (Ritter and Bednekoff 1995). Reduction in effective population size may potentially lead to inbreeding depression. Management intervention to increase the dispersion of the limiting resource may be possible where behavioral observation shows the need.

However, mating skew does not always cause a reduction in genetic variation, particularly where female choice is involved. Female choice is known to be heritable and thus likely to be a character under selection. It appears to have two main benefits: either females gain for their sons those characteristics that make a male attractive or successful in intrasexual competition, or females gain viability genes for all their offspring (Andersson 1994). "Good gene" arguments are often linked to the evolution of display characters since female choice is often based on the size or elaborateness of male sexual ornaments or the intensity of males' displays. These are believed to indicate viability or fitness, but how? The most influential idea is that selection favors the evolution of signals that are costly to the signaler because these honestly reflect the signaler's quality (Zahavi 1975, Grafen 1990). The issue of signal costs will be returned to later in the chapter since the evolution of such "handicaps" may have direct effects on population viability and extinction probabilities. However, the fitness benefits for individuals from mate choice (reviewed by Møller, Christe, and Lux 1999; Jennions and Petrie 2000) are probably concerned principally with the coevolution of parasites and hosts (Hamilton and Zuk 1982). There is currently much interest in choice as it relates to genetic variation in the major histocompatibility complex (MHC), a hypervariable region of the genome concerned with immune function (reviewed by Jordan and Bruford 1998).

Extensive research on mice shows that mates are chosen on the basis of their genetic difference from the subject and that this information is obtained using odors mediated by MHC variation (Potts et al. 1991). The fact that the MHC is a region concerned intimately with immune function suggests that the evolution of dissassortative female choice may be favored by promoting increased disease resistance; for example, through heterozygote advantage. Alternatively MHC variation may simply act as a polymorphic marker to minimize inbreeding (Pusey and Wolf 1996).

A problem for arguments that invoke genetic benefits for female choice is that strong directional selection due to female choice should have depleted any genetic differences among males (as in the case of a reduction in variation due to mating skew, as already outlined). Why is there any genetic variation left among males in the population? This problem, the so-called lek paradox, remains one of the outstanding problems in evolutionary biology and could also have direct consequences for population viability. Theoretically, there should be no additive genetic variance in fitness-related traits (Fisher's fundamental theorem), and where selection is strong, as it is in sexually selected traits, then additive genetic variance should be lower than in nonsexually selected traits. It is therefore surprising that sexually selected characters show higher levels of additive genetic fitness than characters not under sexual selection (Pomiankowski and Møller 1995). What mechanism involving female choice could promote as well as remove genetic variance in fitness-related traits? The answer has practical as well as theoretical significance if the variation produced affects population viability. A possible mechanism (M. Petrie, 2002, pers. comm.) is that female choice could support a higher than normal mutation rate if the mutational load can be revealed in a display character. Simulation modeling by G. Roberts and M. Petrie (2002, pers. comm.) suggests that if females can select males who possess beneficial mutations but who carry fewer deleterious mutations, then mutation rates 10 times those under random mating can be sustained. The idea that female choice can maintain mutation rates provides a self-sustaining solution to the lek paradox and predicts a greater level of evolvability in sexual populations.

This has practical consequences for population conservation since the persistence of lek breeding may thus be important for population viability. Lekking in topi is becoming increasingly rare and persists only in the few remaining high-density populations. In the Mara ecosystem in Kenya it exists only where grassland is lightly utilized (as inside the Masai Mara Game Reserve) but not where it is intensively grazed (as outside the reserve with large densities of livestock). The reason may be that leks form where topi cluster in short-grass patches for antipredator advantage (Gosling 1986); this response occurs only where female topi are forced to avoid surrounding long-grass areas during the resting period of the day and not where the sward is uniformly short. If lekking is influenced by such relatively simple habitat features, it may be possible to intervene to help retain this mating system. Of course it would be desirable to do this in any case because such striking behavior as lekking in topi deserves to be conserved in its own right. But it is also possible that such intervention might conserve behavior that selects for high levels of genetic variation, removes deleterious mutations, and thus promotes population viability.

The possibility that patterns of mate choice may confer such important genetic advantages also has general implications for conservation breeding programs. At present most breeding programs of rare animals in captivity simply attempt to maximize outbreeding to retain as much genetic variation as possible. However, for the reasons already discussed here and by Wedekind (2002a) the benefits of allowing natural choice should be given careful consideration. In practice this could be achieved either by allowing females to choose mates or by artificial selection of mates according to the sort of criteria suggested by recent research on mate choice (e.g., using estimates of MHC similarity). Recent research suggests that not only might natural mate choice prevent inbreeding, it might also be driven by genetic compatibility between potential mates that provides resistance against particular pathogens (Wedekind et al. 1996, Rülicke et al. 1998). Although this latter possibility requires further investigation before its consequences are implemented in conservation breeding programs, there are already grounds for believing that benefits may be derived from allowing natural mate choice. Thus, where possible, and in species with appropriate mating systems, allowing choice should supersede breeding principles based on maximizing outbreeding since achieving high levels of heterozygosity may not outweigh the costs of accumulating deleterious mutations. The possibility that all potential mates will prefer one individual, thus leading to the prospect of severe inbreeding depression, is unlikely because assortative patterns of mating should generally be more common. Direct natural choice should be used wherever possible, but in intensively managed systems this may not be possible. Examples include small declining wild populations where intervention is essential, or captive populations where the financial cost of providing a natural choice is high. In these cases choice of olfactory signals (particularly scent marks, which can be frozen and shipped among cooperating zoos) provides the greatest promise. These odors provide subtle information about genetic variation in potential mates (reviewed by Gosling and Roberts 2001) and they could provide powerful measures of mate preference if used in properly designed choice assays.

Sexual Selection, Speciation, and Extinction

Some recent thinking about the conservation of biodiversity emphasizes the processes that create biodiversity rather than the pattern that happens to be present today (Mace, Balmford, and Ginsberg 1999). Thus, if we wish to maintain the capacity to create future biodiversity, we need to understand the processes responsible. Although the influence of behavior in reproductive isolating mechanisms has long been recognized, it is only recently that substantial support has emerged for the importance of sexual selection in sympatric speciation. Studies of bird speciation show that sexually selected clades (those with greater sexual dimorphism) are more speciose (Barraclough, Harvey, and Nee 1995; Møller and Cuervo 1998). The cichlid species flocks of the African great lakes are the classical example of a group that has shown explosive speciation rates under intense sexual selection (Seehausen 2000).

However, while speciation may proceed rapidly under the influence of sexual selection, the ornamentation or elaborate displays that are generated by male intrasexual competition or female mate preferences may predispose populations that possess them to extinction. Despite the theoretical importance of sexually selected handicaps (Zahavi 1975, Grafen 1990), empirical information on costs is accumulating only slowly. However, signaling intensity and the size of display structures have been shown to have correlated energetic costs in a number of species (e.g., drumming in wolf spiders [Kotiaho et al. 1998]) and to affect life history traits (scent-marking frequency is inversely correlated with growth in mice [Gosling et al. 2000]). The best data on survival are from experiments on barn swallows showing that survival prospects of males are inversely related to experimentally manipulated tail length (Møller 1994). There is also evidence that males carry ornaments at the expense of their resistance to disease and parasites (Folstad and Karter 1992). Although androgens promote the development of male display structures, they may also suppress immune function. Experimental evidence for a trade-off between the sexually selected trait and immunocompetence is now available in birds, including swallows (Saino and Møller 1996; Saino, Bolzer, and Møller 1997) and domestic fowls (Verhulst et al. 1999). Whatever the costs of display traits, all models of sexual selection predict that the evolution of elaborate display traits involves fitness costs that displace males from their survival optimum (Møller 2000).

Evidence that sexually selected traits affect extinction rates includes data supporting Cope's rule (Cope 1896, Eisenberg 1981), which states that body size tends to increase within evolutionary lineages and that the risk of extinction increases with body size. Although Cope's rule does not apply to all taxa, it probably has some general application (McLain 1993), and since larger body size is selected for under intrasexual competition, this effect may be attributed to sexual selection (Møller 2000). Further evidence comes from the probability of survival of introduced bird populations: McLain, Boulton, and Redfearn (1995) found that sexually dichromatic species were significantly less likely to become established than monochromatic species, perhaps because of the demographic consequences of the more costly sexually selected display features. A separate study of introduction success in New Zealand has been variously explained as a result of the degree of sexual dichromatism (Sorci, Møller, and Clobert 1998) or of demographic stochasticity, influenced by the mating system and female choosiness (Legendre et al. 1999).

The loss of biodiversity through an effect on sexually selected traits may sometimes be inadvertent. For example, the processes of sexual selection that produced the rich diversity of cichlid fishes in Lake Victoria may be disrupted by pollution (Seehausen, van Alphen, and Witte 1997). In these species flocks, reproductive isolation is maintained by mate choice using colorful signals. When these are obscured in turbid water, interbreeding between species increases and biodiversity is reduced. Similar arguments involving natural selection have been made by Endler (1997) about changes in the light environment of forests with consequent effects on the ability of cryptically colored animals to escape predation.

Other examples where individual fitness may conflict with population viability occur in cases of sexual conflict where the outcome may be damaging for one or both sexes and thus for population growth. Male bean weevils (*Callosobruchus maculatus*) damage the genitalia of females during copulation, perhaps to help prevent other males from mating with the same female and, as a result, female survival is reduced (Crodgington and Siva-Jothy 2000). In evolutionary arms races between the sexes, an adaptation by one sex that gives it an advantage (for example in mating) is generally matched by a counter-adaptation by the other sex. However, the outcome of such races can some-times favor one sex as revealed in a study of water striders (Heteroptera; Gerridae) (Arnqvist and Rowe 2002). Male water striders attempt to clasp females during mating using clasping genitalia, and since there is a cost to females in being clasped repeatedly after fertilization, females develop counter-adaptations such as abdominal spines. The development of these devices and corresponding behaviors is generally correlated within species, but detailed studies of morphology and reproductive behavior show that the advantage for one sex is greater in some species than in others. This leads to differences in mating rates and thus potentially to differences in population viability.

Sexual conflict is now recognized as being a central process of evolution with the potential to shape both speciation and extinction rates (Parker and Partridge 1998, Arnqvist et al. 2000). Such processes can clearly affect popu-lation viability, but are they accessible to conservation intervention? Direct intervention to prevent animals from damaging each other is possible in con-servation breeding programs (for example, using advanced reproductive technology) but can anything be done in the wild? In general, it depends on the ecological circumstances and whether they can be manipulated. In the example of polygynous antelopes, there is a potential conflict of interest between males that aim to mate with as many females as possible and females that wish to choose between males. Thus females often try to leave territories and males try to herd them back. The ability of males to monopolize females in this way depends critically on the distribution of resources: where resources are concentrated, males can monopolize more females and female choice is more limited. An example is that of male springbok who defend territories near water holes in arid areas (Ritter and Bednekoff 1995). Such behavior potentially leads to reduced effective population size and inbreeding depression, and could be ameliorated simply by providing more water holes. Intrasexual competition among females may also have a negative effect on population viability. For example, in some cavity-nesting ducks, high levels of brood parasitism may result in lower hatching rates in the population because of inefficient incubation of very large numbers of eggs and distur-bance by parasitic females. These effects appear to cause declining populations (Eadie, Sherman, and Semel 1998).

These examples raise the issue of when intervention is ethically accept-able. It is likely to be less acceptable in species in which a sexually selected benefit to one sex is threatening extinction, but more likely to be acceptable when an additional anthropogenic factor is exacerbating the threat. Thus, if the limited availability of water holes artificially increases the benefit to male springbok at the cost of inbreeding depression, it may be sensible to manipulate the distribution of water to ameliorate this effect.

Conclusions and Recommendations

The adaptive behavior of individual animals is not a peripheral issue that somehow embellishes population processes. Instead, adaptive individual behavior is at the heart of these processes. It is impossible to predict accurately population behavior without an understanding of individually based demographic and genetic effects on population viability. Higher-level processes do not equate with those at an individual level, and they cannot be inferred by simply multiplying the effect of average individuals. Selection at an individual level may work against population survival, as in the case of sexually selected handicaps or sexual conflict. Such effects can never be discovered at a population level but only through studies of individual adaptation. Higher-level descriptions of gene–environment interactions are indirect consequences of these adaptations and so must ultimately be less powerful predictors of population viability. A problem does exist in that the route from individual behavior to population process may be complex and difficult to model, particularly where decision rules are used to structure the model throughout. In contrast, higher-level processes are relatively accessible to analysis, and, when their assumptions are clearly understood, they will often have practical utility. We need a clearer understanding of the circumstances under which models using higher-level processes are a sufficient approximation and those in which they are actively misleading. Further advances in both simulation and mathematical approaches to individually based models seem both inevitable and necessary for conservation applications.

Curiously, behavioral theory has still made little impact on conservation practice. One fundamental issue is that interest in the theory of behavior and in practical conservation belongs to two different cultures with different values. One is essentially academic and the other practical. The two cultures are reinforced by patterns of funding and institutional support: academics work in universities and receive grants from government-funded research councils; conservationists work for government or charities with more restrictively defined programs and receive funding from conservation foundations and charities. The two cultures are mutually suspicious. Academics view practical conservationists as narrow and lacking intellectual rigor; conservationists view academics as putting theory before any benefits for their animal subjects. One conservationist has recently remarked that most behavioral ecology is "fiddling while Rome burns." An academic might reply that the rich body of information which is now available for conservation action was achieved by research driven by curiosity, not by practical need. Perhaps most important, practical people whose main focus is on accumulated wisdom through trial and error see science as a radical and disruptive process. In this they are correct, but rather than being a negative attribute, this is the great strength of the scientific process: it is the only mode of investigation that systematically sets out to destroy accepted ideas, and does so because the ideas that survive are generally most useful. But testing ideas to destruction is an uncomfortable process and frequently requires that practice is altered, which is not always popular, particularly in the long and painstaking work that is often required for conservation intervention. However, in the final analysis, practical conservationists and academics interested in animal behavior are both concerned about the conservation of animal diversity, and this goal will suffer until the two cultures reach a working accommodation.

A number of steps can be taken to help achieve this accommodation, but wholesale adoption of a conservation agenda by behavioral ecologists and of behavior studies by conservationists is neither likely nor desirable. Behavioral ecologists are largely curiosity driven and will continue to be curious naturalists, which is in itself important since a fundamental understanding of natural behavior is vital for conservation. Certainly there is no benefit in making good behavioral ecologists into bad conservationists (or vice versa). The body of information that has been, and is being, created is available for use, and the issue is rather one of effective information transfer between the two cultures. This can happen only when the need is recognized, and this is the explicit aim of recent books about behavior's relevance to conservation (Clemmons and Buchholz 1997, Caro 1998b, Gosling and Sutherland 2000), including this one. Hopefully these books will help conservationists and conservation organizations find ways to use fundamental research for practical purposes.

It is also true that some areas of behavioral ecology that would greatly benefit conservation (notably those with links to population viability) do not receive the attention they deserve from academic researchers. Regrettably, this is partly because most academics work on areas that are funded by government institutions. These tend to favor fundamental research so that strategic research on inbreeding depression and population viability is unlikely to be funded. Work of this kind, which proceeds from basic theory and has clear practical benefits, is too practical for a research council but too theoretical for a conservation trust. This problem could be addressed if organizations that fund basic research took a greater interest in conservation aims: some progress has been made in this respect (such as the Wellcome Trusts' biodiversity initiative), but such funding schemes are rare. Similarly, conservation organizations should devote part of their resources to work that aims to understand the basic issues underlying high-priority conservation problems. Last-ditch attempts to save single, highly endangered species are attractive, but they are not enough if we are to make significant strategic advances in the conservation of biodiversity.

Summary

Advances in behavioral ecology have transformed our understanding of animal behavior. Similarly, it has become clear that adaptive individual behavior has consequences for population viability and thus for conservation practice. Individual-based models have made best progress in quantifying the consequences of adaptive individual behavior and in providing techniques that are useful for conservation planning; they have particular utility in providing robust measures of density dependence. Individual behavior becomes particularly important at small population sizes, and mechanisms of the Allee effect, such as failure to find mates, are frequently behavioral. A predictive understanding of behavior is becoming increasingly important as animal populations become fragmented and confined to small protected areas. Sexually selected behavior influences genetic variation within populations and thus the prospects of inbreeding depression and extinction probability. An improved understanding of the genetic consequences of mate choice, particularly in relation to MHC variation, suggest that free choice may be preferable to outbreeding in captive breeding programs. Selection for extravagant displays and evolutionary arms races between the sexes may have negative consequences for population viability. Where these are conditional on environmental variation, conservation intervention may sometimes be possible. Despite these many insights into behavioral effects on population viability, they are rarely used to guide practical conservation. This may be because conservationists and behavioral ecologists belong to different cultures. One is essentially academic and the other practical and these differences are reinforced by patterns of funding and institutional support. Thoughtful changes to funding regimes may be the most realistic means of reaching a working accommodation.

Acknowledgments

I am grateful to John Goss-Custard, Claus Wedekind, David Western, Jan Murie, Marco Festa-Bianchet, and Marco Apollonio, whose comments improved an earlier version of this chapter; and to EJ Millner-Gulland who helped with elephant references. Marion Petrie and Gilbert Roberts kindly allowed me to include their unpublished ideas on the role of mate choice and mutators in the maintenance of genetic variation.

Part II
Resource-Use Strategies in Space and Time

Most animals are mobile: some are born and die within a few square meters, others have seasonal ranges separated by a few kilometers, still others may walk, fly, or swim tens of thousands of kilometers over their lifetime. Movements are motivated by a variety of needs: finding food or mates, avoiding predation, seeking nesting sites or shelter from rigorous weather. As animals move, they may come into conflict with humans, encounter humanmade barriers, or simply change "legal" status as they cross state or other jurisdictional boundaries. A major preoccupation of conservation biology is to ensure that those movements can proceed unimpeded, and that populations have access to all required seasonal habitats. The study of ranging behavior is therefore a key component of conservation-related research on any animal species, and it is one area where many traditional conservation schemes, based on protecting specific areas, or on management policies that change at political boundaries, simply do not work.

To conserve animal populations, we need to know where individuals go at different times of the year, how far they can disperse, what obstacles they can and cannot cross, how likely they are to locate new and suitable habitat, what they would do if their habitat were altered, and how they would react if reintroduced into former habitat. Some populations are "sinks": they are not self-sustaining, and they depend on input from "source" populations to persist, but they may play an important role in metapopulation dynamics. Some source populations may be particularly important for supporting a species over a wide geographical area: clearly, such source populations must be identified and protected.

Animal movements underline the need for coordinated conservation actions and management plans over political and jurisdictional boundaries.

Clearly, the first step is to identify those movements: many conservation problems arise because different governments or different sets of people fail to recognize that they are dealing with a "shared" resource, or because animals move outside protected areas. Animal migrations illustrate the need for international agreements for conservation, which present a challenge to human societies. Governments do not like to give up sovereignty, yet that is precisely what is required to foster international conservation. The five chapters that follow explore how animal movements should be taken into account to decrease extinction risk.

3.

Dispersal and Conservation :

A Behavioral Perspective on Metapopulation Persistence

Rosie Woodroffe

The persistence of small, isolated populations has been a major focus of conservation biology since its inception (Soulé 1987). Human activities have fragmented habitats into small, poorly connected "islands" and archipelagos in a sea of modified habitat. The fragmentation process has left many wild populations small and isolated, sometimes doomed to extinction as a result of catastrophes, environmental and demographic stochasticity, loss of genetic diversity, and inbreeding depression. Dispersal of animals between patches, however, has the capacity to remedy these population problems, increasing numbers, replenishing lost genes and genders, and even recolonizing patches vacated by extinct populations (Brown and Kodric-Brown 1977, Hanski and Gilpin 1997). It is hardly surprising, then, that conservation biologists have seized upon dispersal as something approaching a panacea, and metapopulation biology—the study of small populations connected by dispersal—as the area of research most likely to assist future conservation planning (McCullough 1996, Hanski and Gilpin 1997).

Against this backdrop, in this chapter I review the effects of dispersal on the viability of populations and metapopulations. I also consider the effectiveness, in conservation terms, of attempts to manage dispersal behavior, and discuss the behavioral, ecological, and evolutionary bases of dispersal behavior that may determine why particular management interventions succeed or fail.

Positive and Negative Effects of Dispersal

The fundamental predictions of metapopulation biology are best explained by outlining the "classical" model of Levins (1969). Habitat exists as patches, scattered through a matrix of unsuitable habitat. A proportion, P, of these patches is occupied by subpopulations; thus $(1-P)$ patches are vacant. Subpopulations become extinct with probability e, and vacant patches are (re)colonized with probability c. P therefore changes over time at rate

$$\frac{dP}{dt} = cP(1-P) - eP \qquad (1)$$

The equilibrium value of P, $\hat{P}$, is given by

$$\hat{P} = 1 - \frac{e}{c} \qquad (2)$$

A metapopulation can persist ($\hat{P} > 0$) only if the probability of colonization exceeds the probability of local extinction. Given a constant extinction rate, therefore, metapopulation persistence may depend upon efficient dispersal between habitat patches. If dispersal is hindered, either because of limited innate dispersal abilities or because patches are too isolated from one another, fragmented populations may be doomed to extinction. Consequently, species that are poor dispersers should be more extinction-prone than other species (e.g., Nee and May 1992).

Colonization processes are expected to occur on a much longer timescale than within-patch population dynamics. Metapopulations may therefore take a long time to reach equilibrium. Many species that appear to be persisting relatively well in fragmented habitats may in fact be declining slowly to extinction; this phenomenon has been termed the extinction debt (Tilman et al. 1994, Hanski 1997).

Failure to colonize suitable patches has indeed undermined the viability of world populations of some species. In 1982 the world population of Seychelles warblers (*Acrocephalus sechellensis*) was limited to the 0.29 km² Cousin Island, even though suitable habitat was available just 1.6 km away on Cousine Island (Komdeur et al. 1995). Translocations of warblers to Cousine and to Aride Island (9 km away) allowed the population to almost double its size (Komdeur et al. 1995), suggesting that the species would have been less vulnerable had it been capable of dispersing to other islands naturally. By contrast, species that are good dispersers may be capable of impressive recoveries from population perturbations. Wolves (*Canis lupus*) were eradicated from most of western Europe and large tracts of North America but have staged a remarkable recovery, reoccupying many thousands of square kilometers of their former range.

The positive effects of dispersal on recolonization have prompted much research on how to encourage dispersal between patches, including optimal spacing of habitat patches, construction of corridors, "stepping stones," and artificial translocations. Concerns have likewise been raised about conservation measures such as fences (meant to keep wildlife inside protected areas), which might have some beneficial effects but could also hinder dispersal.

Although dispersal behavior forms an important component of colonization, it may also contribute to local extinctions. Emigration from patches of habitat has the potential to undermine the viability of small, isolated populations, particularly when habitat fragments have high perimeter:area ratios (Stamps, Buechner, and Krishnan 1987).

Emigration from small, isolated patches appears to be an important cause of local extinction in some butterflies (Thomas and Hanski 1997). Thomas and coworkers (unpublished; reported in Thomas and Hanski 1997) modeled the population dynamics of two British butterflies, *Hesperia comma* and *Plebejus argus*, both of which have stringent requirements for habitats that are patchily distributed. They predicted that populations of *H. comma* would collapse in habitat patches below about 0.67 ha because emigration would outstrip breeding. *P. argus*, being less mobile, is less likely to disperse from small patches and should persist in patches down to a minimum of 0.05 ha (Thomas, Baguette, and Lewis 2000). The distribution of the two species across patch sizes gave a good fit to the model predictions.

A similar process operates, on a larger spatial scale, among large mammalian carnivores. For many of these species, which come into conflict with people, "suitable habitat" occurs only inside reserves where they are protected from persecution (Woodroffe and Ginsberg 1998). Animals ranging beyond park borders, however, suffer high mortality due to conflicts with local people, and such mortality has been a major cause of extinction in small, isolated populations (Woodroffe and Ginsberg 1998). Whereas most of this mortality involves resident rather than dispersing animals, mortality of animals dispersing across the inhospitable matrix is likely to be as high or higher, and may effectively halt exchange of animals between reserves.

The Behavioral Ecology of Dispersal

Because dispersal affects both colonization and extinction rates, effective conservation may demand management to encourage or discourage dispersal. Determining whether such management is needed and identifying strategies most likely to be effective demand information on dispersal behavior. Unfortunately, detailed data are lacking for most species. However, insights into the ecological and social factors that have shaped the evolution of dispersal behavior could help to predict little-known species' responses to habitat fragmentation and to management. This section therefore outlines, and attempts to explain, some variation in dispersal behavior.

WHO DISPERSES?

Successful dispersal has two consequences for an individual's reproductive fate: it may alleviate competition for resources, and it may increase offspring fitness by reducing the probability of inbreeding. These two factors have different effects upon the evolution of dispersal patterns.

Abundant correlational data suggest that dispersal may reduce competition for resources needed to breed. Gray-tailed field voles (*Microtus canicaudus*) that disperse grow more rapidly than those that remain in their natal areas (Davis-Born and Wolff 2000). Likewise, female European badgers (*Meles meles*) breed at an earlier age following dispersal (Woodroffe, Macdonald, and da Silva 1995). In a few social and territorial species, however, dispersal may bring few benefits: female lions (*Panthera leo*) that dispersed bred later and died sooner than females that remained in their natal groups (Pusey and Packer 1987).

If dispersers can escape resource competition, dispersal might be expected to occur most frequently at high population densities. This relationship has been found in some species, especially small mammals (Myers and Krebs 1971). In some social and long-lived species, however, precisely the opposite may be found, with dispersal virtually ceasing at high densities (Komdeur et al. 1995; Woodroffe, Macdonald, and da Silva 1995). In territorial species at high densities the chances of finding an unoccupied territory are so low that individuals who remain at home—with the possibility of inheriting the natal territory (and sometimes, though not always, assisting in the care of their younger relatives)—have higher fitness than those that disperse.

Dispersal reduces the chances that mating partners will be close relatives and thereby helps prevent the reduced reproductive success associated with inbreeding depression (Pusey 1987). Avoidance of inbreeding seems to explain why dispersal is often biased toward one sex. In mammals, females are usually philopatric, whereas males disperse; the opposite pattern occurs in birds (Greenwood 1980). This general pattern, however, hides a wealth of variety. The usual mammalian dispersal roles are reversed in great apes, for example, in which females disperse between groups of philopatric males (Harcourt, Stewart, and Fossey 1976). Dispersal may occur in both sexes

(McNutt 1996), or, exceptionally, in neither (Faulkes, Abbott, and Mellor 1990). The proportion of each sex dispersing, and the distance traveled, may also vary substantially within species (Cheeseman et al. 1988; Woodroffe, Macdonald, and da Silva 1995).

Dispersal behavior may also be influenced by social status. Gese and Crabtree (1996) showed that coyotes (*Canis latrans*) that dispersed tended to be low-ranking group members, whereas higher-ranking siblings remained in the natal territory. Harris and White (1992) showed a similar result for red foxes (*Vulpes vulpes*). Woodroffe, Macdonald, and da Silva (1995), by contrast, found that it was the largest (and presumably most dominant) male European badger cubs that dispersed.

PHASES OF DISPERSAL

In considering the metapopulation implications of dispersal, it is helpful to distinguish the phases involved in colonization of new sites. These are (1) emigration from the natal patch, social group, or area; (2) travel to the new area; and (3) immigration to the new site or social group. Colonization may fail at any of these three stages.

Emigration and immigration may be uncoupled: wolves and coyotes, for example, may spend weeks or months traveling over long distances before establishing new territories (Fuller 1989). This uncoupling reaches its extreme in species such as termites, which can be considered obligate dispersers— reproductive offspring can breed only if they emigrate away from the natal colony, and are morphologically adapted for this dispersal.

Other species are more circumspect in their dispersal decisions. Emigration may be contingent upon immigration, with dispersers moving opportunistically to neighboring territories that have been vacated by breeding competitors (Woodroffe, Macdonald, and da Silva 1995). Some species may make temporary forays away from the natal territory but do not truly emigrate until they locate a breeding opportunity (Waser 1996). Such forays may be prolonged; golden jackals (*Canis aureus*), for example, may return to their natal territories several months after initial dispersal, presumably when they have failed to occupy territories elsewhere (Moehlmann 1987). Similar behavior has been recorded in cougars (*Puma concolor*) occupying highly fragmented landscapes with little available habitat (Maehr 1997).

DISPERSAL DISTANCE

Dispersal distances vary substantially and unpredictably among species. Although Van Vuren (1998) found a general relationship between median dispersal distance and body size, this conceals substantial variation in dispersal abilities among related species. Gray wolves of both sexes may disperse long distances (median 58 km, maximum 886 km; Fritts 1983, Van Vuren 1998), whereas the closely related (and critically endangered) Ethiopian wolf (*Canis simensis*) shows much more restricted dispersal with all males remaining in the natal territory (Sillero-Zubiri, Gottelli, and Macdonald 1996). Likewise, median dispersal distance for the American badger (*Taxidea taxus*) is 12 km, whereas European badgers disperse over much shorter distances (Waser 1996).

Dispersal distance might also be predictable from ecological insights. For example, Waser (1985) was able to predict dispersal distances for deer mice (*Peromyscus maniculatus*) from measures of intraspecific competition for vacant territories. The same model, however, failed to predict dispersal distances for either male or female great tits (*Parus major*). Clearly, dispersal distance is shaped by a variety of species- and gender-specific factors.

Correlates of Colonization Success and Failure

Population models predict important effects of immigration and emigration, which may lead to differences in the vulnerability of "good" and "poor" dispersers. Definitions of dispersal ability are, however, inconsistent. Population models tend to consider a good disperser to be a species exhibiting a high colonization rate, but, as already discussed, this variable contains elements of emigration probability, dispersal distance, survival during dispersal, and capacity to breed on arrival. In this section I therefore discuss various behavioral traits that characterize "good" and "poor" dispersers, and consider their possible effects upon population dynamics.

WILLINGNESS TO CROSS PHYSICAL BOUNDARIES

Behavioral responses to arriving at the edge of suitable habitats vary between species—some "bounce off" the edge and remain largely in suitable habitat, whereas others readily enter matrix habitats (Desrochers and Hannon 1997, Schultz 1998, Haddad 1999b). Such variation may have important population consequences.

Some species are unable or unwilling to move between patches of suitable habitat. Consequently, such species' geographic distributions may be limited by comparatively minor physical barriers such as rivers and ranges of hills (Haffer 1997). In natural landscapes and over the very long term, this can promote speciation and may generate high regional biodiversity. In artificially fragmented habitats, however, such cautious behavior can be problematic. For example, dispersing spotted owls (*Strix occidentalis*) may starve at a forest edge rather than disperse to distant patches of suitable habitat (Arcese, Keller, and Cary 1997). American martens (*Martes americana*) are likewise reluctant to cross gaps in forest cover and are rarely found in even moderately fragmented forest landscapes (Hargis, Bissonette, and Turner 1999).

By contrast, some species appear undeterred by geographical features that might be considered likely barriers to dispersal. Coyotes cross rivers up to 100 m wide (Harrison 1992), and field voles (*Microtus agrestis*) commonly swim up to 620 m to disperse between islands in Finland (Pokki 1981). Seven cougars successfully dispersed across a four-lane highway (Sweanor, Logan, and Hornocker 2000), although two others were killed trying to cross this same road after it had been expanded to six lanes.

Willingness to leave suitable habitat may have negative as well as positive effects. For example, a study of forest songbirds' willingness to cross open areas to investigate call playbacks found that nuthatches (*Sitta canadensis*) were less reluctant than other species to leave forest cover, even when an alternative and only slightly longer route was available through forest cover (Desrochers and Hannon 1997). This willingness to enter unsuitable (and potentially hazardous) habitats could explain why fledglings of the related European nuthatch (*Sitta europaea*) maintain high rates of emigration from habitat fragments, leaving high-quality territories unoccupied (Matthysen 1999). Dispersers appear to suffer high mortality in fragmented habitats, making population persistence unlikely in the absence of immigration from larger source populations (Matthysen and Currie 1996, Matthysen 1999). Willingness to enter unsuitable habitat presumably also underlies the mass emigration from small patches observed in the butterflies studied by Thomas, Baguette, and Lewis (2000).

NATAL PHILOPATRY

In some highly social species, reluctance to leave the natal group can limit colonization in the absence of any obvious habitat boundary. These species are said to exhibit a high degree of natal philopatry. For example, female spotted hyenas (*Crocuta crocuta*) very rarely leave their natal groups; virtually all recorded female dispersals have involved fission of clans to occupy neighboring areas (Holekamp et al. 1993, Waser 1996). Males move only to join groups of females (Frank, Holekamp, and Smale 1995). Probably as a consequence, hyenas are extremely slow to recolonize areas subject to predator control (Smuts 1978) and are unlikely to disperse between habitat patches. By contrast, other species disperse readily from their natal areas or groups and may recolonize vacant habitat rapidly (e.g., wolves, African wild dogs

[*Lycaon pictus*]; Ballard, Whitman, and Gardner 1987; Hayes and Gunson 1995; McNutt 1996; Mills et al. 1998).

DISPERSAL DISTANCE

The distances covered by dispersing animals vary substantially between species and may have a powerful effect upon colonization success. Animals often move rapidly when they are dispersing across unsuitable habitat, and can cover distances greater than they would within their normal home ranges (e.g., Schultz 1998).

Measures of dispersal distance may be biased by both the size of the study area within which dispersers are sought, and the structure of the habitat they occupy. Acorn woodpeckers (*Melanerpes formicivorus*), for example, disperse very short distances within contiguous oak woodland and might be considered to have limited dispersal abilities. Extensive study, however, shows that dispersers can cover up to 200 km and that dispersal between patches occurs relatively frequently in fragmented habitats (Lidicker and Koenig 1996).

In mammalian carnivores, Woodroffe (2001) found no significant relationship between species' dispersal distances and their tendency to become extinct outside protected areas, though the trend suggested that long-distance dispersers might be more, rather than less, vulnerable. Thomas (2000) classified butterflies by a combination of dispersal distances and dispersal probabilities and found that species of moderate dispersal ability were most vulnerable to local extinction. The relationship between dispersal distance and vulnerability appears more complex than that assumed by simple metapopulation models.

INTEGRATION ON ARRIVAL

Colonization will not occur if dispersers fail to survive and breed when they reach new patches of suitable habitat. Such settlement often occurs readily: Seychelles warblers translocated to Aride and Cousine Islands, for example, located suitable habitat and established territories within days (sometimes hours) of arrival (Komdeur et al. 1995). Other species may fail to recognize unoccupied habitat as suitable: this has been documented in dispersing Belding's ground squirrels (*Spermophilus beldingi*) and yellow-bellied marmots (*Marmota flaviventris*) (Van Vuren 1998).

More starkly, human activities may thwart immigration. For example, illegal shooting of immigrating wolves has hindered their recovery in Scandinavia, despite the availability of prey and suitable habitat (Yalden 1993).

Where animals enter patches already occupied by conspecifics, they may experience problems integrating into the local social organization. Aggression from resident territory-holders appears to be an important cause of mortality among dispersing lions and wolves (Waser 1996). Likewise, black-tailed prairie dogs (*Cynomys ludovicianis*) often fail to immigrate into established colonies because they are driven away by residents (Hoogland 1995).

Many species appear unwilling to disperse without information on where they are dispersing to. Targets for dispersal may be unoccupied patches of suitable habitat, or social vacancies in existing groups. Male European badgers tend to remain in their natal groups until the death or disappearance of a male in a neighboring group creates a breeding vacancy that they can fill (Woodroffe, Macdonald, and da Silva 1995). The need to locate such breeding vacancies may underlie the forays that frequently precede dispersal in a variety of species (Waser 1996).

Managing Dispersal

It is clear that effective conservation often depends upon managing dispersal, though such management may entail limiting or encouraging dispersal, depending on the circumstances. Propensity to disperse—or not—depends upon a complex array of behaviors that are often species-, gender-, or population-specific. In this section, I discuss approaches to population or environmental management that have been used to manage dispersal.

MANAGEMENT TO ENCOURAGE EMIGRATION

A great deal of attention has been paid to promoting dispersal of animals between patches. Indeed, according to Lidicker and Koenig (1996), "The greatest challenge for land managers and conservation biologists is species that are reluctant to venture out of their preferred habitats at any time" (p. 88). Habitat management to overcome animals' reluctance or inability to disperse includes the construction of corridors and "stepping stones" as well as some other measures.

Corridors

The most widely researched measure for promoting dispersal is the construction of movement corridors. Although corridors have been described as holding "more promise for the conservation of the diversity of life than any other management factor except stabilization of the human population" (p. 493), their value has also been called into question (Simberloff et al. 1992).

Species' use of corridors varies. Experimental linkage of isolated habitat patches by corridors has been shown to increase interpatch movements for a variety of butterfly, small mammal, and bird species (Aars and Ims 1999, Haddad 1999a). However, many studies in nonexperimental environments have failed to establish the importance of corridors because they have sought to detect dispersers only in corridors, not in the surrounding matrix (Simberloff et al. 1992). Certainly some species, such as the wood mouse (*Apodemus sylvaticus*), and several butterfly species, will move between patches by crossing the matrix, as well as by using corridors (Zhang and Usher 1991, Haddad 1999b).

Species' use of corridors appears to be related to their responses to habitat edges. Two butterfly species that tended to "bounce" off the edges of forest clearings were found to use corridors to move between clearings, flying straight ahead rather than diverting into the forest matrix (Haddad 1999b). A third species, with less specialist habitat requirements, did not turn back into clearings when encountering forest edges, and left clearings along corridors only at the rate expected by random movement (Haddad 1999b). Responses to habitat edges may therefore be used to assess the likely value of corridors. For example, corridors were dismissed as a management strategy for the Fender's blue butterfly (*Icaricia icarioides fenderi*) because its tendency to "overshoot" habitat edges suggested that its movements might not be effectively directed by construction of corridors (Schultz 1998).

Behavioral responses to habitat edges presumably also influence the effectiveness of different corridor widths in channeling movement between patches. In the study just described, butterflies turned away from forest edges at distances of up to 8 m, and corridors were 32 m wide (Haddad 1999b). Thus half the width of the corridor was "edge," perhaps explaining why butterflies tended to fly straight forward rather than linger in corridors. Relatively wide (3 m) corridors failed to channel root vole (*Microtus oeconomus*) movements effectively between patches because the animals tended to cross the width of the corridor as well as move forward (Andreassen, Halle, and Ims 1996). Narrower (1 m) corridors were much more effective in encouraging dispersal between patches, although voles were reluctant to enter very narrow (0.4 m) corridors (Andreassen, Hall, and Ims 1996).

Corridor use may vary within as well as between species. Experimental corridors constructed between habitat patches increased natal dispersal of female root voles but had no effect upon the (markedly higher) dispersal rates of males (Aars and Ims 1999). Dispersal is generally male-biased in root vole populations, possibly reflecting a higher motivation for dispersal among males. Interestingly, the presence of corridors increased the rate of secondary (postbreeding) dispersal in both sexes (Aars and Ims 1999). Intraspecific variation in corridor use was also found in one of the butterfly species studied by Haddad (1999b). Resident *Phoebis sennae* responded strongly to habitat edges and tended to leave forest clearings along corridors. Butterflies of the same species studied at a time of year when they were migrating showed similar responses, but the effects were much weaker.

All of these data suggest that, although corridors may sometimes be effective in encouraging movement between patches, their value and optimal design vary substantially within and between species. Crucially, it appears that it would be difficult to predict corridors' likely contribution to the conservation of particular species in the absence of detailed behavioral studies.

Stepping Stones

"Stepping stones," disjointed patches of habitat arranged between larger habitat patches to provide a route for dispersal, have been proposed as an alternative to corridors for more motile species. Stepping-stone arrangements have, for example, been established to protect migratory waterfowl en route between breeding and wintering grounds (Piersma and Baker 2000). Butterflies have naturally recovered by recolonization along stepping stones (Hill, Thomas, and Lewis 1996), and such arrangements have been proposed to link fragments of remaining habitat for other threatened butterfly species (Schultz 1998).

Other Conservation Measures

The widely perceived need to encourage dispersal may also influence decisions concerning other conservation measures. For example, concern was expressed that fencing of Lake Nakuru National Park in Kenya to promote rhino conservation would prevent recolonization by African wild dogs (Frame and Fanshawe 1990).

MANAGEMENT TO LIMIT EMIGRATION

In species where emigration represents a serious threat to the persistence of local populations, management may, paradoxically, be needed to hinder animals' propensity to disperse. Where emigration from isolated patches occurs because patches are small relative to individual home ranges, a simple solution (in theoretical if not in practical terms) is to enlarge the size of habitat patches, or to alter patch geometry to minimize perimeter:area ratios (Stamps, Buechner, and Krishnan 1987; Woodroffe and Ginsberg 1998). For some large-bodied species, fencing may be an effective means of deterring animals from crossing park or habitat borders (Thouless and Sakwa 1995), though the fences needed to enclose large carnivores, in particular, may be substantial and expensive. More subtle management of habitat boundaries might help to deter emigration of smaller-bodied species.

An alternative to reducing individual emigration from habitat patches is to manage surrounding areas to maximize the chances of dispersal success. Such management might involve measures to limit mortality in the matrix; for example, enforced legal protection of species such as wolves and wild dogs, which suffer persecution in areas occupied by people. Alternatively, suitable habitat patches might be created or protected close to source populations to reduce isolation and "catch" dispersers. Such measures have been planned for black-footed ferrets (*Mustela nigripes*), where optimization procedures (a form of modeling) have been used to investigate how the creation of new habitat patches (through localized cessation of rodenticide use to allow prey populations to recover) can best be structured to encourage growth of ferret numbers (Bevers et al. 1997).

Measures to limit emigration may be more direct in some circumstances, especially in the conservation of species that come into conflict with people. Lions, gray wolves, and red wolves (*Canis rufus*) dispersing out of some designated conservation areas have been captured and returned to parks (Stander 1990, Phillips 1995, Phillips and Smith 1996). Such measures have not always resolved conflicts with people but appear to have reduced the numbers of animals having to be killed.

ARTIFICIAL DISPERSAL—TRANSLOCATION

Circumstances may not always permit management to encourage natural dispersal between habitat patches. For example, where patches are very remote from one another, or where they are fenced, measures such as corridors and stepping stones may be inappropriate or uneconomic. In such circumstances, dispersal may have to rely upon artificial translocation. Translocation is widely used to augment populations of game species and is often highly successful (e.g., Castle and Christensen 1990). However, a number of behavioral problems may confound translocation attempts.

Dispersal following Translocation

Release at a new site sometimes seems to elicit elements of natural dispersal behavior. For example, social groups of wolves and African wild dogs translocated together have tended to break up, with animals sometimes moving over thousands of square kilometers following release (Fritts, Paul, and Mech 1984, 1985; Woodroffe and Ginsberg 1999). Translocated wolves settled, on average, 87 km from their release sites (Fritts, Paul, and Mech 1984).

Likewise, about half of the first 37 black-footed ferrets reintroduced to the wild moved rapidly away from their release sites, traveling 4.1 to 17.1 km (Miller, Reading, and Forrest 1996), and translocated sea otters (*Enhydra lutris*) dispersed over 100 km (Estes, Rathbun, and Vanblaricom 1993). Such wide-ranging behavior often involves "homing" to the animal's original home range: 9 of 32 wolves, and 31 of 139 sea otters "homed" (Fritts, Paul, and Mech 1984; Estes, Rathbun, and Vanblaricom 1993). One captive-bred wolf traveled 280 km back to the breeding colony where she was raised (Henshaw et al. 1979), and a translocated leopard (*Panthera pardus*) traveled 540 km back to its home range (Nowell and Jackson 1996). Similar problems of wide-ranging and homing were encountered in experimental transloca-tions of California ground squirrels (*Spermophilus beecheyi*) (Van Vuren et al. 1997).

Problems of postrelease dispersal have been solved, to some extent, by modification of release procedures. Captive-bred black-footed ferrets given prerelease training in survival skills (including access to realistic habitats and live prey) ranged less widely and were less likely to disperse away from the release site (Miller, Reading, and Forrest 1996). Likewise, wolves and wild dogs held in enclosures at the release site for several months have remained in roughly the intended areas (Phillips and Smith 1996, Woodroffe and Ginsberg 1999). Careful structuring of release groups may also have helped to inhibit dispersal; successful wolf and wild dog reintroductions have mostly involved the release of newly formed groups established in captivity by intro-ducing sexually mature males to females (or groups of female kin) (Phillips and Smith 1996, Woodroffe and Ginsberg 1999).

Failure to Survive or Reproduce on Arrival

Recolonization will fail if translocated animals cannot survive or breed in the target area. Managing this possibility demands a realistic assessment of potential release sites. For example, African wild dogs translocated to Tsavo West National Park in Kenya left the reserve following a series of encounters with lions, major predators of wild dogs, and were subsequently killed by people in the surrounding areas (Kock et al. 1999). These interactions with known threats to wild dog populations suggest that the reintroduction site did not represent suitable habitat (Woodroffe and Ginsberg 1999).

Suitable habitat can be recognized by the presence of a breeding population. Indeed, the purpose of translocation is often to augment existing populations and to promote gene flow between them. Behavior patterns may, however, limit the possibilities for successful translocation, especially for social or terri-torial species. A tiger (*Panthera tigris*) translocated to the Sundarbans Tiger

Reserve in India was killed by a larger tiger just 20 m from the release cage (Seidensticker et al. 1976). Likewise, survival of translocated leopards is low, partly because resident territory holders respond aggressively to immigrants (Hamilton 1986).

Despite these concerns, one must be cautious in extrapolating from patterns of behavior in undisturbed populations. In unexploited wolf populations, for example, unrelated individuals are rarely accepted into packs, and intraspecific strife is an important cause of mortality (Fritts and Mech 1981). In contrast, in a harvested population 22% of dispersers were able to join established packs (Ballard, Whitman, and Gardner 1987). This raises the possibility that social integration might sometimes be a comparatively minor problem, especially when population density is low. Indeed, rates and causes of mortality were similar among resident and translocated wolves in Minnesota (Fritts, Paul, and Mech 1985).

Problems of intraspecific strife may be limited by careful choice of animals to be translocated. The wild dog population of Hluhluwe-Umfolozi Park was successfully augmented by releasing a group of related females, designed to mimic the natural immigration of a dispersal group (M. Somers, 1997, pers. comm.).

Conclusions and Recommendations

Although dispersal behavior may have a profound effect on (meta)population persistence, this effect is not always a beneficial one. Managers must not, therefore, assume that dispersal is always to be encouraged.

Management may be necessary either to encourage or to discourage individuals to disperse between patches. Unfortunately, dispersal behavior is, at present, somewhat unpredictable. Species and genders vary in their tendencies to emigrate from natal groups or habitat patches, in their dispersal responses to changing population densities, their dispersal distances, their habitat choices during dispersal, and their capacity to survive and breed on arrival at new sites. All of these factors influence whether dispersal needs to be managed, and, if so, what measures are most likely to prove successful. Conservationists must be cautious, therefore, in adopting management approaches, such as construction of corridors, without the support of behavioral evidence to indicate that such measures will be necessary and effective.

Interspecific variation in dispersal behavior also means that measures which are beneficial for one species may be pointless, or even deleterious, for another. For example, optimal spacing of habitat patches for a vagile species may effectively preclude successful dispersal by a sympatric species with more restricted movements.

Insights into natural dispersal behavior may also be important where habitat fragmentation and isolation are so extreme that dispersal must be effected through artificial translocation. Although mimicking some aspects of natural dispersal (e.g., structure of dispersal groups) may increase the success of translocations, other features of natural dispersal (e.g., traveling long distances) may be counterproductive and need to be discouraged.

All of these observations indicate that behavioral research must be a key element of conservation planning for species inhabiting patchy environments. Where possible, studies need to focus on the actual (meta)population to be protected because dispersal behavior may vary substantially in response to different spatial and social environments. Where habitat management is considered to conserve particular species, research should also consider potential negative effects on other species occupying the same landscapes. Clearly, research will be vital if a wide array of species are to be conserved in the world's increasingly fragmented landscapes.

Summary

Dispersal behavior has important effects on animals' population biology. Colonization of vacant patches and the "rescue" of declining or genetically depauperate populations are both brought about by dispersal. In addition, dispersal away from occupied patches may undermine the viability of isolated populations. These positive and negative effects mean that conservation measures may often involve managing dispersal events.

The relationship between dispersal ability and vulnerability to extinction appears more complex than is assumed by simple models. The few empirical attempts to investigate and test this prediction have established no coherent, and certainly no positive, link between the two. Part of the explanation for this may be the coupling of immigration to one patch to emigration from another: the negative effects of emigration may outweigh the positive effects of immigration, especially when dispersers suffer high mortality in the matrix.

Habitat management may be necessary to encourage—or to discourage—dispersal to improve the viability of metapopulations. Such management includes ensuring the optimal size and spacing of habitat patches, construction of corridors, and, in extreme cases, construction of fences. The effectiveness of these approaches depends primarily on individuals' dispersal behavior, particularly their willingness to cross habitat boundaries. Unfortunately, such behavior is not yet predictable, making it difficult to offer general prescriptions for the management of particular groups of species.

In extreme cases, movement of individuals between habitat patches can be effected only by artificial translocation. Detailed behavioral information may be needed to maximize the success of such translocations, especially for territorial and social species.

All aspects of conservation involving dispersal of animals between habitat patches, from understanding its effects on population viability to recommending the measures most likely to provide effective conservation, require information on the ethology and behavioral ecology of natural dispersal. Unfortunately, there is currently no underlying theory that can reliably predict all aspects of dispersal behavior needed for conservation management, placing a high priority—for the moment—on species- and population-specific research.

4.

Migration and Conservation:

The Case of Sea Turtles

Paolo Luschi

Migration is a widespread behavior by which an animal periodically moves from one region to another that better satisfies the requirements for a phase of its life cycle (Baker 1978, Dingle 1996). Migratory behavior is usually exemplified by animals that regularly shuttle between feeding and breeding areas. Animal migration, however, comprises a broad spectrum of possible patterns, which often makes it difficult to classify a given movement as migratory (Dingle 1996). In this chapter, I use recent research on sea turtles to illustrate the importance of behavior in the conservation of migratory animals.

Migratory movements range from the short-distance changes of habitats of some insects to the spectacular long-distance journeys of some vertebrates. Many migrants moving over large geographical areas traverse a variety of habitats, including some greatly affected by human activities. As a result, migrants are exposed to a number of threats, either along their migratory journey or at their destination, or both. These threats include hunting, habitat loss, and habitat degradation, which can be particularly harmful for migrating animals that are most vulnerable in such physiologically stressful and challenging situations. Migrating birds, for instance, have to meet high energetic demands, often including long stretches of nonstop flight to cross geographical or ecological barriers such as the open sea, deserts, or mountains. To face these demands, many birds spend long periods foraging to replenish fat deposits in stopover sites along migratory routes (Alerstam and Hedenstrom 1998). Consequently, it is important to protect stopover areas of migratory birds. Because bird migrations usually cross political borders, often spanning immense distances, the protection of en route stopover sites requires coordination between different countries, which is often difficult to achieve. Migration is thus a critical phase in the life cycle of migratory species (Dingle 1996), and any anthropogenic threat can be expected to have particularly deleterious consequences on migrating animals. Conservation measures should therefore take into account the animal's behavior and physiology during migration.

Conservation biologists have long shown special attention to migratory animals. An example of a concrete conservation measure is the 1983 Convention on Migratory Species (also known as the Bonn Convention). That convention, signed by 70 countries, aims to conserve migratory species *throughout their range*, and so highlights the specificity of migrants as animals moving across national boundaries and thus needing conservation on a global scale (Hykle 2000). It places special emphasis on the fact that migrants are to be considered a resource shared by different countries and that protection measures have to be agreed upon internationally. Under the auspices of the Bonn Convention, many conservation and management research activities have been undertaken in favor of migratory species ranging from birds to bats and from cetaceans to marine turtles.

Obviously, suitable conservation measures for a given species need to rely on the scientific knowledge available for that species, and a close interaction between conservationists and scientific research is needed to support any well-planned conservation effort. Scientific research on migratory behavior has greatly benefited conservation of migratory animals. Ethological research on animal migration has a long history and has provided valuable insights on various aspects of animal migrations, such as their general extent and phenology, or the physiology and energetics of migrants (e.g., Berthold 1993, Alerstam and Hedenstrom 1998). The recent advent of satellite telemetry (French 1994) has provided migration researchers with a new and powerful tool that makes it possible to track animals' movements anywhere on Earth and so reconstruct migratory pathways with considerable accuracy. This possibility is most valuable for the conservation of the many migratory species for which the actual geographical areas visited during migration remain poorly known or entirely mysterious. Lack of specific knowledge about the animals' whereabouts prevents the planning of any conservation measure aimed at protecting their habitats. The findings obtained with satellite telemetry are thus progressively filling a critical gap in the conservation biology of migratory animals.

Marine turtles were one of the first animals to be tracked by satellite (e.g., Stoneburner 1982), and now a large body of satellite tracking data on their migrations is becoming available. These data, however, do not yet provide a complete view of the entire range of movement patterns of sea turtles. Large gaps remain, such as for juveniles or for many populations of adults (see below). Sea turtles are also threatened animals. Some species of sea turtles are near extinction (Limpus 1995, Pritchard 1997), so their conservation has the highest priority. Proper protection of sea turtles must be based on reliable and detailed knowledge of their biology and especially of their movement patterns, which for the most part remain poorly understood. Satellite telemetry now has the potential to elucidate these issues. Indeed, information derived from tracking experiments is often discussed in view of the implications for the conservation of the specific turtle population studied (Morreale et al. 1996, Polovina et al. 2000, Mortimer and Balazs 2000).

Sea Turtles

Marine turtles are truly migratory animals. Some species travel over hundreds or thousands of kilometers while shuttling between nesting and feeding grounds (Carr 1984, Papi and Luschi 1996). Sea turtles travel large distances not only as adults but in nearly all phases of their life cycle, beginning a few hours after hatching.

A generalized life cycle of the seven recognized species of sea turtles (Fig. 4.1) (Carr, Carr, and Meylan 1978; Miller 1997) highlights how turtles move between different, spatially distant, habitats during all stages of their life. Only the flatback turtle (*Natator depressus*), never migrates away from Australian reef areas (Walker and Parmenter 1990).

Females lay eggs in the sand of a tropical or subtropical nesting beach. After 40 to 80 days, the hatchlings emerge from the underground nest and crawl to the sea where they are transported away by sea currents. In this way, they reach their pelagic nursery habitats, which are usually thousands of kilometers away from the nesting area. These nursery habitats are most probably areas in the ocean (such as oceanic frontal systems or convergence areas), where organisms concentrate and therefore food is abundant. Juvenile turtles remain for about 5 to 10 years in their feeding areas. During this long period turtles will not necessarily remain in the same location, as they continue to be transported by large-scale current movements (hatchling developmental migration; Carr 1987). After some time the young turtles, now larger and close to sexual maturity, will leave their pelagic nursery habitat and settle in a

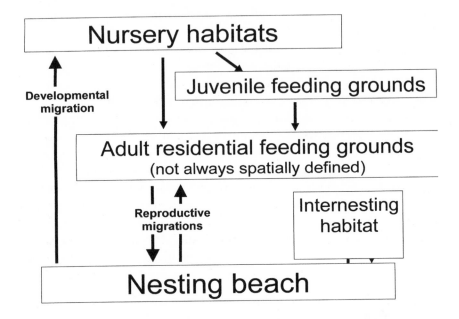

FIG. 4.1. A generalized sea turtle life cycle (modified from Carr, Carr, and Meylan 1978; Miller 1997).

resident feeding ground, which can be either shared with adults or exclusive to juveniles (see Fig. 4.1; Musick and Limpus 1997).

For most species, feeding grounds are specific, spatially defined coastal waters. Leatherback turtles (*Dermochelys coriacea*) and olive ridleys (*Lepidochelys olivacea*), however, have no well-defined feeding areas, and both adults and juveniles simply wander in the open sea without settling in any specific site. At sexual maturity, turtles of both sexes leave the feeding areas and move toward their nesting beach, where females lay eggs several times within a single season, generally every 10 to 14 days. At the end of the reproductive season, both males and females return to the feeding area (see Fig. 4.1). The extent of the reproductive migration can range from a few kilometers (like in many hawksbill turtle [*Eretmochelys imbricata*] populations) to thousands of kilometers, like in green (*Chelonia mydas*) or loggerhead (*Caretta caretta*) turtles. Further short-range movements are made by females between successive egg-layings, when they usually remain in waters close to the nesting beach (internesting habitat). Miller (1997) suggested that some populations also have a specific mating area, but in most cases courtship and mating are thought to occur close to the nesting beaches a few weeks before laying (Godley et al. 2002).

SEA TURTLE CONSERVATION

Many sea turtle populations are critically endangered. Some populations have declined substantially and others have gone extinct (Limpus 1995, Spotila et al. 1996, 2000, Pritchard 1997). Many international conservation organizations have acknowledged this situation. All sea turtle species, excepting the flatback, are regarded as endangered migratory species by the Bonn Convention (listed in Appendix I). They are considered threatened with extinction or endangered by the United States Endangered Species Act and by the Red Book of the World Conservation Union (IUCN). Finally, the Convention on International Trade of Endangered Species (CITES) lists all marine turtles in Appendix I, which prohibits all international trade in turtles or turtle products.

Awareness of the turtles' critical status has led to a number of conservation measures to prevent their decline (Godfrey 1996). These measures have focused on the protection of the nesting areas, which are the only terrestrial habitats used by marine turtles. A series of initiatives have been taken to protect nesting beaches (National Research Council 1990, Lutcavage et al. 1997), including restricted access, especially at night; beach patrols to limit disturbance to nesting turtles (especially natural predation or human harvesting of eggs or adults); and monitoring of egg hatching and hatchling survival, in some cases through the installation of hatcheries where incubation is kept under controlled conditions. All these activities, together with an educational effort to increase awareness of turtle conservation within the local human populations involved, have produced some good results (Hughes 1996; Godley, Broderick, and Hays 2001) and have certainly benefited turtles.

Sea turtles, however, do not live on beaches. Beaches are important for their survival because that is where the delicate and vulnerable nesting activity occurs, but turtles spend the vast majority of their time in the sea. Therefore, any conservation strategy must consider the protection of marine habitats. Beneficial effects of at-sea protection have been recorded, for example for the Ascension Island green turtle population, whose currently favorable status has been attributed to the reduction of harvesting or incidental catch of turtles at their Brazilian feeding areas (Godley, Broderick, and Hays 2001). On a broader perspective, however, little has been done so far, mainly because the scarcity of scientific studies of turtle behavior at sea meant that there was little reliable information about where the different turtle populations stayed after leaving their nesting areas.

Scientists are now filling this gap in our knowledge of turtle behavior. Population genetic studies using molecular markers have joined flipper tagging in providing clues to the geographical origin of turtles found in various locations (Bowen and Karl 1997). Satellite telemetry techniques to monitor marine animals have provided new insights on the actual turtle movements and behavior between the start and end points of migratory journeys.

EXAMPLES OF SEA TURTLE MIGRATORY JOURNEYS

Satellite telemetry (French 1994) has substantially improved our knowledge of turtle behavior at sea, both in the feeding areas and during migration (Papi and Luschi 1996). In most cases, observations were carried out on females, which can be approached while they are nesting, and satellite-linked transmitters can be attached to their large carapace. Most of the available information therefore regards the postnesting migrations of adult females.

The results of satellite telemetry reveal some variability in the migratory strategy both between and within populations of the same species, but some general patterns in the postnesting movements of marine turtles, can be outlined.

1. Female turtles migrate over very large distances. Their travel routes are often straight and directly oriented toward their destination (Balazs 1994, Luschi et al. 1996, 1998). Males most probably do the same (Beavers and Cassano 1996, Balazs and Ellis 2000, Hays et al. 2001).
2. Although turtles nesting in a given area have similar migratory pathways (Fig. 4.2a) (Balazs 1994, Morreale et al. 1996, Luschi et al. 1998), variations exist even within the same species (Fig. 4.2b) (Luschi et al. 1996, Cheng 2000).
3. After leaving the nesting beach, some species (or possibly populations) move directly toward a specific site, reach it quickly, and remain there, presumably until they are ready to breed again. The herbivorous green turtle is the best example of such a pattern (see Fig. 4.2), which is shared with the hawksbill (which usually makes shorter movements, Mortimer and Balazs 2000) and with many loggerhead populations (Papi et al. 1997).
4. Some turtles continue to wander, probably for the entire interreproductive period, generally in the open sea. This is the case of the olive ridley and, especially, of the leatherback turtle (Fig. 4.2c) (Morreale et al. 1996, Hughes et al. 1998)

It is not only adult turtles that migrate. As I suggested earlier, hatchlings and juveniles also perform migratory movements, often over large distances. Satellite data are limited to movements of juveniles or subadults because it is not yet possible to fit radios on hatchlings, which are only a few centimeters long. Despite the technological challenges, a clearer picture of hatchling movements is beginning to emerge. This is especially true of the loggerheads that hatch in the east coast of Florida. The general course of their developmental

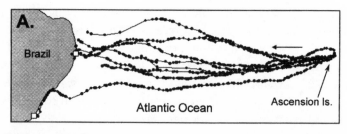

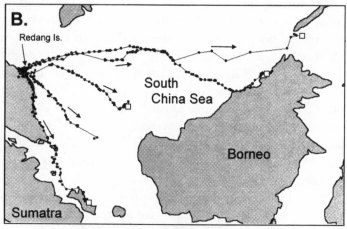

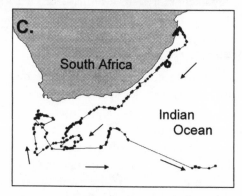

FIG. 4.2. Marine turtle migratory journeys reconstructed by satellite telemetry. (a) Paths of eight green turtles that left nesting beaches at Ascension Island heading for feeding grounds off Brazil (Luschi et al. 1998, Papi et al. 2000). (b) Paths of five Malaysian green turtles dispersing toward different feeding grounds in the South China Sea after nesting on the small Redang Island close to the coast (Luschi et al. 1996). (c) Wandering movements of a leatherback turtle leaving her South African nesting grounds. The journey lasted about 4 months (Hughes et al. 1998). In a and b, white boxes indicate the spatially limited feeding grounds where turtles were localized over a long time.

migration was hypothesized some years ago from hatchling recoveries over many years (Carr 1987), and was recently confirmed by genetic studies (Bolten et al. 1998) and by computer simulations of the hatchling routes (Hays and Marsh 1997). Soon after leaving the Florida shores, these hatchlings enter the Gulf Current, which transports them northeast toward the Azores. From there, other major oceanic currents transport them south toward the Canaries and then westward back onto the Caribbean coasts. It is assumed that the same pattern, including a reliance on oceanic currents to transport the hatchlings and several years of growth in pelagic habitats, also applies to other turtle species and populations (Musick and Limpus 1997). For instance, molecular genetic studies provided evidence that loggerheads born in Japan are transported eastward by oceanic currents to the waters off California, from where they are thought to return to Japan to nest (Bowen et al. 1995).

For older turtles, there are some satellite tracking data available. In most cases, they show slow or wandering movements (e.g., Gitschlag 1996, Polovina et al. 2000, Dellinger and Freitas 2000), with the notable exception of the long-distance movement of a captive loggerhead female of Japanese origin, which was released in Baja California, Mexico, fitted with a radio transmitter, and then tracked while she crossed the entire Pacific Ocean to reach Japan (Nichols et al. 2000). Although this was an adult-sized turtle raised in captivity for years, her journey likely represents a return from her Mexican developmental habitats to her adult feeding grounds in Japanese waters.

SEA TURTLE MIGRATION AND CONSERVATION

These examples demonstrate how research on turtle migration provides a great deal of information about turtles' spatial behavior. The picture emerging from satellite telemetry findings has a number of implications for the conservation of migrating sea turtles.

Turtles move readily across political boundaries; only in rare cases can they be considered the exclusive resource of any one country. On the other hand, their conservation cannot be the responsibility of a single state. The recommendations of the Bonn Convention are therefore particularly appropriate for sea turtles: the conservation of migrating sea turtles must be planned globally, and protective measures have to be agreed on internationally. Our research group already underlined this requirement in 1995 after our first satellite results on Malaysian green turtles (Liew et al. 1995). The IUCN Marine Turtle Specialist Group elaborated in 1995 a global strategy for the conservation and recovery of declining turtle populations. The concept of migrating turtles as a regionally shared natural heritage turned out to be particularly important during a recent legal controversy within the 2000 CITES

Convention of the parties (Richardson 2000). Cuba, supported by Japan, submitted a proposal to reopen the trade in hawksbill shell (downlisting the Caribbean hawksbill from CITES appendix I to II), a proposal that was strongly opposed by many experts in turtle conservation, precisely because hawksbills are migratory animals. Therefore Cuban turtles are not only Cuban, coming from or moving to other Caribbean beaches and waters, as was most clearly demonstrated by a genetic survey of Caribbean hawksbills (Bowen and Karl 1997). These data supported lobbying pressures that led to the rejection of the Cuban proposal, and information on turtle movements substantiated the views of conservationist biologists.

Turtles spend the vast majority of their time at sea, and their movements can be highly variable (see Fig. 4.2b) or totally unpredictable (see Fig. 4.2c). Thus conservation strategies to protect nesting sites are not always effective for migrating turtles and can be used only in specific cases, such as for turtles that migrate to specific, well-defined feeding areas, possibly along similar migratory routes. Satellite telemetry can identify the location of feeding grounds and migratory routes, thus providing the basic information required to substantiate calls for the protection of these important sites. For instance, having identified a migratory corridor in leatherbacks leaving Costa Rican nesting beaches, Morreale et al. (1996) proposed concrete measures to limit fishing activity in that area. More recently, fishing restrictions have been suggested for an oceanic area north of Hawaii that satellite telemetry has shown to be frequented by loggerhead subadults (Polovina et al. 2000). Enforcement of such measures in offshore areas, however, is a challenging task. Finally, satellite telemetry could be used to monitor the fate of turtles caught by fisheries and subsequently released.

The protection of defined geographical areas is not, however, a feasible solution for the protection of turtles that wander over large areas, as is the case of young individuals of most species. Instead, the focus should be less on protection of important sites than on the protection of the turtles themselves. Threats to which turtles are subjected while migrating or staying in the sea must be identified, and their impact must be limited. While at sea, turtles are exposed to many anthropogenic disturbances. The risks of debris ingestion and oil pollution are especially high in the frontal systems that are inhabited by hatchlings for long periods (Eckert 1995, Lutcavage et al. 1997).

The most serious threats, however, are certainly those caused by commercial fisheries (National Research Council 1990, Lutcavage et al. 1997). In particular, shrimp trawling is responsible for about 90% of deaths of adult turtles at sea. A review by American researchers estimated that incidental capture of turtles in shrimp trawls accounted for more deaths than all the other sources of human activities combined. Those researchers estimated that up to 50,000 loggerheads and 5000 ridleys drown every year in shrimp trawls in U.S. waters alone (National Research Council 1990). For the Kemp's ridley (*Lepidochelys kempi*), estimates of fishing-related mortality exceed the estimated number of nesting females in the world. Pelagic fishing gears such as longlines and gill nets used for swordfish and tuna are also harmful, especially for leatherbacks during pelagic movements (Eckert 1995). Fishing devices can be harmful even when abandoned or lost, because turtles may become entangled in them and drown. The combined effect of incidental exploitation on turtles by fishing can be dramatic, as is clearly illustrated by the leatherback populations nesting in Mexico and Costa Rica. A sharp decline was attributed to high mortality away from the nesting beach, mainly due to pelagic fishing activity (Eckert and Sarti 1997, Spotila et al. 2000). Monitoring through satellite telemetry allowed researchers to identify oceanic areas west of South America where leatherbacks moved to and were incidentally captured after the nesting season (Morreale et al. 1996, Eckert and Sarti 1997).

The attention of turtle conservationists has for a long time focused on the effects of fishing activity and on ways to prevent turtle bycatch. Now some remedies are available or under study, the best known of which is the Turtle Excluder Device (TED), a sort of escape hatch placed in a trawler net that allows turtles to escape (National Research Council 1990). The general idea was first conceived in the 1980s and there are now several models available, which are all inexpensive and thought to reduce turtle mortality by up to 97%. Fishermen, however, tend to dislike TEDs, claiming that they reduce the trawl effectiveness in retaining shrimp, although investigations have recorded only negligible losses of shrimp. TEDs are even thought to have positive effects on fishing efficiency by decreasing the mass of bycatch and reducing the vessel's fuel consumption (National Research Council 1990). In 1989, the United States made the use of certified TEDs mandatory for U.S. shrimp trawlers, thus recognizing that TEDs are effective in protecting turtles without impeding shrimp fishing (Crouse 1999). These regulations were then extended to ban shrimp imports to the United States from nations with an indigenous sea turtle population whose fishing vessels do not install TEDs, an action that was taken in view of the migratory nature of sea turtles. Again, international coordination of the countries fishing in the same area is essential to maximize the positive effects of this device.

Conservation of Migratory Animals

Most of the problems faced by migrating sea turtles are common to other migratory species that use habitats in different geographical areas at different times (Dingle 1996). Migratory marine animals like fishes, pinnipeds, or cetaceans move in the same environment as sea turtles, and so are affected by specific problems linked to the marine environment, including pollution, habitat degradation, or fishing activity. But any migratory animal, be it a flying bird or butterfly or a walking mammal, is exposed to many of the threats I have outlined for turtles. The very fact that migrants visit many different areas multiplies the possibility of interaction with harmful human activities, from hunting to destructive agricultural practices. Even apparent harmless actions can have detrimental outcomes: dams to supply water to agriculture or for hydroelectric power, for instance, prevent salmon from performing upstream breeding migrations (Dingle 1996).

Most migrants cross national borders, so their conservation in one country might not guarantee them protection. They are linked (sometimes very faithfully) not only to specific breeding and nonbreeding areas, but also to some intermediate transit areas, such as stopover sites. Habitat degradation could affect any of these sites. For migratory animals, each one of these sites is a potential "weak link" in a chain. Lack of protection in one site overrides protection elsewhere. Migrating birds are often disturbed by humans while fueling at stopover sites, for instance, when they are thought to produce crop losses (e.g., in the case of ducks or geese; Greenwood 1993). The detrimental effect of habitat degradation is best exemplified by the migratory populations of monarch butterflies (*Danaus plexippus*), which are now threatened because of adverse climatic changes of anthropogenic origin in the few Mexican forests where they gather to winter after a 3600 km migration from much of North America. Interestingly, nonmigratory populations of the same species are not threatened (Dingle 1996), which shows how migration is a costly behavior even in evolutionary terms.

Conclusions and Recommendations

This chapter has shown how sea turtle migrations provide many examples of fruitful cooperation between fundamental research on migration and conservation biology. Researchers have provided interesting insights into turtle spatial behavior that have been useful in documenting the interaction between sea turtles and human activities such as fishing. The basic knowledge acquired has greatly benefited turtle conservation and facilitated the planning of appropriate protection measures, which should be maintained and possibly increased. Enlarging the number and extent of protected marine areas, including those far from nesting grounds, and enforcing fishing restrictions in different countries (especially making usage of TED mandatory) are two management actions that are likely to greatly improve the conservation status of sea turtles.

Many aspects of the biology and behavior of migrating turtles are still largely unknown, and scientific research has just begun to investigate them. That research should continue because it will increase our knowledge of fundamental turtle biology and provide valuable contributions for planning effective conservation. Studies on the orientation systems underlying the turtles' migratory performances are expected to be of special importance. Knowledge of orientation mechanisms would improve understanding of the navigational machinery that guides turtles during their migratory journeys, by identifying the environmental cues used. Those cues may help us identify other possible, and perhaps unsuspected, threats to migrating turtles. The navigational mechanisms of turtles remain a subject of speculation, with little or no experimental evidence to support or refute various proposed ideas (Papi and Luschi 1996, Lohmann et al. 1997). The availability of reliable telemetry methods to track turtle movements now makes it possible to test these hypotheses explicitly (Papi et al. 2000), and progress in this field is likely in the near future.

A good example of the practical importance of knowing what stimuli are used by turtles to orient their movements comes from studies of the behavior of hatchlings crawling across the beach immediately after emerging from the underground nest (Godfrey 1996). Hatchlings rely on visual stimuli to orient themselves, crawling toward the brightest horizon of the beach, which in natural conditions indicates where the sea is (Lohmann et al. 1997). These basic research findings have been very useful to evaluate the effects of artificial lighting near turtle nesting beaches. Hatchlings orient in the wrong direction if there are lights at the back of the beach: their natural orientation toward the brightest horizon leads them to move away from the surf, with obvious negative consequences (Lutcavage et al. 1997). The identification of this problem prompted greater care in building new human settlements close to sea turtle nesting areas, together with the proposal of simple modifications of artificial lights to reduce beach lighting or at least its negative impact on hatchling orientation (Witherington and Martin 1996, Lutcavage et al. 1997).

This example shows how basic research findings can be extremely useful in answering to conservation needs. Integration between basic research and conservation is the most powerful tool we have at our disposal, if we are to allow turtles and other animals to continue to migrate across our planet.

Summary

Like many other migratory species, marine turtles visit a variety of different habitats during their long-distance movements and are therefore exposed to threats both along the migratory route and at their destinations. Protecting migrating sea turtles is a challenging task, especially because little is known on turtle behavior at sea. Satellite tracking techniques are progressively filling this gap in our knowledge, and recent findings highlight a number of points of great importance for the conservation of migrating turtles. Turtles usually move across political boundaries and so their protection requires international agreements and global strategies. The turtles' large-scale movements render strategies aimed at preserving specific geographical areas, often successfully used to protect nesting areas, unfeasible or of limited applicability. The most harmful threat to migrating turtles is fishing activity, which, even if it is not targeting turtles, is responsible for the vast majority of turtle deaths at sea. As a tentative countermeasure, actions restricting fishing activity in certain areas or periods have been proposed and sometimes implemented. Specific remedies to limit bycatch captures (especially using modified nets with an escape hatch) are also available, and their use is mandatory in some countries. Also in this case, an international coordination between countries is a key factor to make these solutions most effective. Future conservation efforts should integrate more closely with basic research, which is expected to provide valuable insights on the many poorly known aspects of the biology and behavior of migrating sea turtles.

5.

Bridging the Gap:

Linking Individual Bird Movement and Territory Establishment Rules with Their Patterns of Distribution in Fragmented Forests

André Desrochers

Birds and other terrestrial vertebrates are generally sensitive to habitat changes occurring at the landscape level (Turner, Gardner, and O'Neill 2001). Many papers on landscape management and conservation for birds and other organisms end with a statement on the need to better understand the underlying processes. Yet this message has not elicited much response by ethologists, despite the potential relevance of their work to conservation. Lima and Zollner (1996) illustrated the importance of a "behavioral ecology of landscapes" to provide new ecological insight as well as guidance to landscape managers. In this chapter I examine the potential and realized contribution of ethology toward a theory of habitat fragmentation and, possibly, toward reducing negative effects of habitat fragmentation on wildlife conservation. I define ethology here as the study of animal behavior through direct and detailed observation of individuals. The focus of this essay is forest fragmentation and forest birds, but most issues addressed here should apply to a variety of habitats and organisms.

Forest fragmentation is just one result of the many ways by which humans not only reshape landscapes but also threaten certain wildlife species, regionally or globally. Forest fragmentation is a phenomenon distinct from forest loss; it involves the isolation of habitat patches from one another (Fahrig 1997). In many ways, forest fragmentation is synonymous with forest isolation. Given that birds are among the most vagile terrestrial organisms, they may seem inappropriate models to study in relation to forest fragmentation. There is evidence, however, that woodland animals respond negatively to forest isolation. Most of this evidence takes the form of lower abundance or infrequent occurrence of species in more isolated habitat patches, in species ranging from songbirds (e.g., Opdam, Rijsdijk, and Hustings 1985) to grouse (Åberg, Swenson, and Andrén 2000) and possibly owls (Redpath 1995). More limited evidence points to lower reproductive performance of birds in isolated forests, through unpaired birds (Gibbs and Faaborg 1990; Villard, Martin, and Drummond 1993; van Horn, Gentry, and Faaborg 1995), lower food availability (Zanette, Doyle, and Trémont 2000), increased nest predation or brood parasitism (Robinson et al. 1995), and, possibly, increased fluctuating asymmetry (Lens et al. 1999). Despite growing evidence on landscape use and associated nesting success, no solid theory has emerged to propose a general picture explaining *how* fragmentation leads to the above patterns. It is increasingly clear that the study of habitat fragmentation will not make significant steps forward unless we understand better how wildlife behaves with regard to various aspects of forest fragmentation.

So far, behavioral studies pertaining to the fragmentation issue mostly addressed dispersal and limitations to movements (reviewed in Desrochers et al. 1999). Despite the emphasis on forest fragmentation as a habitat-isolating process, there is not enough evidence to argue that landscape use is only, or even mainly, a result of movement constraints. Therefore, this chapter focuses not only on isolation; it also briefly reviews the main hypotheses that have been proposed to explain landscape use by birds and presents a broader perspective of how behavioral rules of decision may affect landscape use patterns. Specifically, I address the roles of conspecific attraction on reproduction, habitat edges on reproduction and foraging, and movement constraints on dispersal and the search for territories. Although treated separately, these three aspects are not viewed here as exclusive nor independent; in fact it is likely that they do interact, thus posing additional challenges to our understanding. I conclude with some thoughts on ethologists' role in making wildlife conservation not only solid on scientific grounds but also relevant for those who make or break conservation: landscape managers and the public.

Conspecific Attraction

Habitat selection studies of birds generally consider vegetation as the main, if not the sole, factor of importance, even though conspecifics are recognized as an important part of a bird's habitat. The presence of conspecifics is a critical piece of information for birds, yet it is seldom incorporated in empirical or modeling studies by conservation biologists (Reed and Dobson 1993), especially when fragmentation is the subject matter.

The role of conspecifics is highlighted in situations where nesting birds are found in aggregates. During the breeding season, spatial aggregations of individuals are not limited to so-called colonial species. In "typical" songbirds, aggregations may also occur, with no apparent link to patchy resources per se (Stamps 1988). Spatial aggregations apparently originating from neighbors have been noted as early as the 1930s with song sparrows (*Melospiza melodia*), a territorial, socially monogamous songbird (Nice 1937:73). In a study of landscape use by songbirds, Drolet, Desrochers, and Fortin (1999) provided indirect evidence for clusters of territories of songbirds independent of habitat clusters, but these clusters were treated as a statistical nuisance rather than investigated as the result of a potentially important process. Tarof and Ratcliffe (2000) provided a detailed assessment of relationships among individuals in another socially monogamous species found in clusters, the least flycatcher (*Empidonax minimus*).

Why should we expect social aggregations in apparently monogamous songbirds? It is unlikely that such loose aggregations provide much benefit in terms of nestling survival because of the assumed large surface area of the aggregations relative to search patterns by nest predators. However, loose aggregations may provide opportunities for extra-pair copulations as well as provide useful information about the prospects for nesting success (Desrochers and Magrath 1993, Doligez et al. 1999) or possibly factors linked with adult survivorship.

So far, very few field experiments have attempted to single out the effect of conspecific songbirds on territory settlement (see Alatalo, Lundberg, and Bjorklund 1982 for an experiment based on song recordings to entice birds to establish territories). Furthermore, no study has investigated in detail the possible contribution of conspecific attraction in explaining landscape use by forest songbirds. Recent work highlights the influence of landscape structure to extraterritorial movements by hooded warblers (*Wilsonia citrina*) soliciting extra-pair copulations (Norris and Stutchbury 2001). In that study, male hooded warblers did not include in their "copulation neighborhood" females that inhabited woods separated from their own by open areas more than 500 m wide. Given that extra-pair copulations are the rule rather than the exception in songbirds (Petrie and Kempenaers 1998), the hooded warbler story may uncover a widespread phenomenon. But to understand the role of interactions among neighbors and how they are affected by habitat fragmentation, we will have to understand not only the nature of the relationships among neighbors but also the spatial extent of bird neighborhoods in a variety of species and landscapes.

If conspecific attraction is an important cause of avoidance of fragmented forests, then one should expect a positive relationship between species' tendencies to establish clusters of territories and their avoidance of fragmented forests. In such a case, management of landscapes for the accommodation of single territories would tend to overestimate the quality of landscapes for certain species. To prevent this problem, a more realistic approach would attempt to reflect a naturally occurring frequency distribution of territory cluster sizes. Managing for territory clusters would entail the preservation of substantially larger habitat patches, even for species with small territories such as the least flycatcher (Tarof and Ratcliffe 2000).

Responses to Habitat Edges

Depending on the degree of forest cover and associated fragmentation, the amount of forest edges will vary enormously (Fig. 5.1). It is difficult to be far from a forest edge in fragmented forests, compared to contiguous forests, and as a result, we need to address the edge issue when addressing the fragmentation issue. Birds and other terrestrial vertebrates often respond to the amount and proximity of forest edges (reviewed by McCollin 1998, Yahner 1988). For decades, forest edges have been considered as positive, or even essential, landscape elements, particularly for game species (Leopold 1933). Seminal papers by Gates and Gysel and Wilcove (1978, 1985, respectively) initiated a new and darker vision of forest edges as ecological traps where naive nesting songbirds suffered high nest depredation levels.

As a result of Wilcove's work and that of many others (including pioneers Gates and Gysel 1978), ecological relationships occurring near forest edges are the most frequently assumed cause of avian responses to forest fragmentation. Response to edge, however, is often confounded with response to forest fragmentation, even though these two processes are very different. According to the edge hypothesis, forest birds will respond to fragmentation because it leads to an increase in exposure to forest edges. Birds may be attracted to (if nest predators) or avoid (if prey) fragmented landscapes where edges tend to be abundant.

Whether forest edges act as ecological traps to which birds respond remains unsure, however, because the vast majority of studies that documented

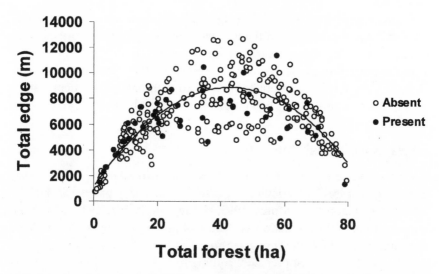

FIG. 5.1. Gray jays are not associated with landscapes dominated by forest edges. Each point represents one 1 km² landscape surrounding a bird point count station (see Drolet, Desrochers, and Fortin 1999 for details). The solid line represents a quadratic curve fit. Points above the line represent point counts made in landscapes with high amount of edge. Open circles: jays present; solid circles: jays absent. Data from J. Ibarzabal and A. Desrochers (unpublished ms).

such effects used artificial nests and therefore had no information relevant to birds' responses to risks associated with forest edges. Even in studies based on real nests, no formal effort has been made to determine whether birds respond to the various threats posed by the proximity of forest edges. We will need input from ethologists and evolutionary ecologists to determine the relative importance of lags in adaptive response and evolutionary trade-offs in explaining how and why forest birds should continue nesting near risky forest edges. Meanwhile, researchers have to contend with broad (and mostly unverified) classifications of species as "edge-associated," "edge-independent," or "forest-interior," as found in Whitcomb et al (1981), all of which are based solely on the pattern of establishment of *territories* relative to forest edges.

These classifications represent almost all we know about bird responses to edges, yet these classifications have to be treated cautiously (Villard 1998) for they are confounded by the diversity of field and analytical methods as well as possible variation in bird responses to edges among localities. Furthermore, these classifications only explore territory settlement patterns, even though responses to edges by songbirds may occur at other scales, such as nest site choice, as illustrated by Kuitunen and Mäkinen (1993).

Rather than seeking unrealistic generalities such as species-wide classifications as "edge-specialists" and the like, we may gain a better theory of avian responses to edges by comparing edge responses among species sharing a similar habitat. But first we must determine whether birds respond adaptively to forest edges and problems associated with them—nest depredation or parasitism, poor foraging habitats, and movement barriers.

Edges as Risky Habitats

According to a large number of studies based on artificial nests, it is clear that forest edges, at least in agricultural landscapes, are risky places in which to place a nest (Paton 1994, Murcia 1995, Hartley and Hunter 1998). Given the importance of nest depredation as a cause of nesting failure (Ricklefs 1969), there has been much attention given to edges as risky nesting habitats. It is usually inferred that nest predators coming from habitats adjacent to forests are the prevailing cause for increases in nest depredation near forest edges (DeGraaf 1995). Nest predators inhabiting forests, however, may also play an important role (Andrén 1995, Hannon and Cotterill 1998). Although the victim side of nest depredation near forest edges is well documented, the nest predator's side of the process is virtually unknown, as is the response of birds to nest depredation risk near forest edges.

Nest depredation is not the only threat that birds nesting near edges face. Brood parasites such as the brown-headed cowbird (*Molothrus ater*) can travel into forests from their foraging habitats (fields and other nonforested areas) to lay their eggs in nests of forest songbirds. Again, the fact that cowbirds generally live outside the forest means that birds nesting near forest edges will be more exposed to their undesirable effects, as demonstrated in numerous studies, including the detailed study by Donovan et al. (1997). As with nest depredation, our knowledge of forest songbirds' responses to brood parasitism is incomplete at best. Forest songbirds show a variety of responses to cowbird parasitism (Rothstein 1990), but responses documented to date generally refer to egg, nestling, or adult recognition, rather than the avoidance of particular nest sites, such as those near forest edges. Hobson and Villard's work (1998) is an interesting exception; they demonstrate that the behavior of adult hosts changes in relation to levels of fragmentation and parasitism risks.

Whether birds respond or not to risks associated with forest edges should affect the way we manage these habitats. Therefore, experimental and observational studies should be designed to (1) determine whether birds prefer to place their nest away from edges (relative to the location of their territory) and (2) determine whether that tendency is modulated by how the risk of either nest depredation or brood parasitism increases near edges.

Edges and Foraging

Interspecific relationships such as those described in the preceding section are not the only reason why birds may avoid forest edges and associated fragmented forests. Food availability has been shown to decrease near forest edges (Jokimäki et al. 1998), and patch size (Burke and Nol 1998; Zanette, Doyle, and Trémont 2000) for certain nesting birds may account for edge avoidance by species such as ovenbirds (*Seiurus aurocapillus*). However, food may be generally more scarce at edges only for certain bird species, depending on their foraging requirements. Furthermore, foraging success has never been shown to decrease near forest edges. The lack of examples showing a decline of foraging success near edges is probably due to the limited research effort to date. However, it is possible that more research will not document such a decline if birds know about and therefore avoid poor foraging areas, such as edges. Again, only experimental, behavioral work can provide strong inference about forest edges as poor foraging habitat.

Can birds determine the potential value of forest edges as foraging habitats? Food hoarding provides interesting opportunities to answer this question. Recently, Brotons et al. conducted an experiment on food hoarding by wintering black-capped chickadees (*Poecile atricapilla*) (Brotons, Desrochers, and Turcotte 2001), with feeders placed near forest edges or in the forest interior. Birds taking food from feeders near edges tended to hoard their seeds further into woodland than birds taking their seeds from the forest-interior feeder. This pattern was especially obvious near edges more exposed to strong winds (Brotons, Desrochers, and Turcotte 2001). This result seems contradictory with Desrochers and Fortin's (2000) finding that wintering chickadees tend to forage near forest edges. However, the latter study was done over 2000 km from Brotons's study area, and was done during clement weather (little or no wind), conditions during which chickadees may not rely as much on food caches. It may thus be that forest edges are suboptimal chickadee foraging habitats only when cached seeds are most needed, such as windy days in late winter (see also Dolby and Grubb 1999).

To conclude with forest edges, it must be noted that associations between forest edges and bird occurrence will come from qualitatively different processes, depending on spatial scale, because of the wide array of spatial scales associated with different bird activities such as nest site choice, foraging, and territory establishment. For example, birds seeking territories may respond to the amount of forest edge within 1 km^2, while paying little attention to local edges (say, within 100 m) while foraging. The converse can also be true, as we found with family groups of gray jays (*Perisoreus canadensis*). Even though gray jays clearly respond to edges at close range when foraging (Fig.

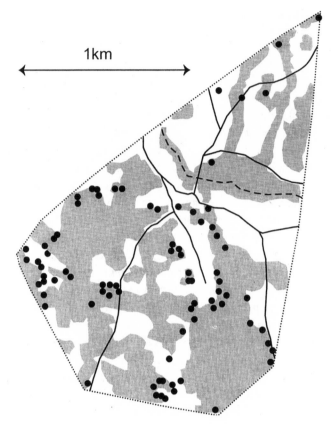

FIG. 5.2. Home range use by a family group of gray jays. Shaded areas represent mature woodland inside the convex polygon delimiting the home range. Data from J. Ibarzabal and A. Desrochers (unpublished ms).

5.2), their response to edges is not observable over whole landscapes of sizes comparable to their home ranges (see Fig. 5.1). This point has major implications, given the abundance of studies adopting a landscape approach, useful for territory establishment patterns but with no fine-grained information on bird movements. Surprising as it may seem, no study has yet attempted to document associations between response to edges at close range versus in the landscape as a whole for given species. It would be useful to test whether species with large home ranges (> 100 ha) are more sensitive to amounts of edge over whole landscapes, and whether species with small home ranges (typically < 10 ha) respond mainly to edges occurring within a few tens of meters. More evidence of the contrast between edge responses at different spatial scales for the same species would force a clearer definition of the term *edge avoidance* and remove some of the confusion now prevailing in the literature since the classic work by Whitcomb et al. (1981) was published.

Scaling edge responses would also help provide more specific guidance for forest management, given that abundant edges at one spatial scale do not imply abundant edges at all scales. For example, forests can be managed for large homogeneous patches surrounded by edges convoluted at a small scale (fine-textured edges), thus benefiting species associated with edges at close range, and species avoiding edges at the landscape scale. Conversely, managers could promote a complementary set of bird species by producing a large number of small stands with linear edges, such as in forests harvested in "checkerboard" patterns (small, square clear-cuts).

Constraints to Movements

The hypotheses addressed in the preceding discussion portray landscape use mainly as the result of choices made by birds that had the opportunity to compare many landscape components. However, landscape use may simply stem from birds' reluctance to venture outside forests when dispersing outside the breeding season, thus missing the opportunity to assess, let alone colonize, isolated forest fragments. If reluctance to enter gaps between forests is important, then forest edges will become barriers. The behavior of birds encountering gaps should therefore contain useful information about movement constraints as a potential process leading to avoidance of fragmented forests. The isolation of a forest fragment may be more important than its size in determining whether it will be used by a given species of bird.

How should we approach the study of constraints to movements across landscapes? The rapid development of movement ecology as a framework to analyze responses of animals to landscape configuration has revived interest in the study of edges, particularly sharp ones (Wiens 1995, Turchin 1998). Modeling tools such as cellular automata (Turchin 1998), and empirical tools such as fractal analysis (Wiens et al. 1995, Desrochers and Fortin 2000) may hold the key to a better understanding of how birds respond to forest edges, and provide an important piece in the puzzle of habitat fragmentation. A major remaining obstacle is the lack of knowledge of the direction (aim) of travel of individual birds through landscapes, as well as the strength of the motivation (Desrochers et al. 1999).

Despite the fact that bird territories or home ranges are sometimes found throughout forested parts of a landscape, irrespective of edges, birds may respond to edges either by "bouncing" on them, following them, or simply passing through them. Forest edges do sometimes act as filters or movement conduits, as has been shown in invertebrates (Wiens, Schooley,

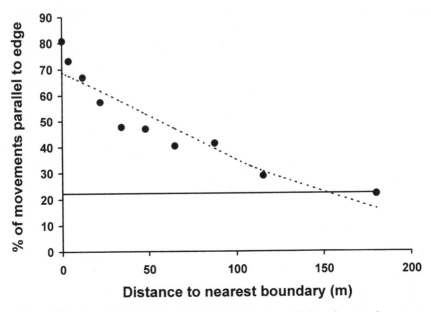

FIG. 5.3. Flocks of black-capped chickadees move parallel to forest edges upon their encounter. Reproduced from Desrochers and Fortin (2000), with permission from Oikos.

and Weeks 1997; Haddad 1999b), mammals (Kozakiewicz 1993), and birds (Machtans, Villard, and Hannon 1996; Desrochers and Fortin 2000). If gaps intervening between forest patches are effective barriers, movements of birds will change abruptly near edges, with consequences in terms of territory settlement as well as dispersal patterns. For example, even though they can be found anywhere in forests, flocks of black-capped chickadees are found disproportionately near edges because when they encounter an edge, flocks tend to move parallel to it rather than reenter the forest or cross a gap (Fig. 5.3). During the dispersal period, when presented with a choice between taking a detour around an open area (field or recent clear-cut), or flying a short distance through the open area, chickadees and other songbirds will often take detours several times longer than the shortcut in the open area, presumably to avoid risks associated with flying into open areas (Desrochers and Hannon 1997). Species-specific responses to forest edges seem to remain fairly stable among seasons, at least for nonmigratory species (Bélisle and Desrochers 2002).

With clear evidence that forest birds avoid traveling into the open, even through distances less than 100 m, one has to ask whether those fine-grained decisions translate into processes operating at the landscape scale. After

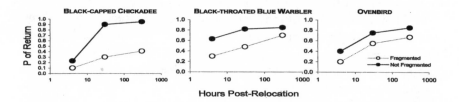

Hours Post-Relocation

FIG. 5.4. Homing time and success of territorial males of three forest songbird species relocated over distances of 1 to 4 km. Sample sizes of 41, 72, and 92 from left to right. Solid lines: birds relocated across contiguous forests (> 70% forest cover); dashed lines: birds relocated across landscapes with 10 to 30% forest cover.

relocating ~200 territorial forest songbirds over distances of 1 to 4 km, Bélisle, Desrochers, and Fortin (2001) discovered that birds homed back less often, and after longer delays, when forced to traverse fragmented forests, relative to birds homing through contiguous forest (Fig. 5.4). Although several interpretations of this result are possible, they unquestionably show that landscape composition influences bird movements over landscapes. Given the investment made by males to establish a territory (and, presumably, the associated drive to return home), the process behind those results is likely a constraint on movements, such as a cumulative cost of encountering small movement barriers repeatedly.

Predictions of the movement constraint hypothesis would undoubtedly be more specific if empirical work such as already described was incorporated as measured (rather than guessed) parameters in modeling efforts such as those using cellular automata in real landscapes portrayed on geographical information systems (GIS). For example, real forest patch use could be compared to patch use by cellular automata moving under gap-crossing rules such as those measured by Desrochers and Hannon (1997). If constraints to movements are important enough, then birds are expected to settle for less when seeking territories, which means that isolated forest patches will be underutilized. Rarer opportunities to find conspecifics would be another negative effect of movement barriers created by gaps separating habitats, especially for less abundant species.

Conclusion and Recommendations

One of the symposia of the 1998 International Ornithological Congress assessed progress in bridging the gap between pattern and process studies of the consequences of habitat fragmentation (Desrochers et al. 1999). The emphasis was on how the study of movement could help understand the study of bird occupancy patterns in the landscape. Since then there has been some progress, both empirical and theoretical, but the foregoing discussion makes it clear that wide gaps remain to be bridged between extensive studies seeking patterns, and intensive, behavior-based studies demonstrating (or rejecting) proposed processes.

More behavioral work will undoubtedly show that fragmentation affects not only the mobility of birds but also aspects such as foraging and reproductive strategy, as well as interactions between these factors. For example, because the dispersal of birds from one nesting season to the next is more prominent following nesting failure (Greenwood and Harvey 1982), it is likely that lower nesting success in edge-rich fragmented habitats will only make the impacts of movement constraints greater. We may never understand well the interactions among processes such as nest predation and dispersal, not to mention processes unaddressed here, such as interspecific competition. But perhaps it would be wise to better understand single processes before tackling interactions among processes.

This chapter illustrated some challenges that ethologists could address to bring some order to the current chaos created by ever-increasing amounts of disparate information on landscape use by animals, especially birds. Without unifying hypotheses based on behavior, the mass of information accumulated by landscape ecologists will amount to little more than hearsay that paves the way to useless, unfalsifiable claims such as "birds respond to fragmentation in variable ways in various places." Worse still, the lack of theory may eventually discredit our efforts toward an understanding of fragmentation effects (and possibly our whole discipline) as an aimless accumulation of information.

How do we work toward unifying hypotheses? One possible answer is the comparative approach, to use rather than dismiss, the variety of responses to habitat fragmentation not only across, but also within, species. Comparative analyses may reveal that species less willing to traverse open areas (to be determined by ethologists) are indeed those that are most sensitive to fragmentation (already determined by landscape ecologists). Also, comparing responses to fragmentation between migrant and resident species may prove fruitful, given their contrasted use of landscapes, at least at certain times of the year. Failing to demonstrate this relationship would be a severe blow to the hypothesis that movement constraints is the key problem. Such a rejection (assuming proper statistical power) would constitute progress in the classic sense of what science is (Quinn and Dunham 1983).

Whatever insights are gained from an effective use of ethology, however, will not necessarily show us the way to effective wildlife conservation. A greater challenge awaits ethologists as to how much decision makers, not to mention other ecologists, will incorporate recent findings into their agenda. Increasing the public's interest in the problem of forest fragmentation is another major task. Currently, studies of animal behavior sway the opinions of decision makers and the public not only through the solid knowledge they provide, but equally or perhaps more because of the sense of marvel toward wildlife they inspire. Whether this is satisfactory or not is a matter of opinion. However, if ethologists really believe they can contribute to alleviate the effects of forest fragmentation, hypothesized or real, I argue that they will have to exercise their skills on both fronts, aesthetic and scientific. This will be achieved by addressing their work not only to peer-reviewed journals and land managers but also to popular magazines, elementary schools, and the like. These undertakings should not be seen as competing with one another. Indeed, good science often naturally leads to results relevant and appealing to a wide audience (which, ultimately, decides on research funding), and thus aesthetics and science generally have a mutualistic, rather than antagonistic, relationship.

Getting people's attention is a tall order in these busy times, especially given that ethologists often seem reluctant to have their research agendas influenced by public or political opinion. Nevertheless, it is to be hoped that at least some ethologists will choose to embark on the journey to a theory of habitat fragmentation, before their own study areas become the stage for more habitat fragmentation and regional species decline.

Summary

The sensitivity of animals to habitat loss and fragmentation is of major conservation concern. Despite their great ability to move, birds appear sensitive to landscape fragmentation. Hypotheses proposed to explain well-documented distribution patterns of nesting birds in landscapes focus either on choices made by birds seeking territories or on behavioral constraints. Birds may choose particular landscape structures not because of movement constraints through landscapes but because of attributes of habitat patches that are correlated with landscape structure. For example, opportunities for extra-pair fertilizations and information exchange among conspecifics may be too limited in fragmented habitat patches to make those patches attractive even for noncolonial species. Additionally, birds may respond to food abundance, nest depredation, or brood parasitism risk near forest edges in fragmented landscapes. Alternatively, birds may be constrained in their use of landscapes because of their reluctance to cross gaps intervening between habitat patches, for gaps may present increased energy costs or depredation risk during dispersal and breeding. Although landscape distribution patterns and their hypothesized causes have been well studied in birds, there has been almost no attempt to link these two kinds of evidence formally. I review recent work that addresses fragmentation issues from the standpoint of bird behavioral decisions with respect to dispersal movements, foraging, territory settlement, and reproduction. I propose that comparative analyses should be used to help provide more incisive predictions linking differences in behavior among species with associated differences in their use of landscapes. I conclude by stressing that solid science may not only provide aid to decision making, it may also provide material appealing to the public at large who, ultimately, decide the future of both behavioral research and land management.

Acknowledgments

This chapter would not have been possible without the inspired and effective work of my graduate students Marc Bélisle, Bruno Drolet, and Jacques Ibarzabal, and postdoc Lluis Brotons, which led to most of the results summarized here. Discussions with Julie Bourque, Marcel Darveau, Susan Hannon, Louis Imbeau, Marc Mazerolle, and Marc-André Villard also played a major role in shaping the ideas that the reader will have found interesting here. I accept the credit for the other ideas. The research behind this essay was funded by a variety of sources, including the Natural Sciences and Engineering Research Council (research grant), and the Fonds pour la Formation de Chercheurs et l'Aide à la Recherche (research team and new researchers programs, and through the Centre de recherche en biologie forestiáre).

6.

Knowledge of Reproductive Behavior Contributes to Conservation Programs

Isabelle M. Côté

Behavioral studies have contributed relatively little to the multidisciplinary field of conservation biology. The relative isolation of behavior from conservation biology is odd, given the contributions that other branches of the natural sciences, such as population ecology, population genetics, and systematics, have made to conservation (Caro 1999), and the many recent attempts to highlight the potential importance of animal behavior for conservation (Ulfstrand 1996, Clemmons and Buchholz 1997, Caro 1998b, Sutherland 1998, Gosling and Sutherland 2000).

To understand how behavior might affect conservation, it is necessary to define clearly the goal of conservation. Textbooks commonly cite a twofold goal to conservation: (1) to assess the effects of humans on species, communities, and ecosystems; and (2) to mitigate these impacts to prevent species extinction and ultimately reintegrate affected species into functioning ecosystems (e.g., Primack 1998). A common human impact on species is a reduction in population size (or in effective population size). Measures to counter human impacts strive to prevent further declines or, ideally, increase population sizes. Thus, to show the importance of a behavioral approach to conservation it is crucial to establish a link between individual behavior and population size. There is now a large body of literature that attests to the interaction between individual behavior and population processes (Goss-Custard et al. 1995a, Sutherland 1996, Anholt 1997, Levin et al. 2000).

In this chapter, I focus on a behavior that can influence population size: reproductive behavior. A wealth of data on reproductive behavior, with a well-developed theoretical framework for understanding and predicting the reproductive characteristics of animals, has accumulated over the last few decades. It is difficult to imagine that this information cannot make useful and novel contributions to conservation biology. Reproductive behavior can be linked to changes in population size, and hence to conservation, in two ways. It can be used as a predictor, where inherent differences in behavior lead to different population responses to conservation measures. Alternatively, reproductive behavior may react to conservation actions, with different reproductive responses leading to differences in population sizes. Although insights into behavior can lead, in both cases, to understanding the population consequences of conservation strategies, I believe that the study of animal behavior can make its most significant contribution as a tool for predicting the effectiveness of conservation measures on population sizes. But is this how reproductive behavior is used in conservation? Does ignoring reproductive behavior have an impact on the accuracy of our predictions of the potential effects of conservation actions?

I briefly review the potential role and current use of reproductive behavior as a predictive tool in four areas of conservation biology; namely, predicting (1) the effects of habitat loss, (2) the effects of exploitation, (3) the risk of extinction, and (4) the success of captive breeding, translocations, and reintroductions. I then present two case studies of simulation models to assess how reproductive behavior may affect estimates of population size. Finally, I consider the relative benefits of investing in studies of behavior versus investing in alternative activities to guide conservation action.

Predicting the Effects of Habitat Loss

The problem of predicting the effects of habitat loss and its frequent precursor, habitat fragmentation, on population sizes has been addressed on two scales. At the macrogeographic scale, the metapopulation approach has primarily used behavioral information on ranging and dispersal (Hanski and Gilpin 1997). At the microgeographic scale, behavior-based models have been highly successful in generating testable predictions of the impact of habitat loss on population size (Sutherland and Dolman 1994; Goss-Custard et al. 1995a; Yates, Goss-Custard, and Rispin 1996). These models have relied so far almost exclusively on detailed studies of foraging behavior. Students of reproductive behavior lag far behind in using their findings to predict the impacts of habitat loss or alteration.

There are, nonetheless, a few remarkable studies of the importance of reproductive behavior in predicting the potential impacts of habitat loss. For example, Smith and colleagues (Smith, Reynolds, and Sutherland 2000; Smith et al. 2000) studied the reproductive behavior of the bitterling (*Rhodeus sericeus*), a European cyprinid fish that lays eggs in the gills of live freshwater mussels and that is threatened over much of its range. Female bitterling avoid mussels that already contain many bitterling eggs (Smith et al. 2000), a behavior that appears adaptive because of high density-dependent mortality of fish embryos in mussels. Using a population model incorporating reproductive behavior and demographic parameters, Smith, Reynolds, and Sutherland (2000) showed that loss of nursery habitat caused by cutting vegetation on riverbanks, a common practice in Europe, would yield a 48% decrease in bitterling population size. However, the accuracy of this prediction, which was validated in the field, depended on the inclusion of reproductive behavior in the model. When avoidance of parasitized mussels by bitterling was excluded from the model, population size was overestimated by 6%. This may seem like a rather small percentage, but it is one-quarter as large as the effect of ignoring predation by perch (*Perca flavescens*) in the model (Smith, Reynolds, and Sutherland 2000). Models of this kind are widely applicable to species, such as many salmonids, for which nesting site quality can be accurately quantified.

Similarly, detailed observations of breeding territory preferences of individual ringed plover (*Charadrius hiaticula*) nesting on beaches in Norfolk, England, revealed that birds settled preferentially on the widest and most vegetated stretches of beach (Liley 2000). This preference appeared related to the lower levels of predation and disturbance associated with wide beaches. By combining a knowledge of breeding territory preferences to potential changes in beach structure resulting from sea-level rise, a serious concern in East Anglia (Boorman, Goss-Custard, and McGrorty 1989), Liley (2000) modeled the impact of various climate change scenarios on plover breeding population size. His model can also be used to manage human disturbance to mitigate the impacts of change in sea level.

More often, however, the links between individual reproductive behavior, habitat quality, and population dynamics are examined without attempting to predict the impact of changes in habitat quality or quantity. For example, when examining the optimal territory selection strategy of individuals recruiting into a breeding population, Boulinier and Danchin (1997) concluded that the best option is to sample several breeding patches before recruiting, even if this results in a missed breeding opportunity, only if the environment is predictable and contains a low proportion of good patches. Such a model could easily be extended to consider the effects of reducing the availability of suitable breeding habitat.

Predicting the Effects of Exploitation

Exploitation relies on the assumption that there is a yearly surplus of animals that can be culled without causing a long-term decline in population size. However, the removal of this supposed surplus is not usually random, and more often than not, it is biased toward larger and older individuals. Sex-biased culling may distort the sex ratio of surviving animals. Furthermore, the combination of gregarious social systems and specific mate preferences may exacerbate the effects of exploitation. For example, poachers in the past have selected mainly large male elephants (*Loxodonta africana*), which yield the largest tusks. This has resulted in highly skewed sex ratios (up to 74 females per mature male in some areas; Dobson and Poole 1998) which, combined with female gregariousness and preference for larger males (Poole 1989), may now limit conception rates (Dobson and Poole 1998). In contrast, in animals with indeterminate growth, such as fishes, selective harvest of the largest individuals often results in female-biased catches (Coleman, Koenig, and Collins 1996). Because egg number increases allometrically with female body length (Bagenal 1978), fishing can have a disproportionate impact on population productivity by removing the most fertile females. Moreover, the increased mortality generated by exploitation can select within a few generations for earlier ages or sizes at maturity (Rochet 1998, Law 2001). Whether these shifts in life histories are genetic or the result of phenotypic plasticity, they result in concomitant reductions in female fecundity with potential impacts for populations.

Reproductive behaviors can have a massive impact on population responses to harvesting. Some fish migrate to traditional spawning sites, which are highly predictable in space and time (Sadovy 1994). Fish with breeding aggregations are more vulnerable to exploitation than those with less discrete reproductive outbursts. Thus the Nassau grouper (*Epinephelus striatus*, family Serranidae), which once formed spawning aggregations of more than 100,000 individuals throughout the Caribbean (Smith 1972), has seen more than one-third of its aggregations disappear due to fishing and is now a candidate for the U.S. endangered species list (Sadovy and Eklund 1999). Similarly, northern cod (*Gadus morhua*, family Gadidae) off Newfoundland and Labrador were so efficiently harvested by trawlers fishing down spawning aggregations that this stock is no longer commercially exploitable (Hutchings 1996). The effects of exploitation may also vary depending on patterns of parental care, with species exhibiting care being more vulnerable to fishing than species without parental care (Bruton 1995). If parental care behaviors make parents more susceptible to being captured, the impacts of exploitation will be far greater than catch numbers would suggest because a whole brood fails for each parent removed.

There have been a few attempts to predict the responses of species to exploitation, mainly on the basis of general life-history characteristics (Reynolds et al. 2001). For example, Jennings, Reynolds, and Mills (1998) found that North Atlantic fish stocks that have decreased in abundance in the past century mature at an older age, attain a larger maximum size, and exhibit significantly lower potential rates of population increase than their closest nondeclining relatives. More specific aspects of reproductive behavior have also been linked to specific population responses to harvesting. Greene et al. (1998) showed that mammalian breeding systems could affect how populations responded to hunting. Monogamous and weakly polygynous species are much more susceptible to culling of males than species where a single dominant male typically mates with many females. Other reproductive characteristics, such as infanticide by newly dominant males and reproductive suppression by dominant females, also contribute to reduce a population's ability to withstand exploitation. Studies of animal behavior can thus generate useful rules of thumb to assess the impacts of exploitation.

Finally, data collected on the reproductive behavior of common species for nonapplied purposes may be useful for the management of other species. The breeding success of pied flycatcher *Ficedula hypoleuca*, for example, is highly correlated with the autumn population levels of woodland grouse species, and thus could be used to set bag limits for these game birds the following autumn (Thingstad 1999). It is much easier to estimate breeding parameters for pied flycatcher than population size for grouse.

Predicting the Risk of Extinction

Through exploitation or habitat loss, populations may be reduced to numbers where positive density dependence prevails, a phenomenon known as the Allee effect (Stephens, Sutherland, and Freckleton 1999). If population density decreases further, it may decline past a threshold where the only outcome is extinction. Alternatively, stochasticity in birth and mortality can produce an increased risk of extinction for small populations (Stephens and Sutherland 1999). Can knowledge of reproductive behavior help us predict the susceptibility of species to Allee effects or to stochastic risk of extinction?

Allee effects that arise from impaired social interactions are more likely to be severe when conspecific attraction, for example to breeding grounds, is strong. Colonially or cooperatively breeding birds are more likely to suffer from Allee effects than solitary nesters (Stephens and Sutherland 1999). Similarly, the fertilization success of marine invertebrates that form spawning aggregations to release their gametes in the sea decreases drastically at low density (Levitan, Sewell, and Chia 1992; Claereboudt 1999).

Mating systems appear to have general and consistent effects on the probability of population extinction. Dobson and Lyles (1989) showed that primate social system influenced threshold population densities below which reproduction would fail. Promiscuous primates, for example, may survive at smaller population densities than solitary or monogamous species. In birds, a similar pattern holds. Legendre et al. (1999) found that the probability of extinction was higher for strictly monogamous birds than for polygynous ones, when population size was small. Modest levels of female choosiness can also lead populations to extinction more quickly than random mating because if preferred males are unavailable, female choice effectively removes from the population females that could otherwise have bred (Legendre et al. 1999). Unfortunately, despite these tantalizing taxon-specific studies, there is still no cross-taxonomic, generalized framework for predicting the likelihood of extinction from reproductive behavior.

Predicting the Success of Captive Breeding, Translocations, and Reintroductions

Behavioral studies are clearly important for ex situ conservation by providing information about the most favorable physical, social, and genetic environments for the captive breeding of endangered animals. Similarly, a knowledge of ecological requirements and of basic behavior is required for the successful translocation or reintroduction of individuals into new habitats. But can the likelihood of success of captive breeding, translocations, and reintroductions be predicted from existing knowledge of reproductive behavior?

Some of the problems of captive breeding can be predicted through an understanding of sexual selection. Lack of mate choice imposed on captive females to promote genetic diversity, through exposure to a single mate or artificial insemination, may actually result in an impoverished gene pool. When given a choice among potential mates, a female should choose that which is healthiest, of highest quality, and most compatible with her own genotype (Andersson 1994). Lack of mate choice can result in low offspring survival (Møller and Alatalo 1999) and the spread of undesirable traits within the captive population. In hatchery-reared fish, for example, the random mixing of male and female gametes, stripped manually from adults, has relaxed sexual selection and resulted in a drastic reduction in male traits associated with sexual competition (Fleming, Jonsson, and Gross 1994).

Sexual selection theory is also helpful for predicting the reproductive characteristics that predispose to successful translocations, reintroductions, or even introductions of exotic species. McLain, Boulton, and Redfearn (1995) found that sexually dichromatic birds introduced to tropical oceanic islands were more likely to go extinct than monochromatic species. Sorci, Møller, and Clobert (1998) found a similar pattern for birds introduced to New Zealand. These results are expected since sexually dichromatic species are generally under sexual selection, and intense male brightness relative to female coloration is correlated with reduced male survival, suggesting a cost of male–male competition (Promislow, Montgomerie, and Martin 1992; Owens and Bennett 1994). Moreover, demographic stochasticity is more likely to bring to extinction small populations in which females are choosy than those where mating is random (Legendre et al. 1999). It thus follows that species under intense sexual selection will require larger effective population sizes for successful introductions, translocations, or reintroductions than those under weaker sexual selection.

What Happens When Reproductive Behavior Is Ignored?

This section uses examples from two fishes to show how alterations of the physical and demographic environments can affect the risk of population extinction. Both cases show how knowledge of reproductive behavior is essential for the conservation of these species.

PREDICTING THE EFFECTS OF HABITAT LOSS ON RIVER BLENNIES

The conservation problems faced by river blennies (*Salaria fluviatilis,* family Blenniidae) are typical of northern Mediterranean fishes that live in small, localized populations. These problems include pollution, the introduction of exotic fish, and habitat loss due to physical alterations of watercourses (Maitland 1995). One relatively common form of waterway alteration, at least on the Iberian Peninsula, is the removal of stones and gravel for the building industry.

Male river blennies establish nests under stones and attract females for spawning. Females deposit a layer of eggs on the underside of the stone, and the male then guards the eggs against predators until they hatch. River blennies from four rivers in three separate drainages showed consistent breeding habitat preferences, with males selecting the largest available stones as nest sites (Côté et al. 1999). We found larger clutches under larger nest

TABLE 6.1. Predictions under different scenarios of river blenny reproductive output before and after a reduction in river stone size from 200 cm^2 to 50 cm^2 caused by extraction.

SCENARIO	NEST DENSITY (nests m^{-2})			EGG PRODUCTIVITY (mm^2 of eggs m^{-2})		
	Before	After	Δ_D	Before	After	Δ_E
A "Reality"	0.15	0.08	−47%	85	342	−75%
B No avoidance of small stones by males	0.16	0.63	+ 294%	368	703	+ 91%
C No preferential spawning by females under large stones	0.15	0.08	−47%	511	266	−48%
B and C	0.16	0.63	+ 294%	549	2195	+ 300%

Δ_D: % change in nest density after stone extraction; Δ_E: % change in egg productivity after stone extraction. "Reality" is a scenario that incorporates knowledge of the reproductive behavior of river blennies.

stones. Combining this knowledge of reproductive ecology with a quantification of stone size distribution at exploited and undisturbed sites, we produced a simulation model that allowed a prediction of the impact of stone removal on blenny nest density and population egg production (Fig. 6.1a and 6.2a). Thus a reduction in mean stone size from 200 cm^2 to 50 cm^2, as observed at our Pyrenean study site, should result in a 47% decrease in nest density and a 75% decrease in egg productivity (Table 6.1; Côté et al. 1999).

Rerunning the simulation model without the constraints and assumptions set by a knowledge of reproductive behavior yields very different predictions. Wild males were never observed nesting under stones of less than 96 cm^2. If male preference for larger stones is ignored and no lower limit of nest stone size suitability is assumed, the relationship between nest density and stone size assumes a decaying exponential shape (Fig. 6.1b). One would then predict higher, rather than lower, nest densities when extraction reduces mean stone size in a river (Table 6.1). Similarly, egg productivity would be predicted to increase substantially (Fig. 6.2b, Table 6.1). If the propensity of females to lay larger clutches under larger stones is removed from the model and an average number of eggs is assigned to all nests regardless of stone size, estimates of nest density are not affected (Table 6.1), but egg production is

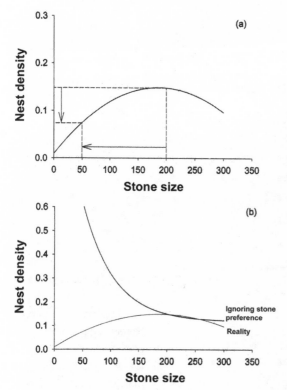

FIG. 6.1. Predicted relationship between nest stone size (cm²) and river blenny nest density (nests/m²), (a) when incorporating male preference for large nest stones and (b) when assuming no male preference, with scenario (a) (= reality) shown in gray for comparison purposes. In (a), the arrows show the concomitant reduction in nest density resulting from a hypothetical reduction in stone size from 200 cm² to 50 cm².

predicted to fall less severely than in the more realistic simulation (Fig. 6.2c, Table 6.1). Finally, if both male and female reproductive behaviors are disregarded, both nest density and egg productivity are predicted to show massive increases in response to stone extraction (Fig. 6.2d, Table 6.1). In fact, the densities of river blennies found in extracted sites were even lower than those predicted by the more realistic simulation, and no nests were ever found in disturbed river sections (personal observations).

It is clear that in the case of river blennies, habitat loss in the form of stone extraction would appear not to be detrimental, and on occasion would seem beneficial, if details of the fish's reproductive behavior were omitted from the predictive model.

PREDICTING THE EFFECTS OF EXPLOITATION ON A HERMAPHRODITIC FISH

Hermaphroditism is relatively common in reef fishes, with sex change from female to male (protogyny) occurring in 15 families and from male to female (protandry) in 6 families (Warner 1984). Individual fishes should change sex

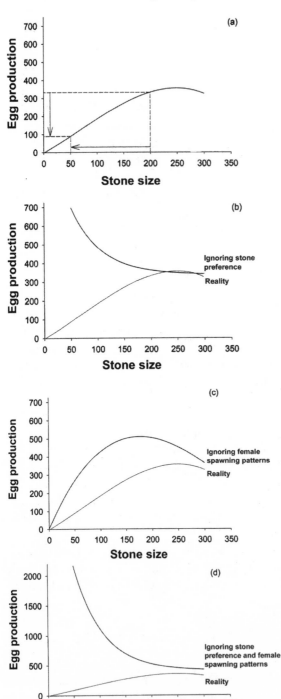

FIG. 6.2. Predicted relationship between nest stone size (cm²) and river blenny egg production (no. eggs produced × 1000) (a) when incorporating male preference for large nest stones, (b) when assuming no male preference for nest stone size, (c) when ignoring female propensity to lay larger clutches under larger stones, and (d) when ignoring both male nest size preference and female spawning pattern. In panels b–d, scenario (a) (= reality) is shown in gray for comparison purposes. In (a), the arrows show the concomitant reduction in egg production resulting from a hypothetical reduction in stone size from 200 cm² to 50 cm².

at a given age or size if sex change increases their reproductive success, compared to not changing sex at all (Warner 1975). The exact trigger for sex change remains unclear, but social and genetic controls have been suggested (Shapiro 1980, Warner 1984). Nevertheless, in sex-changing species, all large individuals tend to be of a single sex.

Current exploitation models (such as dynamic pool models) assume equal growth and fishing mortality for all members of an age or size cohort (Gulland 1977). Species with unorthodox sex determination and sex-specific growth rate, such as hermaphroditic fishes, are therefore overlooked by these models, whose direct application may lead to incorrect estimates of spawning potential or of optimal exploitation levels. Here, I revisit an earlier attempt (Huntsman and Schaaf 1994) to examine the impact of fishing on reproduction of a hermaphroditic grouper, the graysby (*Epinephelus cruentatus*). I ask specifically what are the consequences of ignoring the unusual life history of this species for the assessment of reproductive output.

The graysby is a relatively small (30 cm total length), reef-dwelling, protogynous grouper, ranging from North Carolina through the Caribbean, to Brazil. It is a long-lived, slow-growing species that is fished through most of its range, particularly where larger groupers are now rare. The life history parameters necessary for the simulation model were derived from Nagelkerken (1979). The model itself is based on a simple catch simulation in which the number of individuals alive in each of 10 age/size classes is determined, knowing the initial number in the youngest age class ($N_0 = 1000$) and the total instantaneous mortality rate Z (which is equal to the sum of natural mortality M, initially set at 0.13, and fishing mortality F).

Two sex-change scenarios were investigated: (1) gonochorism, where the number of live individuals in the population were assumed to be mature males or females according to a fixed and nearly even sex ratio (45 males:55 females as found in nature over all age classes; Huntsman and Schaaf 1994), and (2) protogynous hermaphroditism, where the number of mature males and females was determined using the age-specific sex ratios reported by Nagelkerken (1979). For simplicity, I assumed that sex change is under genetic control, occurring at a fixed age/size threshold. The population sex ratio is therefore variable according to the age/size structure of the population.

The reproductive output of the graysby population under each scenario was then approximated as the number of fertilized eggs produced. Egg production was estimated by first relating the number of live, mature females in each age class to biomass, and then female biomass to fecundity using known relationships. The likelihood of fertilization was obtained in two ways: (1) as a ratio of population male biomass to population fecundity (as in Huntsman and Schaaf 1994), which assumes that sperm limitation may occur at low

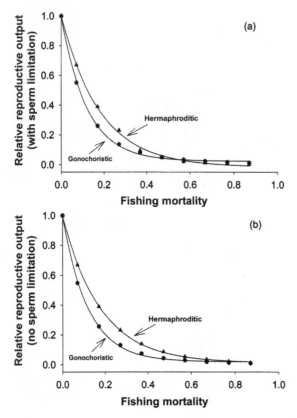

FIG. 6.3. Relative reproductive output of graysby (i.e., fertilized egg production at a given fishing mortality/fertilized egg production at fishing mortality $F = 0$) in relation to fishing mortality for gonochoristic (circles) and hermaphroditic (triangles) stocks when (a) sperm is limited and (b) no sperm limitation occurs.

male numbers; and (2) as a fixed proportion, set at 75%, which reflects recent studies of fertilization rates of hermaphroditic fish in the wild suggesting no evidence for sperm limitation (Petersen et al. 1992). Reproductive output is thus the product of fecundity and likelihood of fertilization, and is expressed relative to reproductive output at $F = 0$.

Not surprisingly, relative reproductive output decreases sharply with increasing fishing intensity (Fig. 6.3). Protogynous graysby tend to lose reproductive capacity more slowly than gonochores at low fishing intensity, but slightly faster at higher fishing mortality. This conclusion depends on whether sperm limitation is assumed to take place (Fig. 6.3a) or not (Fig. 6.3b). In the absence of sperm limitation, hermaphroditic graysby do as well (or as badly) as their gonochoristic counterparts under heavy fishing mortality. These results are largely similar to those of Bannerot, Fox, and Powers (1987), but at odds with those of Huntsman and Schaaf (1994) who suggested that hermaphroditic species lose reproductive capacity more rapidly than gonochores as fishing effort increases and also fail reproductively at lower fishing effort.

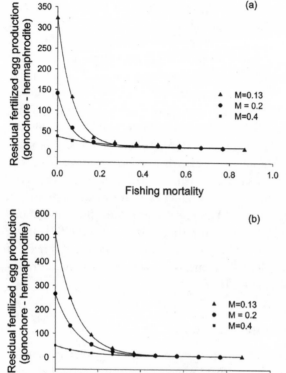

FIG. 6.4. Residual fertilized egg production of graysby (i.e., absolute fertilized egg production of gonochore minus absolute fertilized egg production of hermaphroditic graysby) in relation to fishing mortality at different levels of natural mortality, (a) assuming sperm limitation and (b) assuming no sperm limitation.

The fact that the relative reproductive output of hermaphroditic graysby is generally reduced less than that of gonochoristic graysby is biologically understandable. The exploitation of a gonochoristic fish stock removes from the population large males and large females, with the latter ceasing to contribute disproportionately to egg production. By contrast, exploitation of a hermaphroditic protogynous stock results in the removal of only males from the older size classes. Egg production is therefore not affected to the same extent if sperm depletion does not occur.

Absolute fertilized egg production is predicted to be much higher in the gonochoristic than in the hermaphroditic graysby at low fishing mortalities (Fig. 6.4) because large, very fecund females are assumed to exist under the gonochoristic scenario. However, this effect depends largely on the level of natural mortality: the higher the natural mortality, the lower the apparent initial advantage of gonochores (see Fig. 6.4). Although a relatively low natural mortality ($M = 0.13$) was used throughout these simulations, yearly mortality for species such as graysby may be as high as 0.4 (Pauly 1980, Hoenig 1983).

At high fishing mortalities, absolute egg production for gonochoristic and hermaphroditic stocks converges under all levels of natural mortality (see Fig. 6.4).

Even if gonochores have a higher absolute fertilized egg production most of the time, it is currently impossible to assess whether such an advantage would translate into increased recruitment, which would depend on the strength of density dependence in the survival of eggs and juveniles. Thus, ignoring hermaphroditism when modeling the impact of exploitation on graysby populations will inflate egg production estimates. If density dependence in the grouper's early life stages is weak, then stock size assessment may appear overly optimistic. If density dependence is strong, however, the accuracy of stock size estimates may not be unduly compromised. In fact, weak density dependence is likely to be acting in heavily exploited fish populations because resources are unlikely to be limiting (Lizaso et al. 2000). If so, then heavily exploited hermaphroditic fish stocks, which are most in need of accurate assessment for conservation purposes, will be most inaccurately estimated if sex change is not considered.

Conclusions and Recommendations

Studying reproductive behavior or doing something else: what is best for conservation? This chapter suggests that knowledge of reproductive behavior can, in many cases, alter significantly the predictions made in models of different management strategies. Before jumping to the conclusion that behavioral studies are always essential to conservation, however, we must measure the net benefit of obtaining information on reproductive behavior versus that of engaging in other activities that can also benefit conservation. But how can one compare the value of elucidating the proximate cue for sex change in an exploited hermaphroditic fish and the likely impact that this may have on the predictions on population responses to fishing, to the value of carrying out more field surveys to assess grouper abundance on reefs or in fish markets, or to the value of education programs about coral reef conservation? It may be technically possible to obtain such an answer through sensitivity analyses, but neither behavioral scientists nor conservation managers would deem this a worthy pursuit.

The costs of incorporating behavior into conservation projects can be minimal but the benefits enormous. A lot of the behavioral information necessary to assist in conservation decisions is already available. The details of the mating system or social organization of a rare species in need of conservation intervention may not be known, but often they will be for related species, and behavioral characteristics can be highly conservative. Owens and

Bennett (1995), for example, showed that most of the variability among birds in features such as foraging range and nesting habit occurred at the level of family or above. The behavior of an abundant close relative could therefore be used in lieu of precise knowledge of the behavior of a rare, little-known species to give valuable insights into the latter's requirements.

Several authors have outlined benefits of using a behavioral approach in conservation projects. There are three major advantages:

1. Behavioral scientists generally ask questions about fitness. They therefore identify key factors that affect individual survival and reproduction, and hence population dynamics (Arcese, Keller, and Cary 1997).
2. Behavioral scientists willingly use a predictive approach and have information that makes population models more realistic and more accurate (e.g., Bart 1995).
3. Behavioral scientists simply ask different questions than other members of a conservation team, such as a wildlife manager or even a molecular ecologist (Gavin 1989). They thus offer a different and often valuable perspective on conservation problems.

Given these potential contributions, the question becomes not so much whether behavior should be part of conservation projects but whether conservation projects can afford to ignore a behavioral perspective. As argued by Arcese, Keller, and Cary (1997), when the success of a conservation intervention depends critically on the response of a species to this intervention, the inclusion of a behavioral perspective becomes essential.

My conclusion suggests that the relevance of behavioral studies is limited to conservation projects focused on a single or a few species. This would be a limited sphere of influence indeed, since much of conservation biology today is about identifying and designating areas and ecosystems for protection. As the rate of area preservation slows down, however, issues about the appropriateness of park design in relation to species needs, about edge effects and corridor use, and about human–wildlife conflict management are starting to arise that are fully within the realm of behavioral science. Will behaviorists rise to the challenge of applying their knowledge and perspective to practical purposes? I certainly hope so.

Summary

I review the potential predictive role of reproductive behavior in four aspects of conservation biology: the effects of habitat loss, exploitation and culling, the risk of extinction, and captive breeding, translocations, and reintroductions. Reproductive behavior has the potential to influence the outcome of most of these conservation actions through a number of mechanisms, including mate choice, social facilitation of reproduction, conspecific attraction to breeding sites, and Allee effects on reproduction, and has been used in a limited number of cases to predict the outcome of conservation measures. To examine specifically the effect of ignoring the minutiae of reproductive behavior, I focus on two case studies. The first involves predicting the effects of habitat loss on river blennies, an endangered European bottom-spawning freshwater fish. Considering nest site preferences of river blennies results in a greater predicted decline in population size than when nest preferences are ignored. The second case study investigates the effect of hermaphroditism in marine fishes subjected to exploitation. I compare the effects of fishing on populations of a tropical grouper, comparing predicted reproductive outputs under assumptions of sex change and gonochorism. In this situation, ignoring the fact that the species under exploitation is a protogynous hermaphrodite would lead to overestimates of stock size when the population is at particularly low levels. The details of reproductive behavior may therefore be important when planning conservation action. There may often be little financial cost involved in incorporating a behavioral approach in conservation projects and great benefits to be derived in terms of insights and predictive power.

Acknowledgments

I thank the many people who patiently discussed the various topics considered in this review: Jenny Gill, Phil Stephens, Bill Sutherland, Dolors Vinyoles, and particularly John Reynolds for explaining fisheries models. I thank Marco Apollonio and Marco Festa-Bianchet for allowing me to submit this chapter even though an unforeseen event prevented me from attending the meeting in Erice. I dedicate this chapter to Catherine, the unforeseen event.

7.

Foraging Behavior, Habitat Suitability, and Translocation Success, with Special Reference to Large Mammalian Herbivores

Norman Owen-Smith

This chapter evaluates how an understanding of foraging behavior might contribute to conservation. It is noteworthy that none of three recent books outlining how ethology has contributed to conservation included a chapter on foraging behavior (Clemmons and Buchholz 1997, Caro 1998b, Gosling and Sutherland 2000). Is food-seeking such a minor aspect of population viability that we can relegate it to academic treatments only? I think not. My view is that understanding how resources are acquired and allocated to maintenance, growth, and reproduction is central not only for theoretical ecology but also for conservation applications.

The specific conservation activity that I address here concerns the translocation of animals to new areas in the wild (Kleiman 1989). These exercises may be reintroductions in the sense that the species formerly occurred in the region (also termed repatriation), or introductions to places where the habitat is deemed suitable although historic records are lacking. The animals translocated may either be captive-bred specimens being rehabilitated into the wild, or wild-caught animals being moved to a new area.

Despite large investments of time and money, a substantial proportion of such introductions have not led to self-sustaining populations (Griffiths et al. 1989; Wolf et al. 1996; Wolf, Garland, and Griffith 1998; Fischer and Lindenmayer 2000; Komers and Curman 2000). Failures have been associated with shortcomings in habitat "quality" in the release area (Wolf et al. 1996; Wolf, Garland, and Griffith 1998), and in some cases with high losses to predation (Short et al. 1992). It has been recommended that future translocation protocols be based on a rigorous, quantitative assessment of habitat quality (Wolf, Garland, and Griffith 1998). Subjective assessments of habitat conditions as "good" or "poor," however, are inadequate, given the costs and risks involved. The primary goal of habitat assessments should be to maximize the initial rate of population increase so as to shorten the period during which the population is at risk (Komers and Curman 2000).

Habitat suitability depends fundamentally on the adequacy of food resources. The mere presence of suitable and accessible food is insufficient: the time required to capture this food, and the costs and risks incurred in the process, must be considered. This has been the subject matter of foraging theory since the defining papers of MacArthur and Pianka (1966) and Emlen (1966). For carnivores, potential food items in the form of the bodies of other animals are generally highly nutritious; the problem is how to catch and kill them. For herbivores, vegetation may appear to be available everywhere, but plant species and parts can differ widely in their nutritional value. Decisions about what to eat, and what not to eat, have important consequences for success in meeting physiological demands (Owen-Smith and Novellie 1982), most especially for females that are lactating or in late stages of gestation (Oftedal 1984). Moreover, vegetation growth is usually seasonal, so that the availability and nutritional quality of different food types change widely over time. Foraging behavior must be adapted flexibly to the prevailing circumstances if animals are to meet their nutritional requirements through the seasonal cycle. This encompasses not only diet choice but also food-procuring activities.

Habitat suitability also depends on security from predation and shelter from environmental extremes. The three fundamental habitat requirements—food, shelter, and security—cannot be viewed in isolation. Risk of predation may inhibit animals from using certain areas where suitable food is available, whereas a lack of food in secure habitats may force animals to forage in places that are risky (Sinclair and Arcese 1995a). Animals may avoid extreme cold or heat by forgoing feeding during times when these conditions prevail.

From a wider perspective, the habitat is just one aspect of the ecological niche of a species, its place or "profession" in the environment. The nutritional benefits of different food types, and time and energy costs incurred in foraging, depend on morphological and physiological adaptations. For large mammalian herbivores, features such as mouth width and the capacity of the digestive system relative to body size crucially influence the rate and conversion efficiencies of ingesting and digesting different plant types, such as grass or browse (Owen-Smith 1982, 1985, Gordon and Illius 1994). Grazers and browsers tend to be distinct not only in features of the digestive tract but also in relative liver and salivary gland size (Hofmann 1989; Robbins, Spalinger, and van Hoven 1995). Among grazers, some species are relatively wide-mouthed, whereas others resemble browsers in having relatively narrow muzzles (Gordon and Illius 1988, Owen-Smith 1982, 1989). Anatomical distinctions lead species to partition food resources based on factors such as nutrient concentrations (Jarman 1974), height above ground (du Toit 1990), and leaf size (Cooper and Owen-Smith 1986). The "fundamental niche" may, however, be further restricted by predation risks, as well as by the effects of interspecific competition. These manifold influences were captured by Hutchinson (1959) in his conceptualization of the niche as the region occupied by a species within a multidimensional hyperspace of resources, conditions, and risks.

But this niche concept is misleadingly static. In the real world, individuals must respond to changing conditions over daily, seasonal, and annual time scales. Phenotypic features can be adjusted to some degree to cope with changing needs and constraints. Although modifications of morphology and life history can occur within individual lifespans, the most plastic aspect of the phenotype is behavior. Learning which food types to eat or avoid is an important component of foraging behavior, but so is knowing where to find such foods. Understanding the basic plasticity of foraging behavior of a particular species, as well as its limits, is important for assessing where and when populations will thrive or expire. Selecting the best food is fine when conditions are favorable, but what do animals do when those food types are no longer available? Here the interaction of innate predisposition, physiological tolerance, and capacity to learn at different stages needs to be understood.

My focus here is specifically on large mammalian herbivores. Conservation agencies face the dilemma of how to respond to small or declining populations of such charismatic species within protected areas, and frequently turn to the option of translocation either into or out of the area of concern. Some reintroductions of large ungulates have been spectacularly successful, notably bison (*Bison bison*) in North America, ibex (*Capra ibex*) in Europe, and white rhinoceros (*Ceratotherium simum*) in Africa (Gordon 1991). Yet there have also been numerous failures, most of these not adequately explained (Novellie and Knight 1994).

How could the success of such repatriation operations be more reliably guaranteed? How could the suitability of habitats be identified with confidence before a reintroduction is attempted? How could initial problems that may arise following introductions into novel environments be mitigated? How could an understanding of the basic ethology of food-procuring guide such conservation actions? These are some of the questions I address in this overview.

Case Histories

Particularly illuminating are examples where introduced animals initially performed poorly, only to thrive at a later stage. The following are three case histories, outlined in some detail.

SABLE AT PILANESBERG

My personal confrontation with poor population performance following translocation came from setting up a study to establish why the sable antelope (*Hippotragus niger*) introduced into Pilanesberg National Park in South Africa were doing so poorly (Magome 1991). This park had been created in 1979 by moving people out of an area that had great scenic potential for ecotourism. The 530 km² area was then fenced and stocked with almost 6000 animals, representing 17 ungulate species (Anderson 1986). The understanding was that the people would ultimately benefit from the economic spinoffs from the wildlife park. A few species of large mammal still occurred in the park area, but most had to be purchased and brought in at considerable expense. Among the latter were sable antelope. The park was situated close to the type locality for this species in present-day Northwest Province. The habitat conditions and kinds of grasses available were judged to be ideal for the species. Sixty-seven animals were introduced from nearby game ranches in 1982–83, with special care taken to keep social units intact. Five years later, the population still totaled only about 70 animals, a poor gain for a population that should have increased at 20% per year, if most of the calves potentially produced annually had survived.

Our study examined whether food limitation could be a restriction on the population increase of the sable. We recognized that predation on calves by leopards (*Panthera pardus*) or brown hyenas (*Hyaena brunnea*), could also have limited population growth, but predation was difficult to study. No lions (*Panthera leo*) or spotted hyenas (*Crocuta crocuta*) existed in the park at that time. Hence, habit and food selection were investigated for four sable herds at different spatial scales: the landscapes where home ranges had been established, vegetation communities utilized within these ranges at different stages of the seasonal cycle, and specific grassland types and grass species occurring in feeding sites. However, during the study year most females calved successfully and the population grew to 94 animals, an increase of 34%. By 1991 the sable population totaled 127. Problem solved!

The unexpected outcome of the sable research suggested that 5 years might have been the period required by the sable to establish an effective foraging strategy that enabled them to find nutritionally adequate food throughout the seasonal cycle. However, there were other possible explanations for the delayed population takeoff. The early years postrelease had received below-average rainfall, whereas the study period was a year of high rainfall. The sable depended on green grass regrowth on burns to carry them through the late dry season, and perhaps the prior burning policy had been less effective in making such food available during this critical period.

ORYX IN OMAN

For more enlightenment I turn next to the most thoroughly detailed study of an introduced ungulate population ever undertaken, that of the Arabian oryx (*Oryx leucoryx*) released into the central desert region of Oman in the 1980s (Stanley Price 1990). The species had been completely extirpated in the wild in 1972, the only survivors being animals in zoos. In 1982, 10 captive-raised oryx were released into the wild after a period of acclimation in pens in the chosen area. They were followed by 11 more in 1984. Further releases involved 11 animals in 1988 and 8 in 1989. By 1996 the population had grown to over 300 animals, a shining success. Regretfully, subsequent poaching for live sales of these highly valuable animals has had a severe impact, reducing the wild population to a small remnant (Gorman 1999).

Like the sable, the oryx population grew more slowly during the first few years postrelease than it did subsequently. Over the first 2 to 4 years the inherent rate of population increase, λ, was 1.08 (i.e., 8% per year), estimated from adjusted fecundity and mortality rates for the female segment (Stanley Price 1990); post-1988 it was 1.195 (recalculated from Spalton 1992). There were notable demographic improvements between these two periods: age at first reproduction decreased from 38 months to 24 months, and mortality of calves from birth to 1 month declined from 25% to 14% (Spalton 1992). Although severe drought conditions prevailed during the first 3 years postrelease, animals were supplied with supplementary forage to alleviate the food shortage. Thereafter, supplementary feeding was discontinued so that the oryx had to survive off the natural vegetation alone. Also, one of the dominant males proved to be sterile, which may have reduced the reproductive success of females that associated with him.

Nevertheless, observations revealed fascinating changes in foraging movements as the oryx settled into their new environment. Animals from the first release encountered a habitat devoid of other conspecifics. They made probing movements into new areas following rain, and occupied a sequence of temporary home ranges, each roughly 100 to 300 km² in extent, eventually extending their total range toward nearly 2000 km² (Stanley Price 1990; Tear, Mosley, and Ables 1997). When drought conditions ensued, the animals retreated back to the release area where supplementary forage was available. The second herd was released into drought conditions, and into an environment already occupied by other oryx. They moved less extensively; their total range encompassed 550 km² over 2 years. Observations suggested that individual animals retained a detailed spatial memory of all areas traversed, including the routes that had been followed between different ranges.

Some 6 to 8 years after introduction, oryx in these initial two herds no longer moved widely in response to rain, having apparently established where to find needed resources. In this later stage they made greater use of browse components, including acacia pods as well as leaves and flowers of other tree species, and so became less dependent on the basic *Stipagrostis* grass resource. In contrast, oryx from the two later introductions widened their monthly ranges in response to rain during this same period, just as the previously established herds had done earlier.

Observations indicated how the introduced animals spent additional time and energy getting to know their new environment and the location of resources within it at different times. These animals may have incurred nutritional deficiencies in the process, as evidenced by some mortalities ascribed to botulism due to eating old bones, perhaps to obtain more phosphorus. In later years, animals appeared to exploit the environment more effectively, and the decreased mortality rate, particularly among calves, allowed the population to achieve its full growth potential. If supplementary forage had not been supplied during the initial drought period, the population may have declined. After the first few years, supplementary feeding no longer seemed necessary. However, even after nearly a decade, animals had not located water sources hidden in the foothills on the fringe of the plateau. Although some wolves (*Canis lupus*) and caracals (*Felis caracal*) occurred in the region, predation had no apparent impact on the oryx population.

CARIBOU IN IDAHO

This revealing example involved the translocation from the wild of woodland caribou (*Rangifer tarandus caribou*) of two different ecotypes to augment a remnant native population persisting in the Selkirk Mountains in northern

Idaho (Warren et al. 1996). Animals of the mountain ecotype came from the Canadian extension of the Selkirk range, whereas northern ecotype animals were brought from the Fraser plateau region of British Columbia. Mountain ecotype animals typically depend on arboreal lichens for winter forage, which they can reach because of deep snow accumulations of 2 to 3 m. However, in the plateau habitat inhabited by the northern ecotype, the annual snowfall is much less, permitting animals to access terrestrial lichens by pawing open "craters" in the snow on exposed ridges.

Introduced into the new habitat, the northern animals showed much variability in habitat use during the first year, but only limited dispersal. In contrast, introduced animals of the mountain ecotype showed habitat-use patterns similar to those of the native caribou. The areas that the latter used during fall had more lichen available than those chosen by the northern animals, although by late winter there was no difference. During spring the two ecotypes diverged widely in their habitat choice, with the resident and introduced mountain stock occupying densely forested areas on northern slopes, whereas caribou of northern stock sought open and sparsely forested areas, generally with a southern aspect. Mortality rates over the first 3 years postrelease differed significantly between the two ecotypes, with 64% (14/22) of northern animals dying compared to 33% (6/18) of mountain stock.

In this situation the habitat-use patterns that had been traditional in the original ranges were maintained in the new area, to the disadvantage of one caribou ecotype. However, some modification of habitat selection traditions occurred when animals joined resident caribou and followed the movements of the latter.

OTHER EXAMPLES

In other cases, the initial rate of population increase following introduction was lower than that attained later. Asiatic wild asses (*Equus hemionus*) introduced into the Negev Desert of Israel showed only a 25% overall increase in the breeding female segment (from 12 to 16 animals) over the first 10 years following release (Saltz and Rubenstein 1995; Saltz, Rowen, and Rubenstein 2000). An obvious problem was the male bias in offspring from the prime-age females that predominated among the animals introduced. However, the individual reproductive success of females also remained low during the first 5 years postrelease. This was ascribed to the persistent effects of stress during capture and transport, and a carryover effect from the breeding facility where reproductive performance had been low. However, it is surprising that such effects persisted over 5 years. Foraging problems were not considered.

Similar observations were recorded for Cape mountain zebras (*Equus zebra*) reintroduced within their former range. Population increase over the first 3 to 5 years postrelease averaged only 0.4% annually, compared with a mean of 9.3% subsequently (Novellie, Millar, and Lloyd 1996). Poor performance was ascribed to insufficient numbers being released, together with capture stress, breakup of family groups, and, vaguely, "adaptation to a new environment." Notably, for two cases where populations showed high initial rates of increase, the release site was close to the source population in the Mountain Zebra National Park, in similar habitat.

As already noted, reintroductions of white rhinoceroses have been widely successful. Nevertheless, mean calving intervals somewhat longer than those exhibited by the source population in the Hluhluwe-Umfolozi Park have been documented in several of the new localities (see Table 8.6 in Owen-Smith 1988).

An interestingly different case history is provided by the establishment of a population of mountain gazelles (*Gazella gazelle*) in the Hawtal Reserve in Saudi Arabia (Dunham 1997). The 71 captive-born animals initially introduced had more than doubled in numbers after 3 years. However, mortality over the first year postrelease was substantially higher for gazelles that were older than 3 years when released than for younger animals (54% vs. 19%), whereas among the older group it was greater for males than for females (73% vs. 38%). The direct cause of most mortality was predation by feral dogs and a lone wolf. However, the question remains why these age/sex classes were especially susceptible to predation, and not younger animals of both sexes.

Follow-up of the long-term case histories of ungulates translocated into South African national parks revealed that 85% (17/20) of attempts were successful, where the habitat was deemed suitable and the locality was within the former range of the species (Novellie and Knight 1994). Where either of these conditions was not met, only 13% (2/16) of translocations succeeded. In most instances, animals were moved into small parks where large predators were absent. Of five species reintroductions into the Kruger National Park, where large predators abound (including two cases subsequent to the period covered by Novellie and Knight 1994), only those of white rhinoceros and black rhinoceros (*Diceros bicornis*) have been successful. Notably, for the latter two species, adults are effectively invulnerable to predation. Lichtenstein's hartebeest (*Alcelaphus lichtensteini*) existed only as a small remnant a few years after release from their holding pen, whereas all introduced oribi (*Ourebia ourebi*) and suni (*Nesotragus moschatus*) have disappeared.

Twelve eland (*Taurotragus oryx*) repatriated to the Umfolozi Game Reserve in 1967 failed to establish a population (Brooks and Macdonald 1983). Although some lions were present, tick infestation rather than predation seemed to be the prime cause of this failure. In contrast, all 17 ungulate species translocated into the Pilanesberg National Park between 1979 and 1984 have persisted, despite initial concerns about the viability of some populations (Anderson 1986). Notably, lions and spotted hyenas were absent at the time of the introductions, although lions were later introduced. Ungulates of various species have been translocated successfully to stock private wildlife reserves in South Africa on numerous occasions, although summary data are unavailable. Again, larger predators are usually absent or at least suppressed in abundance in such situations. Likewise, reintroductions of wallabies and smaller macropods in Australia have rarely succeeded in the presence of predators such as introduced foxes, feral cats, and dogs (Short et al. 1992).

OVERALL ASSESSMENT

Some notable features emerge from these case histories. Populations of newly introduced ungulates have frequently increased more slowly initially than subsequently, despite being given special care and despite the absence of predation on adults. Moreover, the restricted population growth initially was generally associated with poor reproductive performance by individual females, although small release number and demographic distortions may have contributed. In at least one instance, mature females also survived less well than young females. Strikingly, I could find no records of viable populations of ungulates established from introductions into areas containing abundant large predators, except for megaherbivores that are largely invulnerable to nonhuman predation.

Foraging Behavior

The implication from the preceding examples is that newly introduced animals perform poorly initially in acquiring the nutritional intake needed to realize their reproductive potential. Animals introduced into a new environment must learn new foraging habits, not only what plant types to consume but also where to find these at different times. Mothers may initially guide their offspring toward favorable plant species (Edwards 1977), but after weaning, young animals must extend this learning further, in particular to secure the key resources needed through the critical periods of late winter or the dry season.

Studies reveal that young ungulates have an innate predisposition to feed on certain food plants, which can become modified through experience. Bottle-raised roe deer (*Capreolus capreolus*) observed from birth to 1.5 months of age discriminated among plant species at first contact, such that more bites were taken from generally preferred than from avoided species (Tixier et al. 1998). Thereafter the naive animals increased their consumption of species that were favored by experienced adults while still eating small amounts of species that wild deer generally rejected, except for two species that produced noxious sensations and potentially toxic consequences. Initial learning appeared to be primarily through taste rather than olfaction, perhaps reinforced by postingestive consequences, but later the fawns apparently learned to avoid the toxic species by odor. By 1 month of age, the fawns distinguished among plant species almost as well as adult deer, despite having had no maternal guidance.

Zoo-bred scimitar-horned oryx (*Oryx dammah*), transferred as adults into an acclimation pen in Tunisia prior to release, approached novel plant species cautiously and smelled the leaves (Gordon 1991). Sometimes the sniffing was followed by tentative nibbling and sometimes by the animals thrashing the bush with their horns. A small woody herb not eaten by sheep or goats was also rejected by the oryx, despite being abundant and green, and a shrub known to be toxic to domestic livestock was also not eaten. Thus food selection was discriminating despite lack of prior experience.

Hand-reared impalas (*Aepyceros melampus*) that had been removed from their mothers when only a few days old accepted without hesitation many of the plant species that were commonly eaten by wild impalas (Frost 1981). Other plant species were ignored, including some that became included in the diet of these animals after they had been released into the Nylsvley study area. During the dry season, our hand-reared kudus (*Tragelaphus strepsiceros*) (also at Nylsvley) expanded their dietary range to incorporate other plant species eaten rarely, or not at all, over the wet season (Owen-Smith and Cooper 1987, 1989). Acceptance ratings for various woody plant species were correlated closely with indices of nutritional value based on relative contents of nutrients, as represented by crude protein, and antinutrients, as represented by condensed tannins (Cooper, Owen-Smith, and Bryant 1988; Owen-Smith 1994). Some forb and shrublet species were never eaten, presumably because they contained unidentified toxins.

Hand-reared lesser kudu (*Tragelaphus imberbis*) and gerenuk (*Litocranius walleri*) likewise showed spontaneous acceptance of certain food species at first presentation, but rejected some species known to be consumed in the wild (Leuthold 1971). Naive white-tailed deer (*Odocoileus virginianus*) fawns selected a diet closely resembling that of experienced fawns and adults in species preference rankings, as well as in bite sizes, biting rates, and intake rate obtained (Spalinger et al. 1997). The notable distinction was for thorny acacia species, for which the bite rate of naive fawns was lower than that of experienced juveniles. Likewise the bite rate of a naive young impala feeding on *Acacia tortilis* was slower than that of experienced animals feeding on the same species because it caught its lips and tongue on the recurved thorns (Dunham 1980).

Sheep that differed in their early nutritional experience, with food sources ranging from hay in pens through sown pastures to semiarid rangeland communities, showed marked differences in grazing preferences that persisted for more than a year despite attempts to change them (Arnold and Maller 1977). Sheep moved from pastoral areas to sown pastures took longer to adjust to the new food source than sheep transferred in the reverse direction, and adults took longer than lambs to adapt to new forages. Sheep that had been reared on hay grazed for 20% more time while feeding on natural rangeland than did sheep reared on pastures, but obtained 40% less food within this time. These large differences show how skills involved in manipulating food types can depend on early experience, with potential consequences for later food selection.

Observations of domestic sheep and goats show that young animals accept novel foods more readily than adults (Provenza and Balph 1987). Dietary learning is based largely on postingestive consequences, both positive and negative, of eating different kinds of forage (Provenza, Pfister, and Cheney 1992). Physiological adaptation to initially noxious chemicals can also occur. For example, sheep became more tolerant to the cyanogenic compounds sometimes present in clover after having been exposed to these chemicals for a period (Harborne 1988). Consequently, foraging behavior must be sufficiently flexible to allow resampling of food types to accommodate changes in noxious properties that may occur over time.

The capacity to learn to avoid poisonous plants may be underlain by a higher innate tolerance for the toxin in native than in alien populations. This has been most clearly documented for kangaroos and other species with distribution ranges either within or outside the range of highly toxic plant genera containing fluoracetate in western Australia (Twigg and King 1991). Eland likewise have some tolerance for this same toxin in a South African shrub called gifblaar (*Dichapetalum cymosum*; Basson et al. 1982). Neither our hand-reared impalas nor the kudus ate gifblaar at any stage while under observation, although they readily consumed similar-looking plants of other species. Cattle, however, cannot resist consuming gifblaar when it presents tempting new leaves in spring while the grass is still brown, usually with fatal consequences. We were unsure how our study goats would respond until one of them ate gifblaar in the presence of my colleague Susan Cooper. She dosed the goat with cooking oil to prevent the toxin from being absorbed and released the goat back into the study area. A few days later this same animal again ate gifblaar when no one was around to help, and died. No learning had occurred on this occasion!

In summary, there is ample evidence that young ungulates can learn which food plants are most nutritious, presumably from subsequent physiological consequences. During the dry season or winter when animals must turn to less favorable species, they cannot depend on maternal tutoring. Animals are vulnerable to being poisoned by certain plant species that are highly toxic unless they have some innate tolerance for the toxin. Learning how to manipulate those plant species that require special handling, such as thorny species, appears to be more strongly restricted by early experience than chemically assessing nutritional value.

Likewise, newly introduced ungulates face similar challenges in learning what not to eat from the novel array of plant species confronting them, and in gaining experience in handling the structural deterrents that many plants possess. Behavioral adjustments and physiological adaptation can take time to become effective, and in the meanwhile animals may be nutritionally disadvantaged.

An animal's performance depends not only on what foods it consumes but also on how efficiently it obtains these foods. We have already noted how efficiency in handling certain food types can affect ingestion rate. Search time is also an important component of foraging behavior. The margin between a daily nutritional intake that is adequate and one that is submaintenance can be quite small. Such differences can be responsible for changing habitat selection over the seasonal cycle.

In the Kruger National Park, kudus used open acacia savanna habitats extensively during the wet season but contracted their foraging range largely to hillslope base regions, or to riparian fringing woodlands, during the dry season (Owen-Smith 1979, du Toit 1995). Foraging efficiency, as assessed by the feeding time obtained per step taken, or more broadly by the proportion of foraging time spent actually feeding, was always higher in the hill base ecotone because of the greater concentration of woody plants in this region. Nevertheless, the greater availability of forbs and creepers, which constitute high-quality food types, in the acacia savanna was probably an attraction. In the dry season, when these plant types became less available and deciduous tree species had mostly shed their leaves, the acacia habitat was largely abandoned by the kudus. Notably, this took place after the feeding time per step had declined to under 2 seconds, and the proportion of foraging time spent feeding to less than 60%.

At Nylsvley, patches of acacia-dominated savanna were likewise favored during the wet season but largely abandoned during the dry season (Owen-Smith 1993). In the Kruger National Park, no resident kudu herds were encountered within a region of largely umbrella thorn savanna that I traversed almost daily during my study period, although impala and giraffe (*Giraffa camelopardalis*) were regularly seen there. Kudus are absent from Tanzania's Serengeti National Park, although both giraffe and impala are common. What subtle vegetation differences distinguish areas that are suitable habitat for some browsing ungulate species and unsuitable habitat for others?

A modeling assessment indicated that kudus would have obtained a sub-maintenance energy intake had they foraged in the acacia patches during the dry season, largely because of the absence of any evergreen or semi-evergreen browse components to provide forage late in that season (Owen-Smith 2002). Impalas can obtain an adequate food intake rate when browsing fine-leaved thorn-trees because of their smaller size and their ability to graze as well as to browse (Cooper and Owen-Smith 1986). Giraffe have a special ability to strip multiple leaves from thorny branch tips (Pellew 1984), and their large size enables them to travel further and to obtain more nutrition from chemically defended evergreen browse during critical periods (du Toit and Owen-Smith 1989). A modeling exercise confirmed how sensitive the nutritional balance of herbivores can be to small differences in bite size and in forage quality, as influenced by morphological differences in body size and oral dimensions (Owen-Smith 1985).

Food sources that may be quite minor in their overall dietary contribution can be crucially important for bridging the critical period through the late dry season into the start of the new growing season (and presumably through late winter in temperate latitudes). For kudus, these foods included certain fruits that ripened toward the end of the dry season, as well as the flowers and foliage that were produced by certain tree species ahead of the rains (Owen-Smith and Cooper 1989). Notably, the latter included mostly species that were otherwise unpalatable due to the high levels of condensed tannins in their leaves (Cooper, Owen-Smith, and Bryant 1988). These bridging resources enabled our study animals to maintain their metabolizable energy intake no more than 10% below their daily maintenance requirement even during the final critical month of the dry season (Owen-Smith 1994). Without such resources, the animals would have starved more rapidly and may not have survived to the new growing season.

In following our habituated young kudus at Nylsvley, it was apparent that, rather than searching at random, they knew where particular resources occurred within the 215 ha pen to which they were confined. For instance, when nearing a tree producing the large "monkey oranges" that were sought out during the late dry season, it became a race as to which kudu reached the tree subcanopy first to find whether any fruits had fallen since the last visit.

Spatial cognitive aspects of foraging behavior have hardly been studied, at least for wild ungulates (Bailey et al. 1996).

Habitat suitability cannot be judged simply from the presence of edible and acceptable browse or graze during the favorable season. It depends also on how the supply of particular resource types persists over the seasonal cycle, and on the effective rates of food intake that these yield. Particularly important are vegetation components providing alternative or reserve resources through the winter or dry season period. Even unpalatable species offering submaintenance nutrition can play a valuable buffering role. Moreover, the use of particular habitats depends on their mosaic juxtaposition with other habitats within regional landscapes.

The chemical basis for food acceptance may be partly innate and readily modified from experience based on postingestive consequences. Less flexible are techniques of manipulating structurally challenging vegetation components to obtain an adequate rate of food intake. Lack of early experience may have more persistent consequences. Even more crucially important is the opportunity to locate resource types or habitat regions that yield an adequate, or at most marginally submaintenance, nutritional gain through critical bottleneck periods of the seasonal cycle. Without such vegetation components being available, what may seem superficially to be suitable habitat becomes unsuitable habitat in its capacity to support a population year-round.

A Current Conservation Dilemma: Roan Antelope in the Kruger Park

I will now highlight some challenging issues concerning the in situ conservation of South Africa's most threatened antelope species, the roan antelope (*Hippotragus equinos*), within its premier national park range. The roan population inhabiting the Kruger Park has always been small, with numbers varying around 300. Because of its rarity it was given special attention, to the extent of immunizing animals against anthrax by darting them with vaccine from helicopters. Nevertheless, between 1986 and 1993 the park population declined from a peak of 450 to a remnant of 45 (Harrington et al. 1999). The problem was recognized as being associated with excessive provision of artificial waterpoints within the roan habitat in the north of the park, which attracted an influx of zebra. The zebra exacerbated grazing impacts during a prolonged drought period, and led to an increase in lion abundance, heightening predation pressure on the roan.

The park managers acted firmly, if belatedly, by closing all boreholes within the core region of the northern plains habitat in 1994. Remnants of roan herds that had been reduced to ones and twos were relocated to a 400 ha fenced enclosure offering suitable habitat within this region. Zebras tended to move off the plains following waterpoint closure, and the lions followed, but not completely because of remaining surface water around the edge of this arid region. Despite these measures, roan numbers continued to decline, such that by 2000 only 23 free-ranging roan remained, concentrated in two to three herds in the area of waterpoint closure. Almost all roan elsewhere seem to have disappeared. However, the 8 roan that were placed in the enclosure had increased to a total of 31.

The dilemma now is how to best preserve this small remnant of Kruger's gene pool of roan antelope. Should animals from the enclosure be released to augment the free-ranging population? Should the remaining free-ranging roan be moved to another enclosure elsewhere in the park? Should all of the roan be moved to captivity outside the park so they could multiply in captivity for later reintroduction? Should other roan be purchased at great expense to increase the Kruger population?

These decisions rest crucially on assessments of likely success, taking into account food supplies relative to predation risks. Animals released into the park face a full gamut of predators, from lions downward. The Lichtenstein's hartebeest that were released from the existing enclosure to make way for the roan have declined to a small, probably nonviable, remnant, probably largely through predation. Roan obtained from captive situations will be naive in their responses to predators and will thus inevitably suffer heightened attrition until the survivors gain experience. The park authorities are reluctant to interfere to the extent of culling predators in a large national park intended to promote natural ecological processes. What fraction of genes would be expected to persist were captive-held animals to be released into an aridified environment containing an abundance of predators? The dilemma is acute and unresolved at the time of writing.

A basic question remains: Why are the roan so vulnerable to predation? Their large body size (280 kg) facilitates digesting the poor-quality forage associated with the nutrient-poor savanna regions they commonly inhabit (Heitkonig and Owen-Smith 1998), but it probably makes them somewhat slow in evading a lion attack, compared with smaller grazers like wildebeest (*Connochaetes taurinus*) and tsessebe (*Damaliscus lunatus*). Also, larger species, who have a lowered surface area to volume ratio than smaller species, have greater difficulty dissipating the internal body heat generated during active foraging (Owen-Smith 1988). This could predispose roan to foraging somewhat more at night than during the day, especially during hot times of the year, which could also expose them to heightened predation risks.

Conclusions and Recommendations

Those responsible for translocation operations involving ungulates need to be more deeply aware of the foraging constraints that may restrict the performance of these animals in their new environment, at least for an initial period. Even when sufficient food may seem to be present, animals may not be able to harvest it at an effective rate. To gauge the success of new introductions, managers must consider the oral and digestive morphology and other relevant phenotypic features of the species concerned. Are the resources that are available adequate to support animals throughout the seasonal cycle, even during extreme years? Will the introduced ungulates be able to readily locate the food types needed to bridge crucial periods? An appropriate computational model is needed to address these leading questions. The model should specifically accommodate the adaptive adjustments of the animals to seasonally and annually changing conditions, in behavior, physiology, and life history features. If there is any uncertainty regarding resource security, supplementary forage should be provided during the crucial periods over the initial few years, then be progressively withdrawn.

The presence of any substantial predation risk, especially to adults, may just tip the balance between a situation where the herbivore population has a small but positive rate of increase initially, and one where a negative population growth leads inexorably to extirpation. Intervention may be needed to eliminate, or at least drastically reduce, the predator threat through the initial postrelease period. Once the herbivores are secure in their resource relationship, they are better able to confront the predation challenge.

Follow-up monitoring of animals postrelease is also crucially needed to identify resource inadequacies before they threaten population viability. Behavioral indicators of foraging efficiency, in particular time allocation both over the day and during foraging spells, should be coupled with spatial assessment of habitat-use patterns. In addition, seasonal changes in diet quality should be assessed from fecal analysis.

Summary

There is quite widespread evidence that the reproductive success and population growth rate of introduced ungulates are often poor in the first few years postrelease, but then tend to improve at a later stage in the same habitat. I suggest that this pattern could be due to the time required to learn where to find and how best to exploit the food resources needed at different stages of the seasonal cycle, and hence to initially inefficient foraging behavior. This is conjectural—only for the Arabian oryx introduction have aspects of foraging behavior been studied in any detail. Other factors could also contribute to the poor population performance initially: physiological adjustments, social disruption, and vulnerability to predation. However, susceptibility to predation can also be heightened by nutritional shortfalls, and inefficient foraging exposes animals to greater predation risks.

Studies of foraging behavior reveal that ungulates are innately predisposed to consume particular plant species but also adjust their diet choice quite rapidly and sensitively to changing resource availability, probably by assessing the postingestive consequences of different forage species. The manipulation techniques initially learned to handle particular food types may be less flexibly altered for mature animals than for younger ones. Animals may require a long time, perhaps several years, to find and efficiently exploit the resources needed to bridge crucial periods of the seasonal cycle. This may result in reduced foraging efficiency and hence curtailed population performance during the first few years in a novel environment. Exactly how managers should cope with such foraging limitations following translocations constitutes the practical conservation dilemma. Recommendations include active intervention through the initial years, in the form of critically ssessed food supplementation coupled with predator suppression where necessary.

Part III
Wildlife Management

Most of the chapters in this section are concerned with sport hunting of large mammals, but the problems considered and the ideas proposed here apply to other groups of animals that are harvested by humans. Sport hunting is the prevailing form of exploitative wildlife management in much of the world, and certainly in Europe and North America.

What does the knowledge of animal behavior have to offer to wildlife management? Wildlife managers are mostly interested in the number of animals available and the consequences of alternative harvesting strategies, so they are mostly concerned with population dynamics. Increasingly, however, managers are interested in individual quality, particularly in the case of trophy hunting, and recognize that the inclusion of species-specific behavior can be a powerful tool for choosing among different management strategies.

The importance of individual differences reverberates through all the chapters in this section. First, Jean-Michel Gaillard and coauthors examine how differences in behavior among species affect the reliability of different census techniques. It may seem incredible that a species could be managed without managers knowing its abundance, yet for roe deer all census techniques fail miserably to provide an adequate measure of population density. The only way to count roe deer is to mark most individuals and then apply capture–mark–recapture methods over several years. For other species, the reliability of censuses may vary with population density. In the second part of their chapter, Gaillard et al. examine how the incorporation of details on population structure may help provide more accurate predictive models of population dynamics. Individuals of different sex and age have different reproductive strategies, and so they have different probabilities of contributing to population dynamics by either reproducing or remaining alive. Information on sex/age composition of exploited populations is therefore particularly useful for managers.

The outcome of predator–prey interactions is affected by the behavior of both prey and predator. As large predators regain some of the ground they lost in the past century, particularly in mountainous areas of Europe and North America, they encounter prey individuals that appear to have forgotten what predators are and how to avoid them, sometimes with dramatic consequences. Joel Berger and collaborators explore some of the consequences of changes in community composition due first to the long-term absence of large predators, and then to their reestablishment, either naturally or through reintroduction programs. The importance of behavior in the planning of reintroduction programs was already underscored in chapter 7 and is further emphasized by Marco Apollonio and coauthors, particularly with regard to the reintroduction of large predators. Few such introductions have been attempted, and the ongoing experimental reintroduction of brown bears (together with the natural recolonization by wolves) in the Alps will tell us a lot about whether modern humans are able to coexist with large predators. The behavior of both species is crucial to their return to the Alps.

Unfortunately, just as some moose in Wyoming do not know how to behave toward large predators, humans in the Alps have forgotten how to behave toward bears and wolves. Much of the current behavior of humans in the Alps, particularly several agricultural practices, is incompatible with large predators. Recovering predator populations are rapidly changing the dynamics of many areas where sport hunting of ungulates has gone on for decades in the absence of any predator-induced mortality on adults. In some of these areas, managers are unfamiliar with the behavior of large carnivores, and results from areas where carnivore populations have persisted may not necessarily be an appropriate guide for how to deal with recovering populations.

Recolonizations by large predators are certain to provide many challenges to conservationists and wildlife managers. At a time when conservation biology mostly deals with losses and extinctions, it is refreshing to realize that some management problems are due to increases in biodiversity. Wolves, bears, and lynx are returning to parts of their historic range in the Alps, the Pyrenees, the Rockies, and Scandinavia. In Sweden, bear populations have recovered to the point where sport hunting of bears has increased. Brown bears are also hunted in Canada, Alaska, and Russia. Normally, one would expect that in a polygynous species it should be possible to harvest a considerable proportion of males because one male can breed with several females. Jon Swenson, however, uses his long-term research on bears in Scandinavia and literature on other large carnivores to suggest that in some cases male harvest may have a greater impact on population dynamics than we may suspect. Individual behavior, once again, is at the base of that suggestion: if surviving male bears kill cubs fathered by bears shot by hunters, the killing of an adult male may have an impact on population growth rate similar to the killing of an adult female.

Managing for sport hunting poses a number of ecological, social, and economic challenges to wildlife biologists. Some of those challenges can only be met by taking animal behavior into account. Inevitably, harvest has a quantitative impact on populations, but over the long term it may also have a qualitative impact. Behavioral ecologists are used to thinking of evolutionary questions, and it may benefit wildlife managers to also think of the potential evolutionary impacts of different harvest strategies, as examined by Marco Festa-Bianchet in the last chapter of this section. Once again, emphasis on individual differences leads one to view the impacts of sport hunting under a broader spectrum than that provided by a simple consideration of numbers counted, numbers shot, and numbers likely to be available next year.

8.

Variation in Life History Traits and Realistic Population Models for Wildlife Management:

The Case of Ungulates

Jean-Michel Gaillard, Anne Loison, and Carole Toïgo

Most populations of large vertebrates are now the target of intensive management or conservation programs (Caughley and Sinclair 1994). These programs usually entail a four-step process (five steps if we include the initial choice of a management goal): (1) assessment of population status using several measurements of population parameters, such as survival and reproductive rates, habitat quality, or animal condition; (2) some measure of population performance to synthesize the different measures performed during the population assessment stage (at this point, managers know whether the population is declining, stable, or increasing); (3) deciding what strategy is most appropriate to attain the goals of a management or conservation plan to balance the observed performance with the desired status of the population; (4) forecasting the population effects of a given management or conservation action so managers can assess the effectiveness of their strategy. At each of these four steps—parameter estimate, population performance assessment, decision, and forecasting—population models play an important role, and the choice of model may affect the outcome of each step.

For population assessment, field data are usually collected through some sampling procedure. Various models can then be used to estimate population parameters from the field. For example, monthly observations of previously marked animals can be collected from a sampling design stratified according to habitat type. Then a capture–mark–recapture (CMR) model (see Schwarz and Seber 2001 for a review) may be used to estimate habitat-specific population size and/or monthly survival.

To obtain a global measure of population performance, information provided by each of the population parameters during the first step has to be combined. Once again, models are required to perform this task. Returning to our example, habitat-specific estimates of population size can be modeled through linear regressions to assess population growth rate (Lebreton and Millier 1982).

To reach their management goal, managers need to identify the target of management actions. Suppose that the goal was to maintain a stable population, but the population actually grew by 30%. Demographic models could simulate different scenarios and suggest which harvest strategy would stabilize the population. Lastly, simulations of the expected consequences of a management strategy could assess whether it had the desired results.

Intensive monitoring programs generally focus on two broad types of populations for which both analyses and currencies generally differ. First, for endangered populations, Population Viability Analysis (PVA) is the most common type of model used. Extinction risk is thus often the currency (Boyce 1992). Population dynamics are also often performed for exploited populations. Here, the currency is the population growth rate or the natural rate of increase (Tuljapurkar and Caswell 1996). This chapter focuses on ungulates, a group that is intensively managed all over the world (Nowak 1991) and whose population dynamics have been intensively studied, particularly in temperate areas (Gaillard et al. 2000).

Historically, two approaches have been used to manage ungulate populations: time series of population counts, which generally consist of yearly estimates of population size, and demographic models based on yearly estimates of fitness components. After briefly reviewing the basic principles of both approaches, we present case studies to highlight the current limits of the models, and then demonstrate how they can be improved by accounting for life history variation among sex/age classes and among individuals.

Using Population Counts: Principles and Limits

Although several methods are available to estimate population growth rate from a time series of population counts (Lebreton and Millier 1982), regressing yearly estimates of population counts (after log-transformation) on time is the most commonly used procedure: the slope of the regression is the population growth rate. Several problems, however, affect the estimation of yearly population size. Despite many improvements in field techniques, sampling designs, and statistical procedures (Seber 1986; Buckland, Goudie, and Borchers 2000; Schwarz and Seber 2001), counts have generally low precision (the coefficient of variation is seldom less than 20%; Caughley 1977) and low accuracy (Strandgaard 1972), and variation in counts is often difficult to interpret (Morellet et al. 2001). Two case studies will demonstrate the magnitude of such problems in ungulate populations.

COUNTING ROE DEER

The roe deer (*Capreolus capreolus*) is a medium-sized, inconspicuous forest-dwelling species. Given these characteristics, we can expect rather large biases in assessment of population size. Indeed, severely biased estimates of roe deer population size have been reported (Strandgaard 1972, Pielowski 1984). In both of these studies, a total removal of roe deer showed that true population size had been underestimated by a factor of three. These experiments, however, do not provide information about whether field assessments of population size over time tracked the real variation of population size over years.

The intensive, long-term monitoring of the roe deer population in Chizé (West France), provides the opportunity for such a test (Gaillard et al. 1998b). Because about 70% of roe deer have been individually marked since 1979, we can confidently assume that the CMR estimates of the population are accurate (Strandgaard 1967, Gaillard et al. 1993). Therefore, we tested whether indices of yearly population size estimated from road counts at night, a census method commonly used for roe deer (Maillard, Gaultier, and Boisaubert 1999), tracked the yearly population size estimated from CMR. Although roe deer population size varied from about 200 in 1980 to 500 in 1984, yearly variation in the number seen per kilometer was not correlated with the yearly variation in population size (Fig. 8.1), indicating that night counts are useless to monitor temporal variation of roe deer population size. Individual variation in behavior likely affects the detection probability of roe deer (Ellenberg 1978), as do observer differences in the ability to detect roe deer from the road (Delorme 1989, Van Laere et al. 2001).

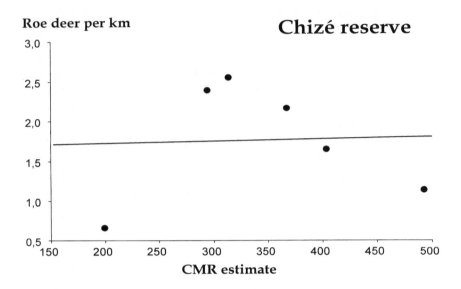

FIG. 8.1. Capture–mark–recapture (CMR) estimates of the size of the Chizé roe deer population (West France) and number of roe deer seen per km of road during night counts over 6 years.

CENSUSING IBEX

Our second example comes from studies of Alpine ibex (*Capra ibex*), a species that contrasts markedly with roe deer: ibex are large, usually live in open mountainous areas, and are very conspicuous (Couturier 1962). Consequently, managers often rely exclusively on ground counts to assess status of ibex populations (Girard 2000). Therefore, we expected ibex counts to be rather accurate. The long-term monitoring of the ibex population of Belledone (France) allowed us to test the reliability of ibex counts. Many ibex at Belledone have been individually marked since 1983, allowing for a CMR estimate of population size. We measured the proportion of underestimation as (CMR estimate–number of ibex counted)/CMR estimate (see Toïgo 1998 for details). Contrary to our expectation, counts consistently underestimated population size by at least 20% (Table 8.1). Moreover, the rate of underestimation increased markedly with population size: less than half the total number of ibex was counted as the population increased beyond 100 (Table 8.1). Differences in ibex behavior according to age and sex class may account for some bias in counts. Thus, although an unbiased count would lead to a

TABLE 8.1. Reliability of total population counts of a colonizing alpine ibex population (Belledone, France).

YEAR	TOTAL COUNT	RATE OF UNDERESTIMATION
1988	39	22
1989	41	37
1990	35	36
1992	35	62
1993	110	43
1994	96	60

Reliability is measured by the rate of underestimation assessed from capture–mark–recapture (CMR) estimates of population size (see Toïgo 1998 and text for details).

sighting rate of 1.00 whatever the ibex category, a study in Les Ecrins National Park revealed between-sex differences in sighting rates. In the colonizing population of Champsaur, sighting rates of 0.84 and 0.92 were estimated for males and females, respectively (Spaggiari 2000). Individual heterogeneities in sighting rates may be higher in well-established populations of ibex, where both nursery groups and male groups are large (Gauthier et al. 1994; Toïgo, Gaillard, and Michallet 1997), increasing the probability that some individuals will be missed.

The case studies of roe deer and ibex underline severe problems with the use of population counts to manage ungulate populations. It is well known that most counts underestimate population size. Our examples highlight two other major pitfalls of population counts that may invalidate them as management tools: inability to track yearly changes in population size (in roe deer) and increasing rate of underestimation with increasing density (in ibex). The latter problem may render hopeless the search for a correction factor, often used to calibrate population indices. Therefore, the main weaknesses of the approach based on counts occurs at the population assessment step. When population counts are applied to the management of ungulates, another major problem occurs at the third step of the process, when a decision has to be made. Population counts only provide a global measure of performance. Even assuming that counts are reliable, however, population growth rate does not tell us how to reach a management goal, and additional information is required. Demographic models built on field assessment of fitness components provide such information.

Demographic Models: State of the Art and Possible Improvements

Demographic models are usually based on population matrices (see Caswell 2001 for a review). Among such models, the simple Leslie matrix model (Leslie 1945), a deterministic, age-structured, and female-dominated model, is most commonly used to manage ungulate populations (Eberhardt 1991). Leslie matrix models can estimate the asymptotic natural rate of increase (λ), the asymptotic stable age structure and the generation time (the mean weighted age of reproducing females). These models require mean yearly estimates of fitness components, including reproductive parameters such as age of primiparity, age-specific proportions of breeding females, age-specific fertility (average number of females produced per female), juvenile survival from birth to 1 year of age, and age-specific survival of yearlings and adults.

Consider an ungulate population where females first breed at 3 years of age, 95% of females 3 years or older breed every year and produce one off-spring (therefore have a fertility of 0.5 assuming an even sex ratio), 65% of newborns survive over their first year, 80% of 1-year-olds survive over their second year, and 90% of adults survive from year to year. That population will have a λ of 1.12 and a generation time of 6.08 years. Leslie matrix models have high heuristic value (Eberhardt 1991) but only account for age varia-tion. Although age is undoubtedly a main structuring factor of vertebrate populations (Charlesworth 1994), life history variation also originates from many other factors. Previous studies have reported that temporal variation generated by density dependence (Fowler 1987) and/or environmental varia-tion (Newton 1998), sex (Short and Baladan 1994), spatial structure (Gilpin and Hanski 1991; Milner-Gulland, Coulson, and Clutton-Brock 2000), phenotypic (Sauer and Slade 1987) or genotypic (Moorcroft et al. 1996, Slate et al. 2000) quality, and infrapopulation structures such as cohort (Albon, Clutton-Brock, and Guinness 1987; Gaillard et al. 1997; Coltman et al. 1999a) and family (Gaillard et al. 1998a) may influence markedly population dynamics of vertebrates. In the following, we will assess whether accounting for environmental variation, cohort variation, and between-sex differences in fitness components will improve our understanding of ungulate population dynamics.

ENVIRONMENTAL VARIATION: ROE DEER AT CHIZÉ

The roe deer population at Chizé has been monitored since 1977. In this fenced forest of 2614 ha, animals were marked either as newborns in May–June or as weaned fawns in January–February, and were thereby of

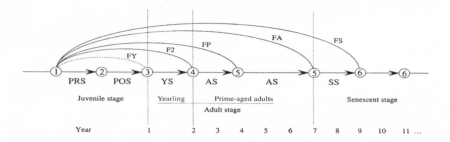

FIG. 8.2. Life cycle of roe deer: (1) newborn, (2) weaned young, (3) yearling, (4) 2-year-olds, (5) prime-aged adults, (6) senescent adults (older than 7 years). Straight lines indicate transitions from one age group to the next, curved lines indicate reproduction. These fitness components describe the development of individuals through the life cycle:

> PRS: summer survival of fawns; POS: winter survival of fawns; YS: yearling survival; AS: prime-age adult survival; SS: senescent survival; FY: fecundity of yearlings; F2: fecundity of 2-year-old females; FA: fecundity of prime-aged females; FS: fecundity of senescent females.

known age. Here we use data collected over 15 years (1985–1998) during which more than 70% of the deer were individually marked, providing reliable estimates of required fitness components from CMR modeling (Gaillard et al. 1992, 1993, 1997, 1998b). As is typical for ungulates in temperate areas (Gaillard et al. 2000), the life cycle of roe deer has three main stages (Fig. 8.2): a juvenile stage subdivided into summer (survival from birth to weaning) and winter (survival from weaning to 1 year), a prime-age stage pooling the nonreproductive yearlings with adults aged 2 to 7 years (Gaillard et al. 1993) that reproduce annually (Gaillard et al. 1992), and a senescent stage (beyond 8 years of age) during which both survival and fertility decline (Gaillard et al. 1998b).

Mean estimates of fitness components (Fig. 8.3) were entered in a Leslie matrix model to calculate the natural rate of increase and the mean generation time. The results showed that between 1985 and 1998 the population had a mean natural rate of increase of 1.188. Generation time was 5.37 years. We then performed a sensitivity analysis to determine which fitness component was the most influential for roe deer population dynamics. A prospective analysis of perturbation (sensu Caswell 2000) based on elasticity (corresponding to the relative sensitivity, de Kroon et al. 1986) showed that survival of prime-aged females had the highest elasticity (0.698), whereas the breeding

CHIZÉ

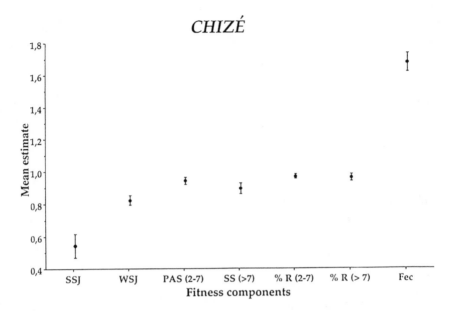

FIG. 8.3. Mean estimate of fitness components obtained from the capture–mark–recapture (CMR) monitoring of the Chizé population (West France) between 1985 and 1998.

proportion of females older than 7 years had the lowest elasticity (0.038). The elasticities of fawn survival (0.191), fertility (0.191), and breeding proportion of prime-aged females (0.111) had intermediate values. These results suggest that the impact of a given variation in survival of prime-aged females on the natural rate of increase is at least 3.65 times higher than the impact of the same variation in any of the recruitment components (juvenile survival, age-specific breeding proportion, and fertility). From the prospective analysis, we may thus conclude that survival of prime-aged females is the critical component of roe deer population dynamics.

Two main limitations are inherent to prospective analyses. First, from a biological viewpoint, the notion that prime-age survival is key to explaining variation in growth of ungulate populations may appear trivial. Indeed, the existence of a "slow–fast continuum" in mammalian life history strategies has been repeatedly demonstrated (Stearns 1983, Gaillard et al. 1989, Read and Harvey 1989, Sæther and Gordon 1994). Ungulates obviously belong to the slow end of the continuum by showing a covariation between large body size, low fertility, and high life expectancy. Therefore, the high elasticity of prime-age survival for population growth of roe deer can be viewed as a simple

TABLE 8.2, Ratios between elasticity of adult survival and elasticity of juvenile survival in ungulate populations.

SPECIES	ELASTICITY RATIO
Pronghorn	8.53
Moose	6.77
Reindeer	6.48
Mountain goat	5.99
Caribou	5.63
Greater kudu	5.30
Red deer	4.81
Mule deer	4.45
Soay sheep	3.83

Ratios indicate by how much more a given change in adult survival will affect the natural rate of increase than the same change in juvenile survival (Gaillard and Yoccoz, unpublished data).

consequence of the covariation between body size and generation time. Supporting this hypothesis, prospective analyses of nine other ungulate species consistently showed that prime-age survival had elasticities between four and nine times higher than those of juvenile survival (Table 8.2). Prospective analyses also have a methodological limit because they do not account for temporal variation in fitness components. Thus the relative importance of fitness components is determined by assuming that all fitness components have a similar level of temporal variation. But is that a reasonable assumption? To answer this question, we used the coefficient of variation (CV) of fitness components calculated from the 15-year time series available at Chizé as a measure of temporal variation. Our results clearly showed large differences of temporal variation among fitness components. Thus summer survival of fawns was highly variable (CV = 0.529), whereas age-specific breeding proportions of females (CV = 0.056 and CV = 0.084 for prime-aged and old females, respectively) and survival of prime-aged females (CV = 0.094) varied only little over the years. Winter survival of fawns (CV = 0.134), yearly survival of old females (CV = 0.135), and fecundity (CV = 0.131, all females combined because we found no differences in litter size among age classes; Gaillard et al. unpublished data) had intermediate levels of temporal variation.

There appears to be a negative correlation between elasticity and temporal variation of different fitness components: components with a strong impact on population growth rate tend to have low temporal variability (see Gaillard et al. 2000 on ungulates, Sæther and Bakke 2000 on birds, Pfister 1998 on a variety of organisms). Thus, in roe deer at Chizé, survival of prime-aged females had high elasticity and low temporal variability, whereas the various components of recruitment showed the opposite covariation of low elasticity and high temporal variability. This demographic pattern seems to be a characteristic feature of ungulates. We previously underlined that demographic analyses of ungulate populations consistently reveal that prime-age survival has the highest elasticity (Table 8.2). Likewise, recruitment parameters typically have high variability (Gaillard et al. 2000). To manage ungulate populations, it is therefore important to account for the differences among fitness components in susceptibility to environmental variation. Retrospective analyses of perturbation (sensu Caswell 2000) are designed for such a task. We thus performed a retrospective analysis on the Chizé roe deer population. Retrospective perturbation analysis involves a decomposition of changes in natural rate of population increase (variance $[\lambda]$) according to fitness components into two parts, the elasticity (e), or the potential impact of a given component on population growth and the coefficient of variation (CV), or the observed variation of a given vital rate (Tuljapurkar 1990, Brault and Caswell 1993, Caswell 2000):

$$\text{Variance } (\lambda) = \Sigma_{\text{ fitness components}} (e^2 \times CV^2) \qquad (1)$$

From Equation (1), it is clear that a fitness component with an elasticity of 0.1 and a CV of 0.5 will contribute equally to changes in population growth as a fitness component with an elasticity of 0.5 and a CV of 0.1 (same value for $e^2 \times CV^2$). Using this method, we can assess whether the differences among fitness components in temporal variation balance the corresponding differences in elasticity. We found that the proportion of variance in roe deer population growth that is accounted for by different fitness components is highly variable. Thus summer survival of fawns accounted for more than 60% of the variation in population growth of Chizé roe deer, whereas adult survival accounted for only 25% of the variation. Other fitness components had a very low influence on roe deer population growth because they had low elasticities and small coefficients of variation. Therefore, retrospective analysis suggests that summer survival of fawns accounts for about 2.37 times more variation in natural rate of increase than what can be accounted for by survival of prime-aged females. Summer survival of fawns is therefore the critical component of roe deer population dynamics in Chizé.

Prospective and retrospective analyses of perturbation therefore appear to provide two very different interpretations of what is driving roe deer population dynamics, because prospective analyses clearly indicate that survival of prime-aged females is the vital rate with the highest elasticity, but retrospective analyses suggest that fawn summer survival accounts for much of the observed temporal variability in population growth. These contrasting interpretations would likely lead to radically different management strategies.

What should a manager trust: prospective or retrospective analyses? Prospective and retrospective analyses answer different questions (see Caswell 2000 for a detailed discussion). Prospective analyses can identify the fitness component that, if it were to vary, would have the greatest influence on population dynamics. For roe deer, that fitness component is the survival of prime-aged females. If some factor, either artificial or environmental, will lead to changes in survival of prime-aged females, that factor will drive future changes in roe deer population sizes. On the other hand, retrospective analyses quantify the respective influences of fitness components from empirical observations. At Chizé, summer survival of fawns drove changes in roe deer population size, and is likely to remain the main driving force as long as environmental variation remains within the range observed during the study. To manage exploited populations, we suggest that retrospective analyses would be better suited for generating decision rules provided that they were based on a monitoring period long enough to be representative of future ecological conditions.

Because our analysis of roe deer was based upon 15 years of monitoring, we suggest that summer survival of fawns is likely to be much more influential in shaping future variations in population size over time than survival of prime-aged females. In that particular case, accounting for environmental variation markedly changes the management decisions. Of course, managers should always assess whether they monitor populations long enough to obtain a reliable picture of the influence of environmental variation. Detailed long-term studies of populations are required at this stage. If, based upon prospective analyses, managers choose to monitor adult survival of females, they will hardly detect any changes in population dynamics that may occur, and will not be able to track changes in population size. On the other hand, if, based upon retrospective analyses, managers choose to monitor recruitment of fawns in the winter population, they will be able to track changes in population size more closely and adjust hunting quotas according to the observed yearly variation in recruitment. By only monitoring recruitment, however, managers will not detect any changes in predation or human activities that may affect mortality of adult females. Under these exceptional conditions, monitoring recruitment can then lead to overhunting and, possibly, extinction of local populations.

COHORT VARIATION: ROE DEER AT TROIS FONTAINES

Trois Fontaines is an enclosed forest of 1360 ha. Roe deer are marked either as newborn in May–June or as fawns in January–February, and are thereby of known age. The sampling design, based on CMR modeling, provided reliable estimates of all fitness components (Gaillard et al. 1992, 1993, 1997, 1998b). We used data from seven cohorts (1976–1982) of females monitored from 1975 to 1999 because to reliably assess between-cohort differences in natural rate of increase, one should wait until all females from all cohorts have died. To avoid the large variation in first-year survival among cohorts (Gaillard et al. 1997, 1998b), we considered only individual performance from 2 years of age onward.

We first estimated the natural rate of increase of the population by applying a Leslie matrix model to survival and reproductive data for 43 females that survived to the age of first reproduction and were monitored throughout their lifespan. That estimate did not account for between-cohort variation. We then estimated cohort-specific fitness components and built Leslie matrix models to estimate the natural rate of increase for each cohort. Lastly, as a measure of cohort variation, we estimated the cohort-specific population doubling time. Using the mean fitness components of females born between 1976 and 1982, we found that the Trois Fontaines population increased over time. The estimated natural rate of increase of 1.385 is typical of colonizing populations of roe deer (Gaillard et al. 1998a) and is close to the value obtained from tranversal age-specific estimates of fitness components ($\lambda = 1.372$; Gaillard 1988). Cohort differences, however, occurred in both age-specific reproductive performance (the number of daughters weaned per female, Fig. 8.4a) and age-specific survivorship (measured by survival curves, e.g., Caughley 1977, Fig. 8.4b). Therefore, natural rates of increase differed sharply among cohorts. Females born in 1977 had the highest λ (1.534), whereas females born in 1982 had the lowest λ (1.265, Table 8.3). Differences in λ, led to marked among-cohort differences in the time required for the population to double (Table 8.3).

It is somewhat surprising that we found evidence for such large cohort variation in λ, because yearly removals maintain the Trois Fontaines roe deer population at low density, simulating a colonizing regime (Gaillard et al. 1993, 1998a). Resources are generally abundant, and very little cohort variation occurs in phenotypic quality (Gaillard et al. 1997). It is likely that cohort-specific growth rates would be much greater in resource-limited populations, such as the one at Chizé, that exhibit strong cohort variation in adult mass (Gaillard et al. 1997, Pettorelli et al. 2002). Therefore, optimal

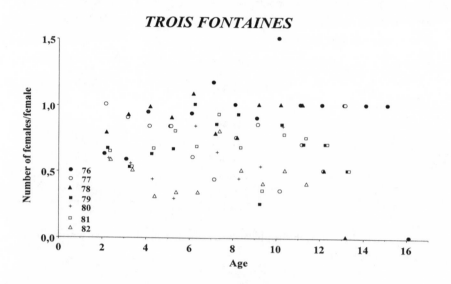

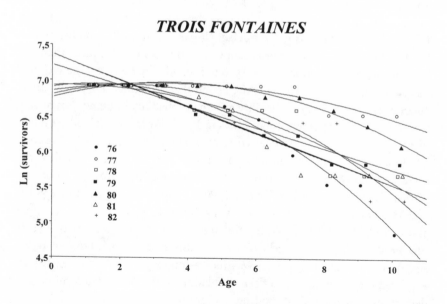

FIG. 8.4. Age-specific (a) reproductive performance and (b) survival curve for seven cohorts of roe deer females monitored from 2 years of age until death at Trois Fontaines (East France, cohorts born in 1976–1982).

TABLE 8.3. Cohort-specific population increase (λ) and cohort-specific time (in years) required for the population to double (TD) for roe deer at Trois Fontaines (East France, see text for details).

COHORT	λ	TD
1976	1.397	2.07
1977	1.534	1.62
1978	1.468	1.81
1979	1.348	2.32
1980	1.347	2.33
1981	1.349	2.32
1982	1.265	2.95

management of roe deer should account for population history and differences in cohort quality.

BETWEEN-SEX DIFFERENCES: RED DEER AT TRONDELAG

To assess whether accounting for between-sex differences in fitness components could make a difference for managers, we used the data collected on red deer (*Cervus elaphus* in Trondelag, Norway, from both monitoring and hunting of known-aged females (Langvatn and Loison 1999). More than 300 deer of each sex (over 600 in total) were individually marked. The red deer data were analyzed by using a CMR design that accounted for yearly differences in both resighting probability and the probability of being shot during the hunting season (Langvatn and Loison 1999). We thus obtained estimates of the fitness components required to estimate the natural rate of increase. We first assessed the potential dynamics of the red deer population by excluding the effects of hunting. Without hunting and irrespective of sex, 75% of fawns survived their first year, 90% of yearlings survived to 2 years, and 93% of females older than 2 years survived each year. Litter size was fixed at one, and 70% of 2-year-old females and 98% of females older than 2 years produced an offspring. A Leslie matrix with these estimates led to λ of 1.191, suggesting that the population was increasing rapidly. As expected for an exploited population, the red deer population at Trondelag had a colonizing demographic regime when hunting was ignored.

We then analyzed the effects of hunting on population demography by using simulations. We set hunting pressure at 10% and estimated the population kinetics over 20 years. The initial population size was set at 1000 red deer. In a first simulation, we accounted for age-structure as previously done for assessing the potential dynamics. The 1000 deer included 160 juveniles,

SEX-BIASED SURVIVAL IN UNGULATES

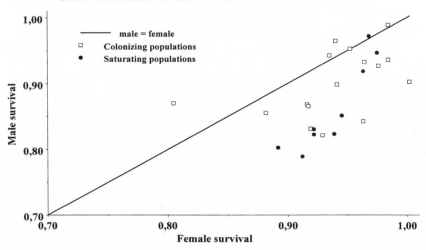

FIG. 8.5. Relationships between adult male survival and adult female survival in ungulate populations. Overall, survival is biased in favor of females because most points are below the line of equal survival between sexes. The magnitude of the sexual differences is higher for populations close to saturation (with density-dependent responses and/or resource-limited, filled points) than for colonizing populations (open squares). Toïgo and Gaillard, unpublished data.

100 yearlings, and 740 adults. We found an asymptotic rate of increase of 1.090 for this red deer population. That first model did not account for between-sex differences, but evidence for sex-biased survival has often been reported in ungulate populations (Clutton-Brock, Guinness, and Albon 1982; Jorgenson et al. 1997; Berger and Gompper 1999).

We obtained mean estimates of sex-specific adult survival in ungulate populations from the literature. Females survived better than males in most populations of polygynous ungulates (Fig. 8.5). Female-biased survival was greater in populations close to saturation than in colonizing populations (Fig. 8.5), suggesting that sex-biased survival is pervasive among populations of polygynous ungulates, especially at high density.

Does sex-biased survival affect management of exploited populations? To simulate the situation where male survival is lower than female survival, we assumed that only 85% of yearling and 80% of adult males survived from 1 year to the next (compared to 90% and 93% in females). The simulation over 20 years with an age- and sex-structured Leslie matrix model

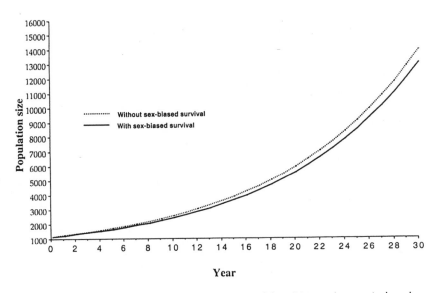

FIG. 8.6. Population kinetics of red deer at Trondelag (Norway) expected under a hunting pressure of 10% with (solid line) and without (dashed line) accounting for between-sex differences in survival (see text for details).

led to exactly the same natural rate of increase of 1.090. Accounting for between-sex differences in survival did not change the estimated population growth, assuming that availability of males does not limit female reproduction (but see Ginsberg and Milner-Gulland 1994). The red deer population estimated from the age- and sex-structured model, however, was consistently lower than that estimated from the age-structured model, the between-model differences increasing over time (Fig. 8.6). Although it did not change the asymptotic λ, low male survival decreased population size and should therefore be accounted for in management decisions. For example, consider a population of 100 deer of each sex with a λ of 1.20 constant over time. Let's assume this population is hunted (with a balanced sex ratio) at the maximum sustainable yield calculated from a model that ignores the lower survival of males. The hunting bag will be 20 individuals of each sex. However, due to higher mortality, male recruitment will not compensate the loss from natural mortality and hunting. The number of males will decrease over time until extinction. The population therefore will go extinct. For species where sex-biased survival is likely to occur, management must be based on sex-structured models to prevent overexploitation.

Conclusions and Recommendations

Our analyses of ungulate case studies highlight three main results that may affect wildlife management. First, although counts appear to be reasonably reliable for monitoring colonizing populations of ungulates, they are most often both inaccurate and imprecise for well-established populations at intermediate and high densities. Many ungulate populations in Europe and North America increased steadily during the last decades (Gill 1990) and are currently at high density (McShea, Underwood, and Rappole 1997; Maillard, Gaultier, and Boisaubert 1999). Consequently, counts can no longer be considered the basic management tool of ungulate populations. Indicators of population performance such as body mass (Gaillard et al. 1996), offspring sex ratio (Kohlmann 1999), browsing pressure (Morellet et al. 2001), or age-dependent survival of radio-tracked animals (Bowden, White, and Bartmann 2000) may be more useful than counts to manage abundant ungulate populations. Measures of population performance such as r or λ are often required by managers, but do not provide sufficient information for efficient management of ungulate populations. A given population growth rate may correspond to different covariation among fitness components and thereby to different population dynamics. For a given population growth rate, environmental variation, cohort variation, and sex-bias in survival all have a strong influence on selecting the optimal management strategy. Indeed, accounting for temporal variation in fitness components often leads to identifying a critical life history stage which differs from that identified by the usual deterministic models. Likewise, cohorts may have markedly different growth rates even in highly productive populations that are rather constant in size over years. Such cohort-specific performance demonstrates that population history plays a determinant role in population dynamics (Coulson et al. 2001) and should be accounted for in management actions. Sampling biases due to cohort effects may bias estimates of population growth rate. Lastly, because most ungulates are polygynous and dimorphic in size (Loison et al. 1999b), males are expected to have lower survival than females, especially in high-density or resource-limited populations. Under such conditions, sex-specific models should be preferred to the female-dominant models that are usually applied. Indeed, even if asymptotic population growth rate is not influenced by low male survival because enough adult males remain to permit females to reproduce yearly (Laurian et al. 2000), lower male survival leads to lower population size in a given year. By neglecting sex-biased survival, managers could therefore overexploit ungulate populations. We conclude confidently that accounting for variation in life history traits provides a first step toward more realistic models of ungulate populations, which will be relevant to management decisions.

Summary

Wildlife managers need information on temporal trends in the size of exploited populations. To obtain this information, one may either analyze a time series of population counts or build a demographic model to estimate the rate of increase (λ or its antilog, the population growth rate [r]). Inaccurate estimates of population parameters, however, often lead to erroneous assessment of population trends. Total counts generally underestimate population size, especially at high density, and have low precision (CV usually > 20%). Some biases in population counts may result from among-individual differences in sighting probability. Census techniques should thus account for behavioral factors that affect sighting probabilities. In the cases examined here, demographic models are used to estimate population trends and to identify the critical parameters responsible for changes in population size. The age-structured Leslie matrix model cannot provide a satisfactory picture of most ungulate populations because it typically does not account for variation in life history traits arising from factors other than age. After reviewing evidence of environmental variation, cohort effects, and between-sex differences in ungulate populations, we show that (1) when temporal variation in fitness components is accounted for, often the key parameter affecting changes in population size differs from the one identified by using a deterministic model; (2) because cohort-specific growth rates may vary substantially, the growth rate calculated at the population level may be biased by sampling heterogeneities among cohorts; and (3) sex-specific models should be preferred to female-dominated models when survival patterns differ between sexes. We conclude that accounting for environmental variation, cohort variation, and between-sex differences would lead to more realistic models of ungulate populations.

9.

Through the Eyes of Prey:

How the Extinction and Conservation of North America's Large Carnivores Alter Prey Systems and Biodiversity

Joel Berger, Steve L. Monfort, Tom Roffe, Peter B. Stacey, and J. Ward Testa

That predators affect the biology of their prey is something that few, if any, people—scientists, writers, naturalists, or laypersons—would dispute. Such processes were obviously recognized by Kipling in describing how prey species shifted in response to Shere Khan's movements.

> *He has no right, he has no right to change his quarters without due warning. He will frighten every head of game within ten miles.*
>> So said Father Wolf of Shere Khan the tiger, Rudyard Kipling, *Jungle Book*, 1894

Aldo Leopold must have believed similarly in 1922, while searching areas of Sonora, Mexico, where jaguars were likely extinct.

> *We saw neither hide nor hair of him . . . (but) . . . no living beast forgot his presence . . . no deer rounded a bush, or stopped to nibble . . . without a premonitory sniff for el tigre.*

And both Kipling and Leopold recognized an ecological role for large carnivores while simultaneously expressing wonderment for how carnivores, as aberrations, may shape prey behavior.

But, to what extent, if any, do large terrestrial carnivores shape prey behavior and ecology? Some evidence may suggest effects are few. Both in southwestern Greenland and on the Svalbard Archipelago (Norway) at 80° north, caribou (*Rangifer tarandus*) have not encountered wolves (*Canis lupus*) for anywhere from 1500 to perhaps 10,000 years, yet it remains uncertain whether group sizes are shaped more by food than by predation (Boving and Post 1997). Where predation once occurred and has subsequently been lost, have antipredator tactics devolved? If so, who really cares other than perhaps a few academics? What can possibly be learned about the role of predation by studying systems where carnivores are now extinct? And, assuming knowledge might be garnered, is it relevant to the biodiversity crossroads that this planet faces? These queries guide this chapter.

Much is already known about this topic, however, and surprises may be few. For example, although confusion exists about mechanisms of possible population-limiting roles of carnivores (Boutin 1992, Krebs 2000), the evidence may often speak for itself. North American bison (*Bison bison*) once numbered in the millions despite coexistence with wolves and grizzly bears. With these carnivores gone from the prairie ecosystem for more than a century, only a neophyte would be shocked to learn that bison in small reserves enjoy rapid population growth when not limited by food. Nor is it surprising that, perhaps due to a small number of founders, the highly inbred population in the Badlands (of South Dakota) contains a high frequency of malformed juveniles; not only do these bison fail to succumb to predation but they survive and may reproduce (Berger and Cunningham 1994). Community-level effects have also arisen as a consequence of the extinction of wolves on North American prairies. In their absence, coyotes (*Canis latrans*) have proliferated (Bekoff 1977, Crabtree and Sheldon 1999). But, where coyotes are reduced by humans, red foxes (*Vulpes vulpes*) have increased (Sargeant, Allen, and Hastings 1987; Sargeant and Allen 1989), and they currently depress the survival of shorebirds and waterfowl (Sargeant, Allen, and Eberhardt 1984). So, on the one hand, although some effects of losing large carnivores appear highly predictable, others are subtle and occur at multiple trophic levels.

This chapter considers questions about interactions between prey species and large terrestrial carnivores, specifically consequences of carnivore loss, maintenance, and restoration. Although these issues can be restricted to prey behavior exclusively, the results in our opinion would be far less interesting if not linked to broader issues involving ecology and conservation. Although behavior in and of itself is clearly fascinating (Tinbergen 1951, Wilson 1975), in many ways its contribution to conservation may be far stronger when interfaced with population processes and community ecology (Caro 1994, 1998b, Gosling and Sutherland 2000). Therefore, we attempt to offer insights at three levels—behavioral, demographic, and ecosystem—by considering spatial and temporal events for each.

Such queries seem relevant for at least three reasons. Large carnivores are being reduced in many regions: for example, tigers (*Panthera tigris*) in Russia and India, jaguars (*P. onca*) in Latin America, and grizzly bears (*Ursus arctos*) in Canada and the United States. They are also being reintroduced or expanding naturally in others: cheetahs (*Acinonyx jubata*) and lions (*Panthera leo*) in South Africa (Hunter and Skinner 1998), wolves in Italy and the United States (Boyd et al. 1994, Boitani 2000), and brown bears in Europe (Breitenmoser 1998, Swenson et al. 1999). Irrespective of whether extermination or expansion is occurring, knowing something about the possible responses of prey populations and ecosystems will help inform decisions about the use and protection of such regions. In the United States, the reintroduction of wolves into the Greater Yellowstone Ecosystem (GYE) has caused great concern about the short-term viability of some prey populations, particularly elk (*Cervus elaphus*) and moose (*Alces alces*) (Phillips and Smith 1996, Boyce 1999), and any knowledge that contributes to wise decision making is often appreciated. Second, mammalian carnivores, whether terrestrial or aquatic, can contribute significantly to ecosystem processes and the maintenance of biological diversity (Ben-David et al. 1998; Estes, Tinker, and Doak 1998; Crooks and Soulé 1999). Both of these commodities are increasingly being viewed as important to the welfare of the environment, and they tend to have economical impacts on neighboring human communities. Finally, the public, whether urban or rural, Botswanan or British, Mongolian or Mexican, is passionate about wildlife in general and carnivores in particular. Gaining societal support for environmental conservation can come about not only through visiting zoos and watching television but also by the popularization of scientific study. Whatever fuels the public imagination and infuses some sense of curiosity about the natural world must be viewed as positive. Understanding animals is but one of the inspirational ways in which the public becomes interested in science and, hopefully, conservation. It is hoped that the enhancement of knowledge about the effects of the presence and absence of large carnivores on prey systems, behavior(s) included, will improve opportunities for conservation measures.

Berger began work on these issues in 1995, focusing on moose, in Alaska where grizzly bears and wolves are still relatively abundant, and in the southern part of the GYE where these two species had been missing for about 60 to

75 years until recently (details to follow). Why moose? Primarily because their dynamics appeared to be inextricably linked to the distribution of grizzly bears and wolves both in systems where these carnivores still existed and in systems where they were absent. Numerous other North American ungulates, of course, fit the binary criteria of large predators present and absent (e.g., caribou, elk), but study situations involving these other species have usually been less than ideal in that observability might be compromised, or females heavily hunted, or habitats strikingly different. For moose, on the other hand, finer nuances of their interactions with bears and wolves had been studied (Peterson 1977; Ballard, Spraker, and Taylor 1991; Schwartz and Franzmann 1991; Gasaway et al. 1992).

Additionally, in 1994, J. Ward Testa and Terry Bowyer initiated independent studies of Alaskan moose and offered to make either their radio-collared study animals (Testa) or components of their data (Bowyer) available to facilitate Berger's impending efforts. Since then, Berger established control (baseline) values of moose at sites where wolves and grizzly bears have never been extirpated and their densities were not radically modified by humans (Miller et al. 1997; Bowyer et al. 1999a; Testa, Becker, and Lee 2000). Experimental treatments were areas where moose had not encountered wolves and grizzly bears for 60 to 75 years irrespective of the causes underlying carnivore losses.

Demography of Moose in the Absence of Wolves and Grizzly Bears

Although grizzly bears and wolves are often major predators of moose in northern circumpolar systems, as ungulate diversity and the availability of smaller-bodied species increase at more southern latitudes, so does prey switching (Boyd et al. 1994, Weaver 1994, Kunkel et al. 1999), and elk and deer become more favored prey than moose (Craighead, Sumner, and Mitchell 1995; Mattson 1997; Kunkel and Pletscher 1999). A consequence of this shift by carnivores to less formidable prey is that moose may be released from predation.

Accepting assumptions about relaxed predation is very different from knowing whether predation release has truly occurred. If the assumption is wrong, then one may conclude that predation has not shaped a system when the converse may be true. In some systems, small-bodied canids such as red foxes become capable predators of young ungulates, a fact that would have remained unknown in the absence of detailed or comparative study (Aanes and Andersen 1996, Andersen and Linnell 1998). But, ignoring long-term effects of historically extinct predators also carries a potential liability and may even preclude an accurate understanding of current prey adaptive responses (Csermely 1996). The "ghost effects of predators past" may shape the biology of a prey species even if predators have been extinct for thousands of years (Byers 1997). Understanding whether recent anthropogenic-induced loss of predators results in an immediate relaxation of predation is important from both demographic and behavioral perspectives.

Consider for example the broad array of sites in North America where wolves and grizzly bears are extinct. Potential extant carnivores include black bears (*Ursus americanus*) and cougars (*Puma concolor*), both of which prey on elk and mule and white-tailed deer or their offspring (Linnell, Aanes, and Andersen 1995; Smith and Andersen 1996). At sites where these extant carnivores exist but not wolves or grizzly bears, is it reasonable to assume that moose have become predator-free?

The simple answer is no. The assumption of predation release would have been a poor one because both in the Canadian Rockies and on Alaska's Kenai Peninsula, respectively, cougars and black bears may be substantial predators of moose calves or yearlings (Schwartz and Franzmann 1991, Ross and Jalkotzy 1996). By contrast, the assumption of relaxed predation is justified in Scandinavia where neither of these two carnivores exist. In their absence an inverse association between juvenile recruitment and large carnivore extinctions is robust (Swenson et al. 1999).

To examine the assumption of predation release in North America, we contrasted neonate recruitment between Alaskan and Wyoming moose populations. Moreover, the comparisons enabled an evaluation of whether mesocarnivores contributed as replacement predators because, despite the loss of grizzly bears and wolves at the Wyoming sites for more than 60 years, black bears and cougars have always been extant. If these latter two carnivora affect neonate recruitment, then differences between Alaska and Wyoming might be less striking than those reported in Scandinavia.

The results of macrogeographical contrasts fail to support this possibility. Juvenile moose survival to 2 months of age was about three times greater in areas of Wyoming than in Alaska or the Yukon (Orians et al. 1997, Berger et al. 1999). Although these results suggest predation may retard recruitment, they fail to account for other possibilities.

Linnell, Aanes, and Andersen (1995), in a provocative analysis, suggested the possibility that the Bambi syndrome (the loss of young cervids to "big bad" predators) has been partially promulgated by socioeconomic concerns about offspring recruitment and the extent to which prey is subsequently available to humans for meat and for trophy. If, however, young are disproportionately fewer in a population irrespective of the presence of larger

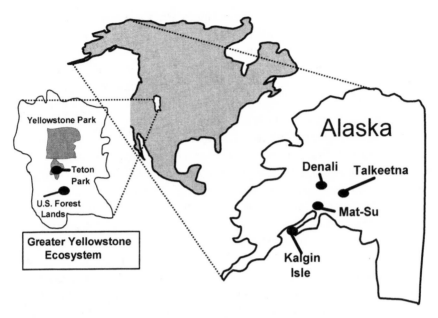

FIG. 9.1. Overview of study areas: carnivores were present at Denali, Talkeetna, and Matunuska-Susitna; carnivores were absent from Teton (until 1990s, brown bears, and 1997, wolves), U.S. Forest Service lands, and Kalgin Island. Brown bears present continuously in Yellowstone Park and a few adjacent (nonpark) areas.

carnivores, the reduction may be due to any number of factors—drowning, light birth mass, low pregnancy rates—some perhaps related to density, others not (Festa-Bianchet 1988b, Gaillard et al. 1993).

To evaluate whether factors other than predation might explain low juvenile recruitment at sites where grizzly bears recently began recolonization (e.g., areas in and adjacent to Grand Teton National Park; Fig. 9.1), we f ocused on a hypothesis alternative to that of predation—that an inadequate food supply limits fecundity and hence neonate production. This hypothesis is amenable to empirical examination because the noninvasive monitoring of fecal steroid hormones permits an unbiased measure of pregnancy (Monfort, Schwartz, and Wasser 1993). The working assumption is that pregnancy rates in food-limited populations will be lower than those in populations with predation. This prediction is anything but novel, having been confirmed empirically for numerous large herbivores (Clutton-Brock and Albon 1989, Sinclair 1989), including moose (Franzmann and Schwartz 1985, Gasaway et al. 1992). But what is important here is that if diminished juvenile representation in a population is explained by food limitation rather than predation, then it becomes possible to test hypotheses about the Bambi syndrome, including the effects of relaxed predation on individual behavior and on community structure.

Predators may kill juveniles, but how did prey fecundity vary in the absence of predation? Because methods have been described in detail elsewhere (Testa and Adams 1998, Berger et al. 1999), only a brief summary is offered here. Pregnancy in restrained females was diagnosed by palpation and follow-up observations of calves in the southern GYE (sGYE) or by analyses of ultrasound, hormones, and observations of calves (Talkeetna Mountains; Testa and Adams 1998; Testa, Becker, and Lee 2000). Assessments in the sGYE were subsequently improved by noninvasive monitoring of fecal progestagen concentration (FPC; Monfort, Schwartz, and Wasser 1993; Schwartz et al. 1995), and with an ecological application involving subsequent births (Berger et al. 1999).

During an approximate 60- to 75-year period, which initially coincided with the local extinction of wolves and grizzly bears, the moose population in the region of Grand Teton National Park experienced exponential growth (Fig. 9.2). During the only period (1963–1966) of this demographic irruption

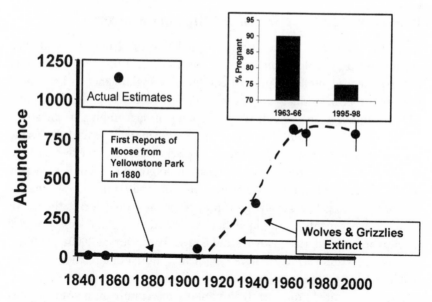

FIG. 9.2. Summary of changes in estimated population size of moose in the Jackson Hole region of the Greater Yellowstone Ecosystem (from Berger et al. 2001). Inset reflects changes in pregnancy rate during the last 30 years (from Berger et al. 1999).

for which data on fecundity were available (Houston 1968), pregnancy rates were 90%, a value consistent with that of other North American populations not limited by food. By 30 years later, pregnancy rates had dropped to 75% (G test for independence; $G_{ad\ j} = 3.36$; $N = 90$; $0.05 < p < 0.10$) during a period in which population growth no longer occurred (Fig. 9.2). These findings are for a population that expanded in an environment free of wolves and essentially free of grizzly bears. The drop in pregnancy rates currently places this population within the lowest fifteenth percentile in North America (Berger et al. 1999).

Although it is unclear how weather, food, or other factors may have contributed to the present relatively low pregnancy rates, populations that are near or below their ecological carrying capacity due to predation (by either natural carnivores or humans) have greater pregnancy rates than those exceeding their food base. Although food quality and availability have not been directly assessed in the sGYE, the decline in pregnancy rates, apparent lack of significant population growth, starvation rate of about 60% (Berger, unpubl.), and general absence of predation on moose are not inconsistent with the possibility of food limitation.

Prey, Scavengers, and Desensitization to Carnivore Loss

Given the marked influence of relaxed predation on demography, to what extent have prey ecology and behavior been affected? In systems that have not experienced predator–prey disequilibria, a major tenet in behavioral ecology is that individuals derive antipredator benefits by group formation, which minimizes an individual's probability of succumbing to an attack (Lima 1987, Dehn 1990). Nevertheless, in a generally asocial, browsing species like moose, it would be difficult to disentangle relative effects of food quality and quantity from those of predation on group formation, especially in the absence of some large-scale field manipulation. An alternative approach to gauge the sensitivity of prey to possible predators is the use of playback experiments, a technique that has proved useful for a variety of taxa (Philips and Alberts 1992, Flowers and Graves 1997, Durant 2000a, 2000b).

Scavengers and carnivores have long held a mutualistic relationship, one that has been conspicuous in the rich folklore of northern boreal zones. Ravens have figured prominently in symbols and culture of the Athabascans (Nelson 1983), and evidence suggests an interdependency involving species such as foxes, bears, wolves, and ravens (Mech 1970, Henry 1986, Heinrich 1989, Peterson 1995). Ravens may be attracted to wolf vocalizations (Harrington 1978). Further south, relationships exist between magpies and predators

(Birkhead 1991), and, perhaps, between prey and scavengers (Stockwell 1991; Genov, Gigantesco, and Massei 1998).

Because scavengers such as ravens are reliant on carnivores to open the carcasses of thick-skinned prey (Heinrich 1989), and are intimately associated with bears and wolves, especially at carcasses (Mech 1970, Mysterud 1973, Craighead 1979), prey species may cue on scavengers to facilitate the early warning or detection of predators. Given the consistent and striking demographic differences in juvenile survival at sites with and without large carnivores (see earlier discussion), we expected deftness in environmental monitoring by prey; that is, vigilance in response to detection of scavengers should be positively associated with predation risk. The converse might also be expected. If recognition of ravens by moose occurs, then following the extinction of grizzly bears and wolves, moose may fail to respond to ravens because the incentive to respond (possible predator detection) is diminished. Neither scenario may, however, be correct. Antipredator responses may be less labile than expected, or the response of moose to the immediate presence of ravens may be independent of the risk of predation.

These possibilities were examined through the use of experimental playback calls in the field. We used six independent study areas, three with intact carnivore communities and three lacking both grizzly bears and wolves (1995–1998). Sites with intact carnivore communities were the Talkeetna Mountains, Denali National Park, and the Matanuska-Susitna Valley regions, areas geographically separated from each other by at least 150 km (see Fig. 9.1). Areas lacking grizzly bears and wolves were Grand Teton National Park and Bridger-Teton National Forest in northwestern Wyoming (each separated by 10–75 km) and Kalgin Island in Alaska's Cook Inlet (a site where moose were transplanted in the late 1950s; Bowyer et al. 1999b). Neither grizzly bears nor wolves occur on the 20 km^2 island, probably due to strong marine undercurrents or a lack of food.

Responses of adult females were assessed during experimental playback of sounds and postplayback periods (Berger 1999). Briefly, sounds were played to adult female moose under four conditions: (1) control (= baseline; no overt or experimental disturbance), and 25 sec playbacks of (2) raven and (3) red-tailed hawk vocalizations, and (4) a dummy sound (running water). Distances between the subject and sounds from a JBL Pro-III speaker connected to a 40 w Kenwood amplifier and powered by a 12 v battery averaged 145 m (SEm = 6.90; N = 203, range 30–800 m). All experiments were on calm days. The response variable was the proportion of time an individual either foraged or was vigilant per 180 sec bout. General linear models were employed for statistical analyses using appropriate transformation (Berger 1999), but data are summarized here as means.

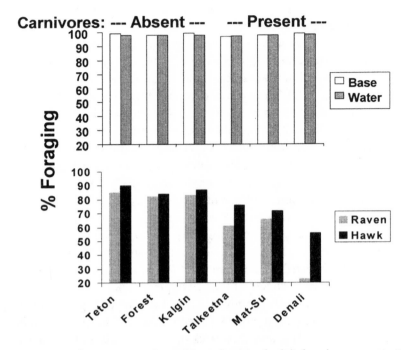

FIG. 9.3. Mean foraging rates (per 180 sec bouts) of adult female moose at sites with and without wolves during two types of auditory playbacks. Sample sizes/site (respectively) are 105, 23, 12, 22, 22, and 25 (modified from Berger 1999).

If predation pressure shaped patterns of foraging, moose from sites with wolves and grizzly bears should feed at greater per capita rates than moose from sites free of wolves and bears. Contrasts among nondisturbed adult female moose involving the six study areas categorized by predation treatment revealed little detectable variation in foraging, nor were differences evident when the sounds of running water were used as a control sound (Fig. 9.3). Therefore, in the absence of disturbance and with exposure to a familiar neutral sound, foraging is independent of predation regime. Hence, observational measures lacking contextual cues associated with predation risk produced no detectable site-specific variation in behavior. If prey foraging was used to assess potential predation risk, then the assumption that differential behavior reflects predation risk would be flawed. Playbacks of familiar sounds of red-tailed hawks and ravens, however, resulted in striking between-site differences that varied by predation pressure. Under relaxed predation, females responded only weakly to hawks or ravens (Fig. 9.3), but at predator-rich sites decreases in foraging were large (both $p < 0.0001$), and greater for ravens than red-tails.

Four additional findings were consistent with the hypothesis that prey were more responsive to scavengers at sites with active predation than at sites with no predators. First, whereas group size had no effect, foraging was less for mothers with calves at sites with predation, a relationship that suggests mothers were more apt to trade-off feeding to survey environments where they coexist with grizzly bears and wolves ($p = 0.034$). Second, distance to the sounds of surrogate predators resulted in differences in moose foraging rates, and these varied by predation treatment but were significant only for ravens ($p = 0.048$); at predator-free sites, the distance of avian sounds had no effect on feeding. Third, vigilance was not related to distance to canopy during raven or hawk vocalizations at predator-free sites but it was in areas with large carnivores, and these relationships differed between predation regime ($p = 0.03$). Finally, patterns of postplayback feeding differed between predator-free and predator-rich sites. With the effects of other variables removed, both ravens and hawks elicited greater reductions in feeding in areas where grizzly bears and wolves still existed (both $p < 0.0001$). These findings affirm not only the recent existence of geographical variation in how mammalian prey rely on cues from aerial scavengers to detect predators, but also that the variation is a consequence of recent ecological instability, namely the extinction of wolves and grizzly bears by humans.

An Experiment: Carnivore Restoration and Reinstillation of Antipredator Behavior

Here we consider how the expansion of grizzly bears and wolves into areas that had been predator-free for many decades may have altered moose antipredator behavior. First it is necessary to frame the ecological context in which carnivore restoration in the GYE has occurred.

In the contiguous United States, both wolves and grizzly bears receive special protection under the Endangered Species Act (ESA). An ultimate goal of the ESA is demographic restoration such that species can be "de-listed" from federal protection once their viability is no longer jeopardized. Although grizzly bears were never extirpated from Yellowstone National Park's 8900 km^2, they were extinct in much of the 100,000 km^2 that constitute the GYE, including areas south of the park. Beginning around 1990 and still continuing, perhaps as a consequence of the 1988 fires that burned nearly half of Yellowstone Park, grizzly bears naturally recolonized portions of the northern Tetons and areas to the east and southeast. Unlike grizzly bears, wolves were extirpated from all of the Rocky Mountain regions during most of the twentieth century. Immigrant wolves from Canada have recently moved into northern Montana (Boyd et al. 1994) but not in the GYE. The U.S. Fish and Wildlife Service restored wolves to Yellowstone Park in 1995, and the first dispersing wolves from there arrived in the Tetons in late 1997.

Between 1996 and 2001, grizzly bears killed at least 12 adult moose in and adjacent to the Tetons, and in March and April 1999 wolves were responsible for the loss of at least eight 9- to 10-month-old moose calves. Some level of predation by both bears and wolves has been restored in a system where it had been lacking for 60 to 75 years. If previously predator-naive moose are quick to learn about predators, it may not be unexpected that they too learn to associate ravens with carnivores, as has occurred in Alaska (see Fig. 9.3).

One model of learning posits the rapid development of predator recognition through individual experience. To examine whether Teton moose who were initially free from predation changed their responsiveness both to wolves directly and to ravens, we contrasted mothers whose calves were killed by wolves and those who did not lose calves (both in the sGYE), and then compared these to Alaskan females (Fig. 9.4).

The responses of sGYE females to the calls of ravens did not vary temporarily with respect to predation events or by maternal status (e.g., whether calves survived or not; see Fig. 9.4). These findings indicate that over short periods of time, moose mothers either fail to associate ravens with carnivores

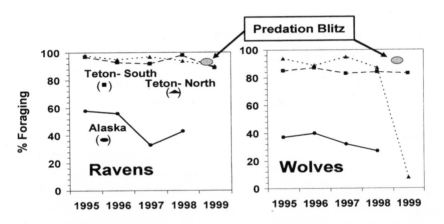

FIG. 9.4. Summary of comparative mean annual postplayback foraging responses of adult female moose to calls of ravens and wolves. The "predation blitz" occurred in early 1999 only in the Teton-South region of Grand Teton National Park (Wyoming). Females of Teton-South that lost their young to wolves had heightened responses in contrast to Teton-North females ($F = 24.625$; $p < 0.0001$). Sample sizes for wolf playbacks are as follows: Tetons–215; Alaska–82.

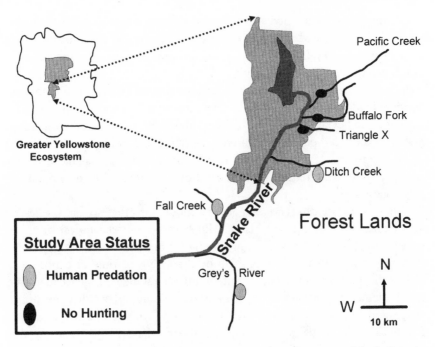

FIG. 9.5. Locations of riparian sites for studies of vegetation and bird diversity within and adjacent to Grand Teton National Park (no hunting—moose at high density) and on adjacent Bridger-Teton National Forest.

or, if they do, they fail to respond. Unlike Alaskan mothers, moose in environments that have been predator-free for 60 to 75 years are less wary of ravens even when predation on their offspring has occurred.

Wolf howls, in contrast, produced massive changes in the behavior of mothers who lost calves. They decreased feeding rates by a magnitude of five, and their reduction in feeding became similar to that of Alaskan moose (Fig. 9.5) only after the predation blitz ($F_{1,\ 89} = 32.516$, $p < 0.0001$). That such dramatic changes were caused by wolf predation per se, rather than other factors, is supported by two comparative analyses (Berger, Swenson, and Lill-Persson 2001): mothers whose calves died due to starvation or vehicles ($N = 3$) had greater feeding rates per bout (x = 71%) during playbacks of wolf howls than mothers with wolf experience (x = 8%) ($F_{1,\ 16} = 24.625$, $p < 0.0001$), and differences between these "control" mothers (who had not lost offspring to wolves) and mothers in areas without wolves (x = 83%) were not detectable ($p = 0.772$).

These findings are important because they (1) suggest that variation in behavior can occur rapidly, in this case in less than a single generation;

(2) demonstrate heightened sensitivity by mothers who lost calves to predators; (3) support the idea of a developmental hierarchy of predator-detection cues in which prey learn about carnivores; and (4) indicate that mothers developed greater response sensitivity to wolves than to ravens. Although it is well known that varied mammalian taxa learn about, or retain memory of, possible predators (Blumstein et al. 2000) and can differentiate among them (Cheney and Seyfarth 1990), what differs about the moose–wolf–bear–raven system is the rapid pace at which receptivity is lost and may be regained. In ground squirrels (*Spermophilus beechei*), for instance, antisnake behavior may persist for hundreds or more generations despite the loss of some predators (Coss 1991).

Whether moose can serve as a general model for other asocial species, or social ones such as elk and bison, wildebeest (*Connochaetes taurinus*) or gazelle (*Gazella* spp.), is uncertain. Among factors that will inevitably shape prey responsiveness to predators will be the length of time of exposure as well as the intensity of predation. But, in today's fragmented, alien-laden, and less biologically diverse world, the emergent issues that will continue to shape and diminish potential prey populations will also be the degree of familiarity to potential selective agents. Already the evidence is strong that alien species, whether dingoes (*Canis lupus*) or foxes into Australia or cats onto islands, have a relative easier time dispatching prey than in coevolved systems. Although for moose, it is now evident that antipredator behavior is relatively labile, at least under some conditions, it remains unclear if the restoration of native carnivores into many of the other systems where they have been extirpated will elicit a similar return of "normalized" antipredator behavior. Only studies in those specific systems will produce answers.

Possible Apex Carnivores and Community Interactions

The preceding findings indicate that the loss of large carnivores has had dramatic effects on the demography and behavior of one species of prey, including alterations of (1) rates of adult pregnancy, (2) patterns of juvenile survival, (3) reactions to scavengers, and (4) learning about predators. Additionally, the release of predation in the sGYE has resulted in a moose population that must hover near to or beyond the food ceiling because population growth has ceased and pregnancy rates have declined (see Fig. 9.2).

The implications of removing large carnivores is relevant for many ecosystems in areas of Europe and Asia and, indeed, globally, where ecological communities are either missing dominant selective forces or have new ones dependent upon humans. These large-scale manipulations offer unique opportunities to investigate how the loss of large carnivores affects community dynamics, particularly the possibility that interactions may be influenced at multiple trophic levels.

How important are carnivores in terrestrial ecosystems? In the neotropics, sites that otherwise appear virtually intact have been referred to as empty forests due to the depletion of game (Redford 1992). Not only may these areas be lacking in wildlife that is consumable by humans, but, because humans and big cats (jaguars and cougars) compete for the same food, these nonhuman carnivores may also be lacking (Jorgensen and Redford 1993), and the system's past equilibria become radically changed (Novaro, Funes, and Walker 2000). Not dissimilarly, in western North America and montane regions of Europe, some lands that look as if they are "wilderness" also lack brown bears and wolves. But the mere loss of large carnivores from ecosystems says little about their role as apex organisms or the scale at which predator–prey disequilibria operate. These losses do, however, create opportunities to learn more about ecological processes and may shed light on the longstanding debate over the role of top carnivores in regulation of prey populations (Terborgh 1987; Wright, Gompper, and De Leon 1994), and whether top-down or bottom-up effects play a larger role in biological organization (Paine 1966, Polis and Strong 1996). Although most research on trophic cascades has focused on aquatic or marine systems (Power 1992; Estes, Tinker, and Doak 1998) and involved heterotherms and invertebrates (Spiller and Schoener 1994, Carter and Rypstra 1995), the recent losses of large vertebrate carnivores provide options to examine how predation and food shape terrestrial communities and relate to the maintenance of biological diversity.

The work we present next summarizes findings on how grizzly bears and wolves may act as apex carnivores, and it is based on studies that we and our colleagues, Matthew Johnson and Lori Bellis, have published elsewhere (Berger et al. 2001). One reason for expecting that grizzly bears, either alone or in combination with wolves, may be apex carnivores is because of their well-established impacts on population growth in moose, which may in turn have subsequent impacts at the landscape level. Although debate still continues over whether moose are regulated by predation per se (Boutin 1992, Orians et al. 1997, Krebs 2000), a series of manipulative but imperfect field studies suggest lower moose population growth where wolves and grizzly bears are not at excessively low densities (Messier 1991; Gasaway et al. 1992; Boertje, Valkenburg, and McNay 1996).

Moose may have substantial localized effects on ecosystems (Pastor et al. 1993, Connor et al. 2000), partly because they consume large quantities of woody shrubs and young trees including aspen, willow, and cottonwood, and also because they achieve densities (Houston 1968) that, in riparian zones, may exceed 20 individuals per km² for up to 5 to 6 months per year. In Alaska and the Yukon, moose density is affected by predation (Gasaway et al. 1992), although periodic severe winters may set an upper limit on population sizes (Messier 1991, Peterson 1999).

Some native herbivores, including cervids other than moose, have the capacity to attain extraordinarily high densities in the absence of carnivores (Alverson and Waller 1997, Schmitz and Sinclair 1997). And, although domestic species may severely impact sensitive habitats (Knopf and Cannon 1982, Kauffman and Krueger 1984, Saab et al. 1995), little is known about the effects of colonizing native browsers. In arid zones like the American West, this can be a critical issue because riparian habitats may constitute 1 to 2% of the landscape but harbor up to 80% of the local biodiversity (Ohmart 1994; Stacey 1995; Dobkin, Rich, and Pyle 1998). It is in these types of systems that intense herbivory may be expected to affect riparian biodiversity.

RATIONALE, RESEARCH DESIGN, AND STUDY AREAS

We studied interactions among moose populations, the structure of their major winter food supply in riparian zones, and avian species diversity in Grand Teton National Park and adjacent public lands managed by the U.S. Forest Service in the Jackson Hole region of the GYE (see Fig. 9.5). Until recently grizzly bears and wolves had been absent for about 60 to 75 years. Outside the national park, more than 10,800 moose were harvested by humans between 1971 and 1991 (Houston 1992). In the national park, the hunting of moose is not permitted, and coincident mean densities vary by a magnitude of almost five (5.2 vs. 1.1 individuals/km²), with densities being lower on lands adjacent to Teton Park.

During the 1800s moose were rare in much of western North America (Karns 1998). They were virtually absent from the Jackson Hole region about 150 years ago, and their rapid population growth appears to have reached a ceiling (see Fig. 9.2). Moose depend upon willow during winter (Peek 1998). Riparian vegetation may be altered more when moose occur at high densities than when they are controlled by predation or hunting. Alternatively, moose may have little if any impact on riparian vegetation.

We evaluated these possibilities by contrasting effects of moose herbivory on riparian willow communities using three areas within Teton Park (no hunting and no large carnivores) and three similar sites in adjacent forest lands (human hunting but no large carnivores). Thus the treatment was predation (absent or present), with each area containing three replicated riparian communities (Fig. 9.5).

TABLE 9.1. Summary of effects of how types of predation on moose influence subsequent browsing in willow communities in the southern Greater Yellowstone Ecosystem.

WILLOW STEMS	PREDATION ON MOOSE		F	P
	Human	None		
Alive–not browsed	0.53 (0.15)	0.10 (0.11)	18.56	0.0001
Alive–browsed	0.22 (0.15)	0.59 (0.10)	7.73	0.0025
Dead–not browsed	0.12 (0.07)	0.01 (0.01)	13.74	0.0002
Dead–browsed	0.13 (0.08)	0.29 (0.09)	5.04	0.0128

Mean (and standard deviation) reflect proportions of stems of dominant willows (Geyer's *Salix geyerianna*, Booth's *S. boothii*, and Wolf's *S. wolfi* ($N = 360$). Analyses are based on two-way, repeated measures analysis of variance (from Berger et al. 2001).

RIPARIAN WILLOWS AND AVIAN SPECIES DIVERSITY

Moose density affected both mean willow height and density. Where moose were reduced by humans, willows were taller and browsed less than in park areas that lacked predation. The proportion of willow stems longer than 100 mm that were browsed and either alive or dead was also associated with moose density (Table 9.1).

Five parameters were used to assess whether moose density was associated with the distribution or abundance of the riparian avian community. Moose density affected all, with each being greater where moose density was limited by predation: (1) species richness of breeding birds ($N = 23$ vs. 18; $p < 0.003$); (2) nesting density ($p < 0.002$); (3) Shannon's diversity index ($p < 0.004$); (4) Hill's Diversity Measure 1 ($p < 0.008$); and (5) Hill's Diversity Measure 2 ($p < 0.021$). Where moose densities were high, the density of nesting avifauna, including willow flycatchers (*Empidonax trailli*), calliope humming-birds (*Stella calliope*), yellow warblers (*Dendroica petechia*), fox sparrows (*Passeralla iliaca*), and blackheaded grosbeaks (*Pheucticus melanocephalus*) were substantially reduced (Table 9.2), and two other species—gray catbirds (*Dumetella carolinensis*) and MacGillivray's warblers (*Oporornis tolmiei*)—were absent. Overall, approximately 50% of the riparian willow bird species were reduced or absent from sites inside Teton Park where moose were protected from predation and thereby attained high local densities. The fact that typical riparian species were present at all sites (e.g., Yellow and Wilson's warblers; see Table 9.2 for others) substantiates that the samples were derived from the same ecological pool.

TABLE 9.2. Relationships between moose densities as influenced by predation (human versus none) and mean breeding densities/10,000 m² (and SD) of several bird species in Wyoming.[1]

SPECIES	PREDATION BY HUMANS	NONE	F	P
Calliope hummingbird	2.00 (0.87)	0.22 (0.44)	10.67	0.0004
Willow flycatcher	0.78 (0.44)	0.55 (0.53)	2.40	0.0994
Gray catbird	0.88 (0.92)	0.00 (0)	5.12	0.0096
Yellow warbler	3.78 (0.97)	2.33 (1.00)	3.12	0.0453
Wilson's warbler	0.11 (0.33)	0.67 (0.87)	1.86	0.1760
Yellowthroat	0.00 (0)	0.11 (0.33)	1.00	0.4582
MacGillvray's warbler	0.22 (0.44)	0.00 (0)	0.80	0.5705
Black-headed grosbeak	0.77 (0.44)	0.22 (0.44)	3.00	0.0552
Song sparrow	0.22 (0.44)	0.00 (0)	3.91	0.0245
Fox sparrow	0.44 (0.53)	0.11 (0.33)	0.85	0.5405
Lincoln's sparrow	0.22 (0.44)	0.66 (0.71)	0.91	0.5038
White-crowned sparrow	0.00 (0)	0.55 (1.13)	0.82	0.5612

[1]Modified from Berger et al. 2001.

Although these data suggest that moose browsing shaped avian communities at a microgeographical scale, the link between the structural modification of riparian willows at our sites and avian diversity had been uncertain. Nevertheless, a direct relationship between willow volume (X; the product of density and cover using the proportion of live willow stems) and avian species diversity (Y) existed ($Y = 0.51X + 0.33$, $n = 18$, $r^2 = 0.24$; $p < 0.03$). Thus it appears that avian species diversity is partially affected through structural modifications of the willow canopy by moose whose densities, in turn, are controlled by humans, either by total protection (park) or hunting (adjacent forest lands).

LARGE HERBIVORES, CARNIVORES, AND TROPHIC CASCADES

Our analyses, though specific to the Jackson Hole area of the GYE, support the idea that a dynamic chain of interactions involving multiple tiers of biological organization were set in motion 60 to 100 years ago, principally by the removal of large predators. Among the key events were (1) human decisions to exterminate large carnivores, especially wolves (Murie 1940, Phillips and Smith 1996) but also grizzly bears (Craighead 1979) both in Yellowstone Park per se and in adjacent regions; (2) a resultant growth of an apparent low-density moose population (see Fig. 9.2) that, although it began expanding from 1880 to 1910, irrupted partially due to a dampened effect on juvenile mortality; (3) increasing herbivory in riparian willow communities at sites lacking predation or hunting; (4) modification of these communities (see Table 9.1); and (5) decreased avian richness and diversity (see Table 9.2). Although the idea of top-down regulation of communities by carnivores, especially in terrestrial ecosystems, has been controversial (Polis and Strong 1996, Terborgh et al. 1999), our data offer support for its importance, at least in this system. To the north, in Yellowstone Park, where elk densities tend be to exceptionally high, evidence is now accumulating for an inverse relationship between elk abundance and wolf densities, with the effect being that aspen (*Populus tremuloides*) sucker heights may be increasing as elk densities drop (Ripple et al. 2001).

Conservation and the Relevance of Behavior in Predator–Prey Systems

Assuming an initial perspective from that of a pure "animal behaviorist" who desires to understand behavioral processes, it would seem desirable to place large carnivores back into their "natural" ecosystems. Obviously, questions of intrigue will always exist for scientists. Other than the financial cost of carnivore restoration, a purist's argument might go something like this: the reestablishment of prey behavior once predators are reintroduced is important because predator–prey processes are likely to be more representative of balanced and naturally operating systems.

But this argument in favor of restoration is a personal judgment, not one steeped in science. Although behavioral studies have strong scientific components, the issue of relevance to conservation is not always obvious. If predator–prey interactions and the restoration of antipredator behavior can legitimately be considered as part of the complex of processes or phenomena involved in maintaining biological diversity (Wilson 1992, Berger 1999), then such processes should not be lost. Such issues aside, is there a more tangible role for behavior to play in conservation?

Some would argue no because many conservation biologists and wildlife managers are often less concerned about animal behavior in and of itself and more concerned that systems become (or are maintained as) functionally operating (Redford and Feinsinger 2001, Pyare and Berger 2003). Understanding a species' behavior may at times help to detect when (or whether) that goal is achieved, but unless the benefit is substantial, the usefulness of behavioral knowledge for its own sake may be limited. Perhaps the most relevant area where understanding prey behavior has been linked directly to conservation programs has been in situations in which population viability may be jeopardized. These situations tend to occur when prey are naive and unable to cope with predation, whether by native or alien species (McClean 1997; Blumstein et al. 2000; Berger, Swenson, and Lill-Persson 2001), and it is here that an understanding of mechanisms may be more germane to restoration, especially in areas with many alien species (Griffin, Blumstein, and Evans 2000; Short, Kinnear, and Robley 2002).

In shifting from behavioral-oriented approaches to on-the-ground approaches that confront the real management of protected areas, a curious irony is brought forth. In numerous ecological reserves, including (until recently) Grand Teton National Park, the intent has been to enhance biological diversity through "hands off" management, yet the opposite has occurred. Protection, after the localized extinction of large carnivores, has instead resulted in ecological processes such as the decline of taxa within a given park's boundaries; in contrast, lands outside the park where the active management of wildlife has occurred resulted in greater levels of avian diversity (see Table 9.2). Studies of the behavioral implications of such ecological processes may not be relevant here, yet the conservation of biological diversity surely is, irrespective of whether it be birds, bats, or butterflies, ecological processes, or other levels of organization.

One thematic region where behavioral approaches may be combined with ecological ones to help the management of predator–prey systems involves functional redundancy. Can (indeed, should) human predation substitute for predation by carnivores if prey are so abundant that other tiers of biological diversity are lost? It is already obvious that major differences occur in gender- and age-specific predation and killing practices (Ginsberg and Milner-Gulland 1994, Berger and Gompper 1999), but the extent to which prey differ in their behavior in response to human and nonhuman predation is not totally clear. And, the extent to which ecological shifts occur under contrasting predation regimes and whether these have implications for conservation of ecosystems is still something for which data are only emerging (Ripple et al. 2001).

Conclusions and Recommendations

We argued that large carnivores have the capacity to play substantial roles in systems' dynamics independent of whether the level is that of a single species or an entire landscape. The consequences of losing large carnivores can be long lasting. Cheetahs and other fleet carnivores have been gone from North

America since the Pleistocene, yet these "ghost carnivores" affect pronghorn (*Antilocapra americana*) who retain not only morphological adaptations for speed but behavior reflective of pressures derived from extinct predators (Byers 1997).

In today's anthropogenically inflicted world, however, the alterations that we report stemming from loss of carnivores are not ones that have transpired over thousands of generations. Instead, they have occurred in as few as 10 generations, as in the abrupt decoupling of a prey–scavenger recognition system, or even in less than a year as in one-time experience-based learning by moose mothers who lost young to wolves. Although such behavioral changes are specific to the system we worked in, the implications are likely to be broader. The desensitization of prey to both scavengers and carnivores at sites lacking predation indicates that subtle, community-level interactions now no longer occur, at least in some landscapes. It is anyone's guess as to how other interspecific interactions may have been shaped or reshaped by carnivore losses. Sublethal effects of predators can be significant (Sih and Wooster 1994, Lima 1998), and for ungulates may include habitat shifts, group size formation, sexual segregation, and changes in feeding, pregnancy, and many others still in need of testing (Berger 1998). The extent to which a system's equilibria may shift as a result of behavioral alterations has only recently been receiving attention. So if one were set on providing insights on the consequences of predator–prey disequilibria, this may be one way in which behaviorists can contribute. Nevertheless, as the loss and replacement of carnivores is likely to continue, other processes will be unraveled, many of which have yet to be described (Wilson 1987, Crooks and Soulé 1999). How the study of animal behavior contributes to conservation at this level is not so clear.

Science, of course, is different from conservation, and progressive efforts concerning the latter are sources for both optimism and study. In localized regions of Europe, lynx (*Lynx lynx*), brown bears, and wolves may all be expanding, either naturally or through reintroductions, or both (Breitenmoser 1998; Swenson, Sandgren, and Soderberg 1998; Boitani 2000). Such conquests are relatively limited when viewed globally against remaining challenges. A larger issue is how best to discover and combat negative ecological responses of prey to replacement of native predators by aliens—the feral species and translocated native ones that we humans have disseminated across this planet and which now reduce our biological heritage through population reduction and extinction

Restoration options steeped in biological principles will always be important and science will always play some needed role to further sort out ecological dynamics and pursue conservation gains. Yet, a more massive challenge awaits us all—to spread our messages widely, to shape public opinion, and, in doing so, bring government to action. Though Shere Khan and el tigre may no longer exist in the environments where Rudyard Kipling and Aldo Leopold toiled, we need to focus our restoration efforts on the next generation while not ignoring the rich messages of the past.

Summary

The effects of losing large carnivores on the functioning of ecosystems and on individual prey have been uncertain. We offer empirical evidence about how the loss and restoration of wolves and brown bears is likely to affect temperate and boreal ecosystems, specifically on some of the interactions involving wolves, brown bears, ravens, avian Neotropical migrants, and one large herbivore—moose. We show through geographical contrasts of sites in Alaska and the Yellowstone region as well as experimental alterations of carnivore presence that (1) behavior of individual prey changes as a result of carnivore recolonization, and (2) demography is altered in ways in which maternal behavior is linked to changing predator regimes. We also infer that (3) released from predation, the irruption of moose led to a cascade of ecological interactions that diminished avian biodiversity.

Our findings underscore the importance of detailed single-species and community approaches as well as the adoption of experimental protocols set within an ecological framework to help facilitate biological restoration. The study of animal behavior per se has played only a limited role in functional conservation. Nevertheless, such study has increased knowledge about prey adjustments to predation and the extent to which demographic blitzkriegs may be avoided by beginning to understand rates at which effective antipredator behavior may be reinstilled in previously naive and exceptionally vulnerable prey. Research involving the application of animal behavior to conservation may be limited, but where the goal is to contribute to predator–prey systems one might profitably focus on population vulnerability in relation to predation pressure and habitat alterations, especially when predators are aliens.

Acknowledgments

Grants from Beringea South, National Science Foundation, Denver Zoological Society, the Engelhard Foundation, the Eppley Foundation, and Wildlife Conservation Society, and support from Alaska Department of Fish and Game, Universities of Nevada and New Mexico, Wyoming Department of Game and Fish, and the National Park Service (Denali and Grand Teton) facilitated this work. L. Bellis, M. Ben-David, D. Blumstein, D. Brimeyer,

J. Bohne, J. H. Brown, S. Cain, F. Camenzind, D. Craighead, C. Cunningham, D. Finch, J. M. Johnson, B. Kus, D. Ligon, W. Longland, M. Mastellar, B. Miller, R. Reading, R. Schiller, C. Schwartz, K. Stahlnecker, S. Vanderwall, K. White; and especially S. Engelhard, K. Redford, and W. Weber contributed to this study. We thank Marco Festa-Bianchet and Marco Apollonio, who encouraged attendance at the symposium in Erice.

10.

Behavioral Aspects of Conservation and Management of European Mammals

Marco Apollonio, Bruno Bassano, and Andrea Mustoni

This chapter considers a few case studies where knowledge or ignorance of animal behavior led to the success or failure of wildlife management programs in Europe. We also comment on some of the commonest management problems on our continent, and argue that, despite recent strong calls to consider the links between wildlife conservation and ethology (Clemmons and Buchholz 1997, Caro 1998b, Gosling and Sutherland 1999), very often basic aspects of animal behavior are ignored. There are two main reasons why animal behavior is seldom accounted for in European wildlife management: many wildlife managers know little about behavioral ecology, and "traditional practices" persist despite evidence that they are inadequate. We examine case studies where ranging behavior, reproductive behavior, and social behavior have important implications for wildlife management.

Ranging Behavior

Spatial behavior is of great importance to management. Documentation of the ranging behavior of a mammal population, however, can be a demanding task because of the costs and the time necessary to obtain reliable data. This is especially true in the case of wide-ranging species such as most large and medium-sized mammals (Harestad and Bunnell 1979), which unfortunately are also the most difficult to manage without information about their home range size and the frequency and potential speed of their movements. Information on ranging behavior is crucial to choosing the spatial scale of management plans and testing their validity over time. We illustrate this point with examples from two carnivores and one ungulate.

BROWN BEAR REINTRODUCTION AND CONSERVATION IN SOUTHERN EUROPE

The brown bear's (*Ursus arctos*) range in southern Europe was greatly reduced over the last few centuries. A few remain in the Alps and the Pyrenees, with small but viable populations in the Apennines (a few dozen bears) and in the Cantabrian Mountains (50–65 individuals). Larger populations persist in the Dinaric and Carpathian Mountains (Swenson et al. 2000).

Because the Alpine and Pyrenean populations were reduced to only three to four individuals (Swenson et al. 2000), reintroduction programs were planned for both. Information on the ranging behavior of bears played a key role in the planning of these operations. The political complexity of Europe, both between and within countries, required coordination among local and national governments for the reintroduction of a large carnivore that could potentially be in conflict with agriculture. "Old" data on home ranges of Alpine or Pyrenean brown bears were scarce and contradictory (Daldoss 1981). Couturier (1954) suggested that in the Pyrenees, home ranges were only 3 km^2. A pilot radiotelemetry study by Roth and Osti on the small remnant population in the Italian Alps monitored three individuals and found that home range size was about 300 km^2 for males and about 100 km^2 for females (Roth 1983, Osti 1999). Those results were possibly influenced by the old average age of surviving bears and by the small population size, confirmed by more recent studies (INFS 1998), of the remnant group of bears studied. The "recent era" in brown bear ranging behavior knowledge started with the first reintroduction in Austria in 1989. The first bear, a female, had a home range of about 335 km^2; but the second, an adult female, ranged over 4730 km^2, moving up to 67 km from the release site (Rauer and Gutleb 1997; Zedrosser, Gerstl, and Rauer 1999). These results are similar to those obtained in recent reintroductions in the French Pyrenees and the Italian Alps (Table 10.1; P. Y. Quenette, 2000, pers. comm., Mustoni et al. unpublished). Meanwhile, data on ranging behavior of autochthonous brown bears in southern Europe were collected in Slovenia (see Table 10.1; Kaczensky 2000). Their home ranges appear smaller than those of reintroduced bears in

TABLE 10.1. Home range of reintroduced and native brown bears in southern Europe (calculated as minimum convex polygons)

AGE/SEX	ORIGIN	NUMBER	HOME RANGE (km²)	HOME RANGE MEAN (km²)	STANDARD DEVIATION (km²)
Adult/males	Reintroduced	2	286–1230	758	667.5
Subadult/males	Reintroduced	2	1722–2376	2049	462.4
Subadult/males	Native	4	100–516	322	177.4
Adult/females*	Reintroduced	6	33–4730	1780.8	1637.1
Adult/females	Reintroduced	5	870–4730	2130.4	1560
Adult/females	Native	4	39–63	51.5	10.1
Subadult/females	Reintroduced	2	216–355	285.5	98.3
Subadult/females	Native	2	41–287	164	173.9

*Reintroduced bears data are from reintroduction projects in Austria, France, and Italy. Native bears were released near the capture site in Slovenia.

all possible comparisons (all males U = 0, all females U = 3, adult females U = 1, Mann Whitney U test, $p < 0.05$) if the second-year home range of a reintroduced female was excluded. If the small home range of the reintroduced female is included, the home ranges of native brown bears seem to be smaller only for all males (U = 0, Mann-Whitney U test, $I < 0.05$). It therefore seems likely that the reintroduced bears have an exploratory behavior compared with the native ones.

Reintroduced bears also made rapid and sudden movements. In the Italian Alps a male moved about 30 km in 4 days, and 16 km in just 1. In the same area a female traveled over 29 km in 36 hours. These results were also confirmed by the 23 km traveled in 24 hours by one of the "Austrian" reintroduced bears (Zedrosser, unpublished). Autochthonous bears can also make long-distance movements, as illustrated by a large male that traveled 110 km in less than 10 days in 1999 from the province of Belluno to the province of Trento (Wildlife Management Service, Trento Province, unpublished). Some of these movements occurred in December, contradicting the assumption of winter dormancy in brown bears, confirming recent findings on this species (G. Jonozovic, 2000, pers. comm., Swenson et al. 1997b, Kaczensky 2000).

The relevance of these data for brown bear conservation in southern Europe is easy to understand. A successful reintroduction requires three main components organized over a large geographical scale: public information, damage prevention and compensation, and continuous monitoring of released bears. The sudden arrival of a bear in an area can lead to disaster if local people are not informed about the program, including the potential for damages (particularly to beehives). Lack of information may lead to hostility toward the bear reintroduction program. In Europe, administrative power is very fragmented. For instance, in Italy, regions and provinces are in charge of wildlife management. A province can vary in size from 2000 to 7500 km²; it is therefore obvious how important coordination can be. In the recent reintroduction program in the central Alps, conducted in Italy by the Adamello Brenta Natural Park, the success of the operation was due to careful planning of the three steps, previously mentioned, among five provinces. Currently, this is the largest brown bear reintroduction in Europe, with 10 bears successfully released and being monitored. After the first 3 years of the project, the bears have been documented in three provinces (Fig. 10.1), no illegal killings have occurred, and about €30,000 in damages has been paid out.

Another important aspect of bear conservation that can be influenced by the knowledge of their ranging behavior is population estimate. On many occasions wide movements can cause a gross overestimate of the population because a few bears with very large home ranges and very rapid movements are assumed to be many bears with small home ranges. The same bear can be seen tens of kilometers apart over a few days. An inexperienced wildlife manager may conclude that a flourishing population exists, instead of a small and threatened one. This problem may apply to the small Apennine populations where an overestimation of bear numbers is highly probable. In the long run, an erroneous overestimation of population size may delay restocking operations and possibly increase the risk of extinction.

WOLF RETURN TO THE WESTERN ALPS

Wolves (*Canis lupus*) once ranged throughout all of Italy and neighboring countries, but in the nineteenth century they gradually disappeared from the Alps. The last recorded killing was in 1923 in the Maritime Alps (Cagnolaro et al. 1974). The wolf in Italy decreased to its lowest numbers in the early 1970s, when it was reduced to a few areas of the Apennines and the Tyrrenian coast. Beginning in the 1980s, however, the species expanded in distribution and probably in numbers. This reversal of fortune was due both to legal protection and to the rapid recovery of woodland that allowed the development of flourishing wild ungulate populations.

In 1985 a dead wolf was found less than 100 km from the Alps, on a northern Apennines range directly leading to them. Not surprisingly, wolves then appeared on the French side of the western Alps (Fig. 10.2). In 1988 one

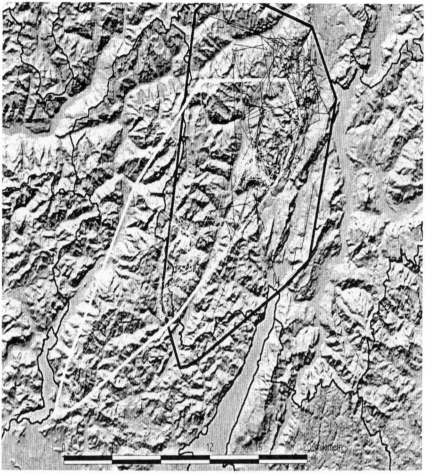

FIG. 10.1. Annual home ranges of reintroduced brown bears in the Italian Alps. In white, subadult male; in dark black, subadult female; in light black, province border.

wolf was found dead, then in 1992 a pack of wolves was observed in the Mercantour massif. Wolves were seen on the Italian side of the Alps in 1994 and in Switzerland in 1995. Genetic analyses showed that these wolves were of Italian origin (Scandura et al. 2001) despite claims of illegal reintroduction.

The recovery of wolves in the Alps was rapid (see Fig. 10.2) but not without consequences. The main conflict arose with sheep breeders, and its intensity varied among countries in relation to sheep numbers and breeding practices. In Italy and France reimbursements were established from the very beginning of wolf recolonization. In Italy the lower level of wolf damages (about €35,000 compared to about €342,000 in France in 2001) was due to a

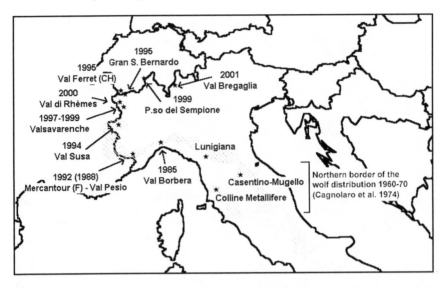

FIG. 10.2. Wolf recovery in the Alps. Each locality is labeled with the date of first occurrence. The year in brackets for the Mercantour is the date of the first recovery of a dead wolf in the Alps.

lower number of sheep and to the smaller size of sheep flocks. The decreasing human population of many alpine valleys in Italy is another cause of this difference. A further reason is likely a difference in sheep-raising practices. In France, flocks of up to 800 to 1000 are not uncommon, whereas flocks on the Italian side are roughly five to ten times smaller. In France, where the amount of damages increased from €8690 in 1993 to €227,234 in 1998 (Poulle et al. 2000), sheep breeders asked for controls on the wolf's distribution (French Ministry of Agriculture and Fisheries 2000) even though the number of wolves was estimated at only 14 to 19 in the period from 1996 to 2000 (French National Hunting Office, 2000, pers. comm). There were also six documented illegal wolf killings from 1995 to 2001. In Italy the low level of conflicts helped to maintain full protection of this species and to reduce illegal killings to only two between 1994 and 2001. In Switzerland, hostile public opinion led to the legal killing of any wolf believed to have killed 50 sheep. Given that it is impossible to properly establish which wolf kills which sheep, this measure opened the door to legal control of any wolf, leading to three legal, one illegal, and two suspected wolf killings. Not surprisingly, no evidence of wolf reproduction has been reported in Switzerland. The reported loss was 36 sheep in 1998, 266 in 1999, and 242 in 2000, out of about one million sheep in Switzerland (Kora unpublished data).

The pronounced tolerance of Italian authorities and citizens toward wolves may also be due to a tradition of considering it a rare species that requires protection. Thirty years of legal protection and of public campaigns for wolf conservation, combined with an effective compensation scheme, likely contributed to acceptance of wolves in newly reoccupied areas. On the contrary, in France and Switzerland the species disappeared completely: there have been no resident wolves in Switzerland for more than a century and in France for more than 50 years. Consequently, the public had neither knowledge of nor tolerance toward these predators. Moreover, sheep-breeding practices in these countries increase the potential for conflicts. The most extreme position was taken in Switzerland, where wolves had been absent for a longer period and where the government heavily subsidizes sheep breeding in the Alps.

When the Italian wolf populations started to recover, much information was available in the scientific literature on the dispersal capacity of wolves (Fritts 1983, Van Vueren 1998) and on the social mechanisms leading to dispersal (Mech 1995). It was therefore easy to foresee the return of wolves to the Alps, but, at that time, nobody warned public administrations about the need to develop programs to increase public awareness about the return of the wolf, or about how to prevent damages. Lack of initiative taken by wildlife managers and scientists obviously worsened the conflicts arising from the recovery of wolves in the western Alps, creating a situation where three neighboring countries have three different attitudes toward the management of a wide-ranging species that ignores national borders. In western Europe wolves are considered a high-priority species by many national laws and by the Habitat Directive of the European Community: failure to prepare in advance for the return of the wolf led to several conflicts and possibly hampered recolonization of some areas.

RED DEER IN THE ALPS AND NEIGHBORING MOUNTAINS

The red deer (*Cervus elaphus*) is the largest herbivore in southern Europe. Its original wide distribution was reduced during the nineteenth century to mountain areas, especially in alpine countries such as France, Italy, Austria, Switzerland, Germany, and Slovenia. In the mountains deer must face difficult environmental conditions, including harsh and snowy winters. The strong seasonality of the Alps encourages traditional movements among seasonal ranges, determined by food availability and snow cover. Gossow and Stadlmann (1985) identified distinct summer and winter ranges on mountain slopes, showing how the availability of winter ranges was reduced by habitat modifications caused by human activities. Radiotelemetry studies, however, revealed a more complex situation: both on the north (Georgii

1980, Georgii and Schroeder 1983, Leoni 1995) and on the south side of the Alps (Luccarini and Mauri 2000) red deer populations include migratory and sedentary individuals. The former made long-range seasonal movements from a winter range at low elevation to a summer range at high elevation, whereas the latter remained in the same general area year-round. As a result, in winter the entire population was restricted to the same low-altitude areas that were used year-round by a substantial number of deer. Low-elevation areas in the Alps are intensively used by humans, reducing the availability of foraging and resting areas for red deer. In addition, artificial feeding stations can concentrate deer in small areas. The winter home ranges of artificially fed populations are smaller than those of unfed populations (Luccarini and Mauri, unpublished).

Supplemental feeding increases winter survival, and it is frequently used in the central and eastern Alps and neighboring mountains. Supplemental feeding is sometimes prescribed by national laws, for example in Austria, but it can result in major damage to trees. The combined effect of browsing and other problems such as acid rain can damage over 70% of the trees (Wotschikowsky 1978). A more extreme management practice consists of fencing all deer into corrals where they are fed over winter and then released. The enclosed area is completely devastated by browsing and the deer are vulnerable to contagious diseases (Wotschikowsky 1978).

Although red deer are now widely distributed over the Alps and neighboring mountains where they are an important game species, they are not taken into account when urban developments are planned. Even if some extreme forms of management such as massive winter feeding in otherwise unsuitable areas have kept some deer populations at high elevations in winter (Schmidt 1993), the forest damage caused by winter concentrations of deer at low altitudes is a widespread problem (Voelk 1998).

Reproductive Behavior

Reproductive behavior and mating systems affect wildlife conservation and management because they can have a profound impact on population structure (Clutton-Brock, Guinness, and Albon 1982), genetic variability (Apollonio and Hartl 1993), and, ultimately, species evolution. Because the rut is a key phase of the biology of a population, any perturbations to its regular course may have strong consequences for population dynamics. The interrelations of wildlife management practice with ungulate reproductive behavior are many. Here we consider the possible consequences of trophy hunting on ungulate males. The practice of hunting large males to obtain prestige is very old in Europe. According to Julius Caesar in his *De bello gallico* (first century

B.C.): "Aurochsen are smaller than elephants; they look like bulls. . . . Germans like to trap and kill them. This kind of activity is useful for young men to exercise themselves and become stronger, those that kill many aurochsen show the horns of their preys in public and receive great honors." The human interest in displaying large horns or antlers can have major effects on the sex/age structure and population dynamics of ungulates. Horns and antlers are often secondary sexual characters, whose size is strongly correlated with age.

The outcome of the search for large trophies is an overkill of mature males. The structure of the male part of the populations can become highly skewed toward young age classes. Data on the age of chamois (*Rupicapra rupicapra*) shot over 7 years of hunting seasons in the free hunting districts of the Aosta Valley record that less than 5% of males were older than 9 years, and less than 10% were older than 7 years. Chamois age distribution can be quite different in areas with a less severe hunting effort. Those harvested over four seasons in private reserves in the Aosta Valley, where hunting pressure is lower than in free access areas, included 12% of males older than 9 years and 21% older than 7 years, more than twice the numbers in free hunting territories ($\chi^2 = 138.26$, $p < 0.001$) (Fig. 10.3A,B). If we consider that harvests were biased by the preference for older males, we obtain a discouraging picture of the management of the most common alpine ungulate, at least where shooting plans do not call for age-specific quotas. The same trend is evident comparing male census data of the unhunted alpine ibex (*Capra ibex*) population of Gran Paradiso National Park (Italy) and those of the Swiss hunted ibex population of the Graubunden Kanton: the differences in age class distribution in the male populations were very marked ($\chi^2 = 358,82$, $p < 0.001$ comparing years with lower presence of males older than 10 years in the populations; $\chi^2 = 457,52$, $p < 0.001$ comparing years with higher presence of males older than 10 years in the populations) (Fig. 10.3C,D). This last example could also indicate that moderate hunting pressure, such as that exerted on the Swiss alpine ibex population, may lead to significant differences in population structure from completely unhunted populations.

The consequences for the ungulate populations of intense removal of mature males in relation to reproductive strategies can be variable and complex. Male age is correlated with reproductive success, and there is often a clear age threshold below which males do not reproduce. In unhunted populations, this age is 5 years for red deer (Clutton-Brock, Guinness, and Albon 1982), 4 years for fallow deer (*Dama dama*) (Clutton-Brock, Albon, and Guinness 1988; Apollonio, Festa-Bianchet, and Mari 1989; McElligott and Hayden 2000), about 3 years for roe deer (*Capreolus capreolus*) (Liberg et al. 1998), 10 years for alpine ibex (Apollonio, Mauri, and Bassano 1997), and probably

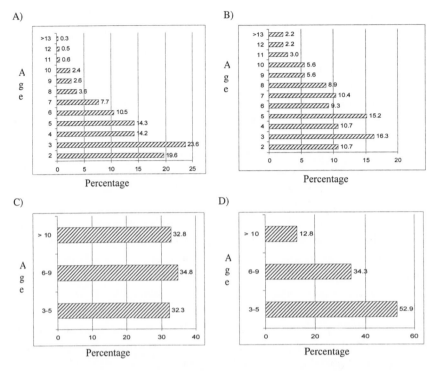

FIG. 10.3. (a) Age distribution of 2721 male chamois shot in free hunting area of the Aosta Valley from 1985 to 1991. (b) Age distribution of 270 male chamois shot in private hunting reservations in the Aosta Valley from 1989 to 1991. (c) Age distribution of 2682 male ibex in the unhunted population of Gran Paradiso National Park (1984 and 1992). (d) Age distribution of 3383 male ibex in the hunted population of Kanton Graubunden Hunting Districts (1996 and 2001).

7 to 9 years for chamois (the minimum age at which males defend territories) (von Hardenberg et al. 2000).

Given that older males potentially have high reproductive success and take an active part in reproductive behavior, their removal may have many serious consequences. One such consequence could be the possibility for younger males to take part in reproduction, including bearing the costs of this activity and therefore experiencing increased mortality (Geist 1971). This process can induce a steady decrease of the mean age of males to extremely low levels. There are also some indications that the removal of some old, highly success-ful males may decrease the mating rate of females, possibly because of the social instability and intense fighting among surviving males (Apollonio, Festa-Bianchet, and Mari 1989). Changes in male age structure can lead to a change in mating system (Byers and Kitchen 1994), possibly forcing females to make suboptimal mate choices that could be associated with lower probabilities of conception.

Mature males are the most fit individuals, able to survive well past the age of first reproduction: the hunter's choice counters natural selection. This is especially true when individuals with high reproductive success are those that live longer, as recently reported in fallow deer (McElligott and Hayden 2000).

Criteria commonly used to assess the "value" of a trophy, such as size and shape, may have little to do with a male's reproductive success. Complex formulae used by the Conseil International de la Chasse to evaluate ungulate trophies are based on various measurements of length and width. On two occasions we obtained these antler measurements for lekking fallow deer bucks of known copulatory success ($n = 6$ and $n = 16$). In only one case (circumference of the right beam and copulatory success, $n = 16$, $r_s = 0.544$, $p = 0.029$), was there a correlation between reproductive success and individual measurements or the trophy score ($r_s \leq 0.676$, $n = 6$, ns; $r_s < 0.425$, $n = 16–14$, ns). It was obvious that differences among lekking bucks were so small that a clear trend relating antler size to reproductive performance was difficult to demonstrate. Although it is difficult to measure reproductive success of males, data from several other ungulate species suggest that antler morphology does not always predict male reproductive potential. For roe deer, age seems to be the major factor influencing mating success (Liberg et al. 1996).

Criteria used to decide which male to remove may not have any selective value. Consequently, any selective harvesting policy may resemble a random removal. In these conditions, the risk of a mistake is higher when shooting an old than a young male because the former has overcome more selective difficulties than the latter. Moreover, in species with polygynous mating systems, such as most European ungulates (Clutton-Brock 1989), male mating success is generally highly skewed (Clutton-Brock 1988; Apollonio, Festa-Bianchet, and Mari 1989; McElligott and Hayden 2000). Therefore, given the difficulties of recognizing high-quality males from morphology, the possibility of killing successful males is always present when mature males are removed.

Social Behavior, Social Structure, and Management Operations

Many mammals have well-defined social units and strong individual relationships. Sociality is often advantageous in cooperative predator detection, caring for the young, and searching for food. Attaining a critical population size is often key to the survival of social species, which often show a metapopulation structure with dispersal among subpopulations. Here we discuss how ungulate reintroduction should be planned in relation to social and metapopulation structure (see chapter 7). Ungulate reintroductions are an important part of wildlife management in Europe: recovery of many species has been based on a series of reintroductions, often planned without a general framework.

ALPINE IBEX REINTRODUCTION IN THE SOUTHERN ALPS

Ibex were almost extinct in the Alps, and their recovery was entirely due to protection of the last population and to reintroduction in six different countries. The only surviving population in the second half of the nineteenth century was on the southern slopes of the Aosta Valley, Italy, protected in a royal hunting preserve. The population estimate in 1879 was of 790 ibex (Passerin d'Entreves 2000). The first reintroductions took place at the beginning of the twentieth century, both in Italy and in neighboring countries. In 1940 there were just four populations in the southern Alps, one autochthonous, two expansions from reintroduced Swiss colonies, and one resulting from a reintroduction in the Maritime Alps in Italy. In 1999 there were about 70 populations in the Italian Alps, with more than 13,000 ibex. The average size of these populations was 192 individuals, but the standard deviation was very high, at 471 individuals, suggesting a very uneven reintroduction success, even if differences in time since reintroduction likely explain some of this variance.

Taking into consideration that the alpine ibex is a social species in which males and females segregate for most of the year (Nievergelt 1967, Toïgo et al. 1998, Villaret and Bon 1998), and that individuals tend to move freely between valleys joining adjacent groups, we examined two possible effects of sociality and metapopulation structure on reintroduction success. We analyzed 14 reintroductions in Italy, for which data were available, to test whether the success of a reintroduction (measured as population size 5 years postrelease) was related to the number of ibex introduced in each release or to the number of males and females. Using stepwise multiple regression analysis we found that only the number of females introduced was a significant predictor of variation in population size 5 years after release ($R^2 = 0.83$, $p < 0.0001$), but the number of males and the total number of ibex introduced had no significant effects. A large number of females in the reintroduction nucleus was obviously important to obtain a large population as well as a rapid growth rate. We also examined the mean annual increase of isolated populations (with no other population within 30 km) and those within less than 30 km from at least two other well-established populations. Preliminary results suggest that connected population may be growing faster 5 years after reintroduction than isolated ones (Mann Whitney U test, $U = 11$, $p = 0.02$). When we controlled for the number of ibex introduced, number of females introduced, and number of males introduced using analysis of covariance, however, the increase of the population after 5 years was not different between isolated and connected populations ($F_{1, 14} = 2.49$, $p = 0.14$).

Even if these results are not unequivocal in presenting evidence for the influence of size and composition of the founding nucleus or of the existence of a metapopulation frame, it is interesting to note that ibex reintroductions in Italy have often been planned with little consideration for the social characteristics of this species. In fact, the most common reason for ibex reintroduction is politically motivated, based on the desire of a given public administration to have ibex on "their" mountain. Consequently, there is no planning on a wider geographical scale to reestablish a viable ibex metapopulation. Moreover the composition and size of the reintroduction nucleus has often been dictated by casual context rather than planned with a scientific approach. A noteworthy exception is provided by some well-planned operations in the central Alps that used about 90 ibex as a starting group. But other difficulties included low financial support for these operations and the lack of postrelease monitoring, a critical aspect rarely planned after a reintroduction (the omission of which also complicated our analysis). Many opportunities remain to establish a wide-ranging plan for ibex reintroduction in the southern slopes of the Alps because more than 80% of all ibex in Italy are currently confined in the western half of the Alps.

Conclusions and Recommendations

Our chapter emphasizes the need for a new approach to management of large mammals in Europe. The economic and social importance of these species requires a consideration of their biology and conservation needs in the planning of many human activities. We should not limit our interest to hunting or recreation. For example, we suggest that it is necessary to include the conservation of large herbivores in the planning of urban areas in mountains or hilly districts. Much research is required to forecast the future demographic and geographical expansion of the large carnivore populations now existing in Europe. Behavioral ecologists are accustomed to formulating hypotheses and making predictions about their research subjects. They must also use their communication skills to advise politicians and land managers about policy options that take into account the current distribution of large mammals and also encourage their future recolonization of suitable habitats.

Summary

This chapter deals with some of the more recent and most relevant management problems for large mammals in western Europe. Brown bear reintroduction in the Italian Alps provides insights about how the ranging behavior of this species makes it necessary to formulate plans for public information, damage prevention, and compensation, and requires intensive monitoring of reintroduced individuals. The recent wolf recovery in the western Alps is a good example of contrasting approaches to the management of the same predator in neighboring countries. Lack of coordination across political boundaries hampers the conservation of large predators. Red deer management is a longstanding problem in western European mountains. Although some deer remain at low elevation all year, others migrate long distances over altitudinal gradients. The provisioning of artificial feeding stations has a negative effect on red deer population because it causes unnatural crowding, thereby reducing home range size. We examined the consequences for ungulate populations of heavy removal of mature males in relation to reproductive strategies. There are major differences in population structure between populations of the same species under different management regimes. Often, the criteria used to assess the "value" of a trophy male may not be related to reproductive success. We suggest that poor knowledge of male reproductive skew can lead to the use of nonselective harvest criteria and ultimately to unwanted genetic or demographic consequences. Finally, using the reintroduction of ibex in the Alps as an example, we examine how the planning of such operations at the metapopulation scale may influence the success of the individual reintroduced herds.

Acknowledgments

We are grateful to Marco Festa-Bianchet and Jean-Michel Gaillard for their comments on the first draft of this chapter. We thank Marco Festa-Bianchet for his editorial assistance. Finally we wish to thank Elena Rossi for her continous help in the preparation of this chapter.

11.

Implications of Sexually Selected Infanticide for the Hunting of Large Carnivores

Jon E. Swenson

When estimating sustainable harvest rates, wildlife managers are typically interested in population-level parameters, such as reproductive rates, survival rates, age distribution, sex ratios, density dependence, and the effect that hunting has on these parameters (Sutherland 2001). The potential effects of hunting on behavior, however, are usually not considered. As one might expect, hunting can affect wariness (Vosburgh and Irby 1998) and short-term habitat selection (Swenson 1982). In addition, effects of hunting on mating system and pair bonds have been documented. For example, pronghorn antelope (*Antilocapra americana*) may shift from resource-defense to harem-defense polygyny in response to a decline in the proportion of older males (Byers and Kitchen 1988). Also, killing male mallards (*Anas platyrhynchos*) after the pair bond has formed on the wintering area lowers subsequent reproduction by yearling females (Lercel et al. 1999). Greene et al. (1998) modeled the effects of hunting in relation to breeding behavior for some African mammals and concluded that the effect of harvest depends, in part, on the interaction between hunter selectivity and breeding system. They found that for moderately polygynous ungulates, high-intensity hunting that focuses on either males or on adults of either sex has a greater effect on populations than hunting of all age and sex classes, and that monogamous and weakly polygynous species are particularly sensitive to hunting. In large carnivores, hunting dominant males can have a population effect through increased juvenile mortality due to sexually selected infanticide (SSI) following male turnover (Starfield, Furniss, and Smuts 1981; Swenson et al. 1997; Greene et al. 1998). In addition, decreased reproductive success has been reported when females choose less productive habitats to avoid potentially infanticidal males (Wielgus and Bunnell 1994, 2000).

In this chapter, I describe the SSI hypothesis and briefly summarize empirical evidence for SSI in many mammalian groups, including large carnivores, although SSI has been well described in many other groups of animals, including birds and invertebrates. I then focus on the evidence of SSI in large carnivores, particularly the lion (*Panthera leo*) in the Serengeti and the brown bear (*Ursus arctos*) in Scandinavia. I discuss the ramifications that SSI can potentially have for sport hunting of large carnivores and the controversy that this concept has generated.

Sexually Selected Infanticide

Infanticide is the killing of dependent offspring. A common formal definition is "any behavior that makes a direct and significant contribution to the immediate death of an embryo or newly hatched or born member of the perpetrator's own species" (Hrdy and Hausfater 1984). Hrdy (1979) classified the potential reasons for infanticide and concluded that infanticide can benefit the perpetrator if it is linked to competition for limited resources. When that limited resource is mates, the competition is intrasexual and the infanticide is termed sexually selected. The requirements of the SSI hypothesis are (1) infanticidal males should not kill offspring they have sired, (2) infanticide should shorten the interbirth period of the victimized females, and (3) infanticidal males should mate with the mother of the dead infant and sire her subsequent offspring (Ebensperger 1998). If loss of part or all of a litter will increase the size or survival of the subsequent litter, however, the interbirth period does not have to be shortened for SSI to occur (van Schaik 2000a). I limit my discussion about infanticide to SSI because it has the most direct implications for sport hunting of large carnivores.

Because infanticide reduces the fitness of the victim's parents, they are expected to evolve counterstrategies to prevent infanticide (Ebensperger 1998). For mothers, these may include (1) pregnancy termination to prevent additional investment on infants that would likely be killed (the Bruce effect), (2) maternal aggression to protect the offspring, (3) female coalitions to repel infanticidal conspecifics, (4) avoidance of potentially infanticidal males, (5) promiscuity to confuse paternity and provide an incentive for males to tolerate the young, and (6) territoriality to keep potential intruders away from vulnerable infants (Ebensperger 1998). Behaviors consistent with these counterstrategies have been reported in various species of large carnivores (Craighead, Sumner, and Mitchell 1995; Craighead et al. 1995; Ebensperger 1998; Wielgus and Bunnell 1994, 1995, 2000; Logan and Sweanor 2001), although one would not necessarily expect to find all of these in every species showing SSI.

Patterns of Sexually Selected Infanticide in Mammals

A recent review revealed that various forms of infanticide have been observed in 91 species of wild or captive mammals (Ebensperger 1998). Van Schaik (2000a) investigated whether infanticide by males is concentrated in species with female life history patterns that are expected to promote SSI, especially long lactation (or any prolonged maternal care that delays conception) relative to gestation, because postpartum mating is unlikely given that it would produce two sets of young of different ages. Van Schaik (2000a) found that the gestation period is longer than lactation in most mammals. Species in which male infanticide has been documented, however, had significantly longer lactation:gestation ratios, as expected from the SSI hypothesis. The few species with short lactation and documented infanticide by males are rodents with large litters, where the loss of one litter either decreases the interbirth interval or increases the size of the next litter (Elwood and Ostermeyer 1984).

Van Schaik (2000a) found reports of infanticide in five taxonomic groups, all with lactation:gestation ratios greater than or equal to 1; primates, fissiped carnivores, odontocetid cetaceans, sciurognath rodents, and perissodactyls. In the other eight groups of nonvolant mammals examined, lactation:gestation ratios were significantly less than 1, and no deliberate infanticide by males had been recorded. The results of this analysis strongly support the SSI hypothesis and show that male infanticide is found primarily in three groups: primates, fissiped carnivores, and sciurognath rodents.

Another way to assess the importance of SSI is to determine whether females of species that are expected to show SSI also show counterstrategies against it. Van Noordwijk and van Schaik (2000) examined four predicted counterstrategies: (1) polyandrous preconception mating (leading to paternity confusion), (2) long mating periods (to promote polyandry), (3) postconception mating (which also confuses paternity because a male mates with a female that is already pregnant), and (4) embryo abandonment in response to the arrival of potentially infanticidal males. They classified species as vulnerable if lactation lasts longer than gestation, reproduction is not strictly annual and seasonal, or implantation is facultatively delayed or the birth interval shortened after a lost litter. Females in vulnerable primate species mate polyandrously more often (62% of 47 species) than females of nonvulnerable species (9% of 11). In carnivores, 87% of 40 vulnerable species mate polyandrously compared with 58% of 26 nonvulnerable species. They found no clear pattern for length of mating periods or post-conception mating, but the latter is common among the vulnerable species in two of the most vulnerable orders, primates and carnivores. Although the data on embryo resorption or abortion are limited, especially for wild mammals, it is observed in many species in circumstances where subsequent infanticide is likely.

Primates are the best-studied group of mammals exhibiting male infanticide. Van Schaik (2000b) examined 54 cases of infanticide to see if they correspond with the predictions of SSI. He found that 85% of the cases of infanticide occurred after a reproductively able male became dominant; 94 to 98% of the infants were not killed by their father; 67 to 100% of the infants, depending on species, were killed at an age that would have shortened the time to their mother's next estrus; and, in 78% of the cases, the male gained mating access to the mother. These results are consistent with SSI. Recent DNA analyses have shown that male langur monkeys (*Presbytis entellus*) killed or attacked only infants they had not sired, and were the likely fathers of subsequent infants (Borries et al. 1999).

Packer and Pusey (1984) predicted that carnivores were more likely to exhibit infanticide than any other mammalian order. These studies provide strong theoretical and empirical support for the occurrence of SSI in carnivore species with female reproductive patterns that make them vulnerable for SSI.

Lions in the Serengeti

Among large carnivores, SSI has been best documented in lions, especially in a 24-year study in the Serengeti (Pusey and Packer 1994a). The presence of resident males in a pride deters infanticide by alien males. Virtually all cubs less than 9 months of age died after groups of males took over a pride of females. There was a clear benefit for the males to kill the cubs after a takeover because the time from cub death to conception is about 5 months, compared with 20 months from birth to conception when the cubs survive. Because the average tenure of a male coalition is only 24 months, the advantage is obvious. Another advantage to the overtaking males is that the mothers terminate their alliance with the former males after their young die, and begin to cooperate with the new males to keep out other males (Pusey and Packer 1994a).

As expected, female lions show counterstrategies to infanticide (Pusey and Packer 1994a). McComb et al. (1993) documented that mother lions can identify the roars of potentially infanticidal males. When mothers heard roars of alien males, they stood up, grimaced or snarled, and stared in the direction of the roars. Most mothers then ran or walked with their cubs in the opposite direction. When meeting alien males, females with cubs threaten or attack them, and may be wounded or killed in the fighting (Pusey and Packer 1994a). Grouping by females appears to be an important defensive strategy; cubs of solitary females were more likely to be killed than cubs of groups of two or more females (Packer, Scheel, and Pusey 1990), even though the typical number of females in a crèche exceeds the optimal group size for female foraging efficiency (Packer, Scheel, and Pusey 1990) and crèche formation leads to milk theft by nonoffspring (Pusey and Packer 1994b). Also, alien males are more reluctant to approach playbacks of three females roaring in a chorus than single females roaring (Grinnell and McComb 1996). Pusey and Packer (1994a) concluded that risk of infanticide by males was an important factor influencing pride size.

Another counterstrategy to infanticide is attracting large male coalitions (Pusey and Packer 1994a). Because the length of tenure increases with coalition size, females associated with a large male coalition suffer lower rates of infanticide. Following takeover, females show heightened sexual activity but reduced fecundity for about 100 days. The heightened sexual activity of many estrus females that have lost their cubs encourages competition among coalitions, and the largest coalition eventually becomes resident in the pride. The reduced fecundity is thought to be a strategy to attract a larger male coalition before conceiving (Pusey and Packer 1994a). Delayed conception would lead to a higher female lifetime reproductive success if it increased the chance of attracting a large coalition by 30% (Packer and Pusey 1983a). Packer and Pusey (1983b) found that females also produced a higher proportion of sons in the first 300 days after a takeover. This was expected because the young males in the pride will then produce potentially larger male coalitions when they mature, increasing their chances of invading new prides.

The research in the Serengeti clearly shows SSI in lions. The three requirements of SSI were met and females showed five of the six counterstrategies. When the effect of SSI after male takeover is incorporated into harvesting models, several authors have shown that killing harem-holding males reduced population growth rates (Starfield, Furniss, and Smuts 1981; Venter and Hopkins 1988; Starfield and Bleloch 1991; Greene et al. 1998). It is important to note, however, that the occurrence of SSI seems to vary among lion populations. The smaller parks of East Africa contain small isolated populations, and dispersing males suffer high mortality outside the protected areas. In these parks male takeovers apparently occur at a much lower frequency than in the Serengeti and there is only anecdotal evidence of infanticide (Packer and Pusey 1984). Also, cub:female ratios were not affected by removal of adult males in a lion population in Zambia, where the mortality rate of adult males was especially high (Yamazaki 1996). This apparently influenced the social organization, allowing males to copulate with females from several prides. I suggest that SSI would be lower in this situation, with uncertain paternity.

Brown Bears in Scandinavia

Bears are a likely candidate for SSI (Packer and Pusey 1984). Female reproductive intervals are 2.0 to 3.0 years in American black bears (*Ursus americanus*) (Garshelis 1994), 2.6 to 4.6 years in North American brown bears (McLellan 1994), and an average of 3.7 years in polar bears (*U. maritimus*) (Derocher and Taylor 1994). During the breeding season, captive female American black bears breed 2 to 3 weeks after their cubs are removed (LeCount 1983), but there is some evidence that black and brown bears in the wild may come into estrus within 2 to 4 days after losing their cubs (LeCount 1983, Hessing and Aumiller 1994). Evidently, a male has much to gain by killing cubs he has not fathered if he can sire the subsequent litter. Observed female counterstrategies include aggressive defense of young, and mothers may be killed or wounded in the fighting (McLellan 1994; Craighead, Sumner, and Mitchell 1995; Swenson, Dahle, and Sandegren 2001b); polyandry (Craighead, Sumner, and Mitchell 1995; Craighead et al. 1995); and avoidance of habitats frequented by males (Pearson 1975; Murie 1981; Mattson, Knight, and Blanchard 1987; McLellan and Shackleton 1988; Wielgus and Bunnell 1994, 1995, 2000).

It is well known that bear cubs, as well as older bears, including females, are killed by adult conspecifics of both sexes (Garshelis 1994, McLellan 1994). Killing of nonkin offspring by females may reduce competition for resources (Hrdy 1979, Lindzey et al. 1986). Although much conspecific predation is clearly not SSI, I will concentrate on male infanticide because it is most relevant to sport hunting.

Unlike lions, male brown bears are not associated with a harem, and neither sex is territorial. Bear home ranges overlap those of other bears of both sexes (Mace and Waller 1997, McLellan and Hovey 2001). During the breeding season (mainly May and June), males and females usually remain together for several days to 3 weeks, although consortships can be as short as a few hours (Craighead, Hornocker, and Craighead 1969; Murie 1981; Herrero and Hamer 1997). Both sexes are promiscuous, with females mating with up to eight males in one season (Craighead, Sumner, and Mitchell 1995). Cubs stay with their mothers until they are 1 to 2 years old in Scandinavia (Swenson et al. 2001), but in North America few cubs leave as yearlings, and they can stay with their mother for up to 4 years (McLellan 1994).

HARVEST OF ADULT MALES AND CUB SURVIVAL

The reported effects of hunting male bears on cub survival are equivocal. Some studies of brown and American black bears have reported a negative relationship between recruitment and density of adult males (McCullough 1981, Stringham 1983, Clark and Smith 1994), counter to the SSI hypothesis. The opposite effect has also been postulated, that increased hunting of adult males can increase cub mortality through SSI by immigrating males (Stringham 1980). Empirical evidence for SSI comes from one population of American black bears and a comparison of two populations of brown bears (LeCount 1987, Swenson et al. 1997). Miller (1990b) concluded that neither a positive nor a negative effect of killing adult males on cub survival has been adequately demonstrated.

My collaborators and I have been studying two populations of brown bear since 1984, one in northern Sweden (about 8000 km²) and one in central Sweden–southeastern Norway (about 13,000 km²); they are described in Bjärvall and Sandegren (1987). The study areas are about 600 km apart and are near the northern and southern edges of the species' range in Sweden. Both populations increased rapidly from 1984 to 1995, indicating they were below carrying capacity. The exponential rate of increase (r) was 0.13 in the north and 0.15 in the south (Sæther et al. 1998). Bear hunting was allowed during the autumn in both areas, but the northern area includes three national parks where hunting is forbidden. Hunting pressure has increased sevenfold since 1995 in the southern area. In the northern area, there is evidence of considerable poaching of bears, about 2.8 times greater than the legal harvest (Swenson and Sandegren 1999). Poaching is less important on the study area than in the surroundings, and it may restrict immigration to the study area (Swenson et al. 2001).

Swenson et al. (1997) examined data from 1985 to 1995 to test the following predictions of the SSI hypothesis: (1) premature loss of cubs would shorten the interval to subsequent estrus, (2) cubs would disappear (presumably die) during the breeding season, (3) cub survival would be lower following the killing of adult ($\geq$ 5 years) males in the south, where hunters killed males during the period from 1985 to 1995, and (4) cub survival would be high in the north, where no males had been killed by hunters. Unmarked cubs were monitored by following their radio-marked mothers. All the predictions were met (Swenson et al. 1997): (1) 8 of 10 females that lost all their cubs gave birth the next year, compared with none of 40 that kept their cubs; (2) 75% of the 20 cubs that disappeared did so during the breeding season; (3) survival of 74 cubs was significantly lower both 0.5 and 1.5 years after an adult male had been killed on the 11,200 km² study area in the south, but not 2.5 years after; and (4) cub survival was significantly lower in the south (0.72, $N = 74$) than in the north (0.98, $N = 50$). The time lags are not whole numbers because males were killed in the fall, and cub loss occurred the following spring.

We concluded that killing an adult male would disrupt the male social organization for 1.5 years and decreased the population growth rate (λ) by 3.4%. Killing an adult male in our southern study area led to a loss of recruitment equivalent to killing 0.5 to 1.0 adult females (Swenson et al. 1997). The time lag we recorded does not seem unreasonable for brown bears if the loss of cubs is primarily caused by infanticide by immigrating males that establish a home range on the study area after the death of a resident adult male. Bears are generally killed during the fall, when fattening for winter denning is important. The breeding season starts in the spring not long after den emergence and continues to midsummer. Thus there is a relatively short time for an immigrating male to become established in a vacancy from the dead adult male and to breed (Swenson et al. 2001). In addition, a young male will probably have difficulty killing defended young, but can become more effective with increasing age.

We have continued our research on SSI and reanalyzed our data using all adult male deaths, not just hunting kills, and extended the period to 1998. We changed the study area definition from a composite area containing all females with cubs for all years to an area containing the home ranges of females with cubs for each individual year. We made spatial and temporal comparisons to examine whether nutritional, social (SSI), or den disturbance factors best explained the observed variation in cub survival (Swenson et al. 2001). Cub survival was 0.96 in the north ($N = 78$) and 0.65 in the south ($N = 126$). The loss of cubs at both the spatial and the temporal levels of comparison was best explained by social factors.

Nutrition did not seem important because cub loss was greater in years when the mass of adult females and cubs was highest (Swenson et al. 1997,

2001). Disturbance was evaluated only in the south and explained some variation in cub survival. In the north, few adult males died and three adult males lost early in the study there were not replaced for many years, presumably due to little immigration of new males. In the south, five times as many males died annually. In years with recorded adult male mortality, an average of 20% died. The estimated number of adult males remained stable, presumably due to immigration. The number of adult males dying 1.5 years previously in the area containing all radio-marked females with cubs in a given year was correlated negatively with cub survival in the south. In the north, no factors correlated with temporal patterns of cub loss, but loss of adult males in these areas 0.5 to 1.5 years previously was the best explanatory variable among those tested. In the north, the few males present were young, and most first bred successfully as 3-year-olds, when they are possibly not large or experienced enough to kill cubs that are defended by their mothers. Swenson et al. (2001) concluded that immigrating males kill cubs, as predicted by the SSI hypothesis.

IDENTIFICATION OF INFANTICIDAL BEARS

It is extremely difficult to observe infanticide or to identify the perpetrator. We have assumed that the perpetrators were primarily immigrant males based on the findings in many other studies of SSI, described earlier, and other evidence. Cub survival was high in the north during a period with no known adult mortality and little or no immigration, and in the south following years when no adult males were known to have died. An increase in the local density of subadult males has been associated with removing adult males in American black bears (Sargeant and Ruff 2001), and the 1.5-year time lag in increased cub mortality suggested that immigrating males could be responsible, as did the stable adult male numbers in spite of adult male mortality in the south. In our earlier analysis, we also found that cub mortality was elevated 0.5 year after the killing of an adult male (Swenson et al. 1997). We did not find this in the second analysis, but it included all dead adult males, not just hunter-kills, and the definition of the study area had been changed. Therefore, these two results are not directly comparable.

We have continued our investigations about this phenomenon, followed females with cubs intensively in 1998–1999, and expanded our studies using DNA fingerprinting. Here I will report some preliminary results. We collected tissue samples from the mother, the infanticidal male, and the killed cub(s) on four occasions. In all cases, DNA analyses revealed that the infanticidal male was not the father of the cubs he killed (E. Bellemain et al.

unpublished). Two infanticidal bears were marked adult resident males, 9 and 11 years old. Also, we observed males with mothers within a few days after cub loss and determined the father of the subsequent litter on six occasions. Four of these males (aged 6, 9, 12, 27 years) were marked and all were the father of the next litter. Two (aged 6 and 9) were unmarked at the time the female was observed with an unmarked male. Our DNA records suggest that one 6-year-old male was probably a first-time breeder, and the other a nonresident. Although we have no proof that these males killed the cubs, it is likely that they did. This shows clearly that the first male with a female just after she lost her cubs has a high probability of siring her next litter. Both of these findings are critical for the SSI hypothesis; the perpetrator is usually not the father of the cubs he kills, and he has a substantial chance to father the next litter. The DNA results also gave some limited support for the hypothesis that immigrants can be infanticidal. In addition, the only year we documented a 3-year-old male mating successfully in the south was 1.5 years after four adult males had been killed there.

These new preliminary results show clearly that resident adult males are also infanticidal in a manner consistent with SSI. Others have also reported that primarily large adult males kill cubs (Troyer and Hensel 1962, Murie 1981, Mattson, Knight, and Blanchard 1992; Olsen 1993; Hessing and Aumiller 1994; McLellan 1994). This is a major difference from the SSI observed in lions. An explanation may be that brown bears are not territorial, and home ranges of adult males overlap. Of course, SSI increases the fitness of a resident male as much, or more, than that of an immigrating male, and nothing in the SSI hypothesis requires that the species be territorial or social. How can we reconcile the relationship between the death of adult males and cub mortality with the evidence that resident males also are infanticidal? We have still not examined the possibilities, but one is that resident males may shift their home ranges after an adult male dies, bringing them into contact with new adult females.

We also looked at the bear-caused deaths of 13 subadult bears (1–4 years old) in relation to the death of adult males from 1984 to 1999 (Swenson, Dahle, and Sandegren 2001b). Most yearlings separated from their mothers in May. We found area differences in the rates of intraspecific predation only for yearling females, which was higher in the south (0.162, $N = 38$) than in the north (0, $N = 28$). Bears killed no subadult females older than yearlings, but males were killed as 1-, 2-, and 3-year-olds. Neither population density nor food abundance influenced rates of intraspecific predation on yearlings, but intraspecific predation on yearling females increased with the number of adult males that had died 2.5 years previously and when any adult male had died 1.5 years previously. Because the pattern for intraspecific predation on

TABLE 11.1. Effects of loss of at least one adult (≥ 5 years) male Scandinavian brown bear, with a 1.5-year time lag, on cub survival, yearling female survival, and population growth rate in the southern study area in Sweden.

	ADULT MALE DIED	NONE KNOWN DIED	SOURCE
Cub survival	0.55	0.92	Swenson et al. (2001)
Yearling female survival	0.70	0.85	Swenson, Dahl, and Sandegren (2001b)
Population growth (λ)	1.078	1.128	

The reproductive rate was increased to account for the shortened interlitter interval due to higher litter loss for the years following adult male death.

yearling females was similar to that for cubs, we speculated that infanticidal males might be prone to kill subadult bears, although this is clearly not SSI (Swenson, Dahle, and Sandegren 2001b). Intraspecific predation on subadults was highest during the breeding season, as it was for cubs and as reported by Mattson, Knight, and Blanchard (1992).

When I combined the results of our studies (Swenson, Dahle, and Sandegren 2001b; Swenson et al. 2001) and calculated population growth using a standard deterministic model (Ferson and Akçakaya 1990), I found that the loss of adult male(s) was associated with a 4.5% reduction in the population growth (Table 11.1).

We also had the opportunity to test whether an increase in harvesting adult male bears would increase cub mortality through SSI. Because the southern population showed a 16% annual growth rate from 1985 to 1995 (Sæther et al. 1998), harvest quotas were increased markedly. In Dalarna and Gävleborg counties the annual number of harvested bears increased sixfold after 1995, the annual number of harvested adult (≥ 5 years old) males increased 35-fold ($U = 6$, $p = 0.001$), and the total annual mortality of radio-marked adult males doubled ($z = 1.12$, $p = 0.26$), as did mortality of cubs accompanying radio-marked females ($z = 2.82$, $p = 0.005$). Thus the results supported the SSI hypothesis (Table 11.2).

We also studied females with cubs to determine whether they showed counterstrategies to infanticide, as would be expected if SSI were an important factor affecting female reproductive success (Agrell et al. 1998, Ebensperger 1998). We followed adult males and females with and without cubs intensively to determine whether the females with cubs exhibited any of the following counterstrategies to avoid meeting males: (1) avoiding males by (a) different activity

TABLE 11.2. Annual legal harvest of all brown bears and adult (≥ 5 years) males in Dalarna and Gävleborg Counties, Sweden, the total annual mortality of radio-marked adult male bears, and the annual mortality of cubs accompanying radio-marked females in the southern study area (1985–1995 and 1996–2001).

	1985–1995	1996–2001
Total number harvested annually	2.8 ± 0.74 (SE)	17.7 ± 2.33
Number adult males harvested annually	0.1 ± 0.09	3.5 ± 0.85
Adult male mortality	0.07 ± 0.04	0.14 ± 0.05
Cub mortality	0.28 ± 0.05	0.47 ± 0.04

SE: Standard Error

rhythms than males, (b) less movement during the breeding season, (c) different use of habitat, and (2) by mating promiscuously.

We found support for the hypothesized counterstrategies: (1a) During the breeding season, females with cubs were less active than males and females without cubs, and most active when adult males were least active (Myre 2000). (1b) Females with cubs moved less than either males or females without cubs during the breeding season (Zakrisson 2000). One could argue that this is because cubs restrict female movement, but home range sizes of females with cubs were negatively correlated with population density (Dahle and Swenson in press). Thus it is not only the cubs causing females to move less. (1c) Females with cubs used different habitats during the breeding season than those without cubs. Bed sites for females with cubs were located in areas with better visibility and large Scots pine (*Pinus sylvestris*) trees (Katajisto 2001, Kristoffersson 2003). We observed several cubs that avoided infanticide by climbing large pine trees (Fig. 11.1). Males killed cubs more often than expected in areas without large pine trees (Kristoffersson 2003). (2) Females mated promiscuously and several litters had mixed paternity (Bellemain 2001), as observed in Alaska (Craighead et al. 1995). During the breeding season, females have one or two estrus cycles of 16 to 27 days each (Craighead, Hornocker, and Craighead 1969). We have shown that the first male with the female often fathers the litter; it is therefore possible that dual estrus cycles are an example of postconception mating.

In conclusion, our results show that the three requirements for SSI are met in brown bears. Females with cubs showed three or four of the proposed counterstrategies: aggressive physical defense (Craighead, Sumner, and Mitchell 1995), avoiding males, promiscuity, and perhaps postconception mating. As far as we know, however, females do not use pregnancy block, group defense, or territoriality as counterstrategies.

FIG. 11.1. Attempted infanticide by an unmarked brown bear male on a cub in Sweden, 1999. The cub saved himself by staying at the top of the Scots pine tree. The male had already killed the cub's two siblings when the picture was taken. DNA from hairs collected from the tree revealed that he was not the father of the cubs. The cub may himself have been a product of sexually selected infanticide (SSI) because his mother lost her litter in 1998, and his father was observed with her just after the litter disappeared. (Photograph by Jonna Katajisto).

EVALUATING THE EVIDENCE IN SCANDINAVIA AND NORTH AMERICA

Our recent research has strengthened our conclusion that SSI is an important factor in bear population dynamics. However, this does not mean that SSI is important in every population. Reviews of North American studies on black (Garshelis 1994), brown (McLellan 1994), and polar bears (Derocher and Taylor 1994) did not reveal conclusive evidence of SSI in any population. Infanticide was an important cause of cub mortality in one population of American black bears (LeCount 1987), but not in another (Elowe and Dodge 1989). In British Columbia, Hovey and McLellan (1996) found high cub survival in a hunted brown bear population adjacent to the unhunted Glacier National

Park, a potential source of immigrating males, and none of 44 cubs died during the breeding season (B. McLellan, 2002, pers. comm.). Miller (1990b) found no change in cub survival rates with increasing harvest rates of adult male brown bears in Alaska. However, there is not necessarily a linear relationship between the loss of adult males and cub mortality due to SSI (Swenson et al. 2001).

Is there a fundamental difference between Scandinavian and North American brown bear males? Perhaps we should remove some of the burden from the males because a high rate of infanticide also suggests that female counterstrategies are not functioning well (Janson and van Schaik 2000). What could inhibit these counterstrategies? North American and Scandinavian brown bears have very different histories. Humans tried to exterminate bears in Scandinavia with all available technology for hundreds of years, and almost succeeded (Swenson et al. 1995). This long history of persecution may have been an important selective force shaping life history strategies (Stearns 1992). After a long period with persistently high human-induced mortality rates and low density, it is not surprising that European brown bears are less aggressive toward humans (Swenson et al. 1999a), more nocturnal (Roth 1983), and more productive (Sæther et al. 1998) than North American brown bears. They are also poorer predators on adult moose (*Alces alces*) (Swenson, Dahle, and Sandegren 2001a). Lowered aggressiveness and increased productivity, perhaps as a trade-off against body growth, may make European brown bear females less able than North American females to defend their cubs from infanticidal males. This may not have been a problem in low-density, heavily hunted populations, but may be a problem in the present higher-density populations.

In contrast to Europe, brown bears in North America were exterminated rapidly after European immigrants arrived; they survived only in inaccessible areas, primarily in the north. In southwestern North America, for example, the brown bear was exterminated about 60 years after the first European settlement (Brown 1985). I am making essentially the same arguments regarding the effect of long-term heavy hunting pressure on the selection for life history traits that Festa-Bianchet discusses in chapter 12. An in-depth comparison of behavioral differences between European and North American brown bears may provide important insight into the mechanisms of SSI, the factors affecting the effectiveness of counterstrategies, and, ultimately, the implications for managing hunting of brown bears.

Evidence from Large Solitary Cats

There is also evidence of SSI in several species of large, territorial, nonsocial cats. Logan and Sweanor (2001) found that male cougars (*Puma concolor*) caused 44% of all kitten mortalities, but females killed no kittens; infanticidal males were "apparently" not the fathers of the kittens they killed; and in 13 cases resident males, including 3 cases of known sires, associated safely with females with kittens. Two males that killed and ate entire litters subsequently sired litters with the mother of the killed kittens. The loss of newborns to 2-month-old kittens may accelerate breeding access to females by 8 to 10 months. Logan and Sweanor (2001) also found several counterstrategies among females with kittens: aggressive defense, avoidance of other cougars, and perhaps "pseudoestrus." Spreadbury et al. (1996) reported that two young kittens were killed by two different transient males. Ross and Jalkotzy (1992) recorded high kitten survival (97%) with a stable male population. Later, with a high turnover in adult males due to trophy hunting, the kitten survival rate declined (I. Ross, 2002, pers. comm.).

Leopard (*Panthera pardus*) males have also been documented siring the subsequent litter after killing kittens (Bailey 1993). A suggestion of SSI has also been reported in tigers (*P. tigris*). Smith and McDougal (1991) estimated 90% cub survival and no infanticide when resident males were stable, but 33% cub survival and much infanticide when new males were taking over the territories of former residents.

Although these results are not conclusive, they are consistent with SSI. They also strengthen the case for SSI in nonsocial carnivores.

The Controversy

Although there are few critics of SSI among behavioral ecologists, it is a very controversial hypothesis, especially among some anthropologists who view infanticide as maladaptive or aberrant (Sommer 2000). In a more extreme criticism of SSI in lions, Dagg (1999) stated that the SSI hypothesis "resonates with Western culture in which many people accept male dominance and aggression and condone in part the control of female sexuality by men." For other reasons, the SSI hypothesis is also controversial among some North American wildlife biologists (Wielgus and Bunnell 2000), although some North American wildlife biologists have proposed its existence (Stringham 1980, LeCount 1987). SSI is also a controversial issue in hunting of cougars, where some managers apparently believe that harvesting males increases kitten survival (I. Ross, 2002, pers. comm.), just as some do for bears (Miller 1990b). Researchers have also speculated that harvesting males may increase juvenile mortality through SSI in cougars (Ross and Jalkotzy 1992; Murphy, Ross, and Hornocker 1999; Logan and Sweanor 2001) and wolverines (*Gulo gulo*) (Landa et al. 2000).

Some critics of SSI in bears believe that SSI should not be expected in nonsocial and nonterritorial species. The SSI hypothesis is not specific to any social organization, and the large number of species exhibiting SSI also argues against it being restricted to any specific social organization. The evidence of SSI in solitary cats supports this. Our preliminary DNA-based results suggest that the pattern of SSI in nonterritorial species may differ from that in territorial species. In nonterritorial species, infanticide by resident males that are not the father to the young may be more common. Such behavior would be hindered in territorial species, where the male has a more exclusive access to the females in his territory. SSI is also easier to observe in social species. In an intensively studied red howler monkey (*Alouatta seniculus*) population, the infant mortality rate was 200 times greater within 1 month after a male status change than during nonchange periods of the same groups. However, less than 5% of the suspected infant killings by newly dominant males were observed (Crockett and Sekulic 1984). Also, infanticide was observed only seven times in the Serengeti and Ngorongoro Crater in Tanzania during intensive studies on lions between 1966 and 1982, but circumstantial evidence suggested that infanticide occurred almost every time a new coalition of males took over a pride (Packer and Pusey 1984).

Paradigm shifts always take time in science in general and in wildlife management in particular, and many wildlife managers remain skeptical. Conservatism was evident in a report that rejected SSI as a possible biological consequence of hunting brown bears, because only one study had suggested it, whereas others showed no clear trend, even though no study had clearly rejected it (British Columbia Ministry of Environment, Lands and Parks 1995). It is well known that many large carnivores can only sustain quite low levels of human-induced mortality, and that adult females are particularly sensitive (Knight and Eberhardt 1985, Miller 1990a). As a consequence, some agencies encourage large carnivore hunters to select males (e.g., Smith 1991).

Wildlife research is usually not optimal to document SSI because it involves marking many animals and monitoring them relatively infrequently to obtain rates of mortality and reproduction. Intensive, individual-based studies are the exception, rather than the rule. It is perhaps not surprising that behavioral ecologists, who work more intensively with their study animals, find evidence of SSI, rather than wildlife biologists. It is very difficult to observe infanticide in wild mammals and many conclusions are based on patterns in relation to hypotheses being examined, including our conclusions regarding SSI in brown bears.

Conclusions and Recommendations

How should managers react to the possibility of SSI in populations that they manage? The SSI hypothesis clearly predicts that many large carnivores should show SSI, but is that enough to accept it for management purposes? A rodent that was expected to show infanticide did not (Ebensperger 2001). Should managers just ignore these hypotheses for now? I would argue that they should not. The review and tests of the SSI hypothesis in addition to the evidence of its existence in lions, brown bears, and some large solitary cats strongly suggest that SSI should be considered when managing harvest of large carnivores. Also, various estimates of the effects of SSI on population growth are large enough to have consequences for management. Given the controversy among researchers, however, it is not surprising that many managers are unsure about SSI.

Male infanticide has been observed in the wild in a context that is consistent with SSI in many species of large carnivores that are commonly exploited by humans: spotted hyena (*Crocuta crocuta*), cougar, Canada lynx (*Lynx canadensis*), bobcat (*L. rufus*), lion, leopard, tiger, American black bear, brown bear, and polar bear (van Schaik 2000). I would add the wolverine. Although it is a seasonal breeder and most adult females mate every year (Magoun 1985), only 40 to 60% of the females raise cubs each year (Landa et al. 1998), and infanticide is a common cause of cub death in some years (Persson et al. in press). If the killing of cubs in 1 year increases the probability of successfully rearing cubs the following year, SSI would be adaptive, as has been suggested in a captive population of another seasonal breeder, the red deer (*Cervus elaphus*) (Bartos and Madlafousek 1994).

The ramifications for hunting are more difficult to predict. Because SSI is associated with the turnover of adult resident males, hunting should have a greater impact on the level of SSI in species in which hunters can distinguish the sexes in the field, such as lions and cougars (and somewhat for bears, Smith 1991). However, hunting should have an effect on SSI whenever it reduces the survival rate of adult males, as it does for brown bears (McLellan et al. 1999). The death of 20% of the adult male brown bears reduced population growth rate by 4.5% (from 1.13 to 1.08) in the local area, and the hunting death of a male reduced population growth by 3.4% (Swenson et al. 1997). Using a very different approach, Wielgus et al. (2001) estimated that a grizzly bear population could experience a 5.7% reduction in growth if females avoided productive habitats frequented by potentially infanticidal males that immigrated after resident adult males were killed. Actual infanticide was not included in their model. Models of lion population dynamics have also showed reduced population growth when adult males are killed (Starfield, Furniss, and Smuts 1981; Venter and Hopkins 1988; Starfield and Bleloch 1991; Greene et al. 1998).

It appears that some populations of lions, brown bears, American black bears, and probably other large carnivores are more susceptible to losses of young than others. Also, in some species both resident adults and immigrants can be expected to exhibit SSI. If many ecological and environmental factors affect the expression of SSI, it will be difficult for a manager to predict the effects of killing adult males in a given population. Obviously, we need more research on SSI in hunted large carnivores, particularly to allow us to predict when SSI should be expected.

Generally, too little is known about the effects of hunting on the behavior of hunted species. Festa-Bianchet (chapter 12) discusses the effects of hunting in terms of selection pressure on life history traits, and notes that there is little research in this important area. This applies equally well to effects on social organization and mating systems. I searched the last 5 years of the *Journal of Wildlife Management* and *Behavioral Ecology* for articles on this subject. I found that no papers in *Behavioral Ecology* and only 0.7% of the species-oriented papers in the *Journal of Wildlife Management* were on the behavioral effects of hunting (Table 11.3). Who will study the effects of hunting on animal behavior, if not behavioral ecologists or wildlife biologists? I reiterate Festa-Bianchet's (chapter 12) ethical concerns about manipulating the morphology and behavior of hunted species by hunting. Is it ethically right to allow hunting when we do not understand its consequences? One could counter that it is usually necessary to hunt large carnivores where the landscape is human-dominated and tolerance to depredations caused by large carnivores is low, but that does not release us from our ethical obligation to understand what we are doing to these species when we hunt them.

Now, back to the manager. A biologist responsible for managing the hunting of large carnivores can either ignore SSI until it has been conclusively demonstrated, or assume a population consequence of harvesting adult males unless SSI has been documented not to occur in a species and area. I recommend the latter, which follows the precautionary principle, and which I believe is only good management procedure. Just as harvesting should be more conservative when population estimates are uncertain (Tufto et al. 1999), it should also be more conservative when the effects of hunting are

TABLE 11.3. Articles published in the *Journal of Wildlife Management* and in *Behavioral Ecology* (1996–2000) that examined the effects of hunting on behavior.

	JWM	BE
Number of articles	698	439
Species-oriented articles:	651	386
about commonly exploited species	65%	10%
effects of hunting reported	3.6%	0.7%
behavioral effects of hunting reported	0.7%	0

uncertain because of factors such as SSI (Boyce, Sinclair, and White 1999; Anthony and Blumstein 2000).

Summary

Sexually selected infanticide (SSI) can occur when a male, who is not the father of a dependent young, may gain increased mating success by killing the young. It is promoted by disruption of the male social organization by killing resident adult males, thus allowing new males into an area or perhaps allowing other resident males to realign their home ranges. It has a solid and well-documented theoretical basis and should be expected in many species of large carnivores. SSI has been well documented in one population of lions, strongly indicated in brown bears in Sweden, and suggested for cougars, leopards, tigers, American black bears, and wolverines. Estimates of the effects of SSI following killing adult males on population growth (3.4–5.7% reductions in r) are large enough to have consequences for management. In species exhibiting SSI, hunting adult males can promote it. According to the precautionary principle, we should consider SSI when managing the hunting of large carnivores. Because there may be geographical or population differences in the occurrence of SSI, however, much more research is required before we can reliably apply knowledge of SSI to carnivore hunting management. The effects of hunting on the behavior of the hunted animals should receive increased attention from behavioral ecologists and wildlife biologists.

Acknowledgments

I appreciate the constructive criticisms of an earlier draft of this manuscript that I received from B. McLellan, S. Stewart, R. Harris, and T. Komberec. The personnel and students of the Scandinavian Brown Bear Research Project have worked hard to gather the bear data presented in this chapter. They are a great group, and I thank them. The Scandinavian Brown Bear Research Project has been supported by the Swedish Environmental Protection Agency, Norwegian Directorate for Nature Management, Norwegian Institute for Nature Research, Swedish Association for Hunting and Wildlife Management, WWF–Sweden, the Research Council of Norway, Orsa Besparingsskog, and several private foundations.

12.

Exploitative Wildlife Management as a Selective Pressure for Life-History Evolution of Large Mammals

Marco Festa-Bianchet

This chapter explores the usefulness of behavioral ecology when sport hunting is either a component or the major objective of a wildlife management strategy. I examine the potential selective effects of different management practices, and argue that wildlife managers' ignorance of those effects could have long-term negative ecological and economic consequences. Knowledge of the selective pressures caused by sport harvest could help define harvesting programs that avoid or reduce artificial changes in the genetic makeup of harvested populations. I will assume that the main goal of sport hunting is to provide recreational opportunities, not to maximize meat production or the number of animals killed. Within that framework, I suggest that minimizing the impact of sport hunting on the evolution of the hunted species should be a major preoccupation of wildlife managers.

Until recently, most wildlife management was concerned with numbers of animals within a hunted population, and their relationships with their habitat. Hunting regulations and harvest quotas are typically based on population goals. Managers seek either to harvest enough animals to prevent some type of habitat or health degradation (such as allowing forest regeneration or decreasing the risk of epizootics), or to avoid overharvesting and thereby maintain the ability to harvest the population in the future or to increase long-term yield (Caughley and Sinclair 1994). Consequently, much management-oriented research has focused on population dynamics, particularly questions of density dependence and of time lags in population and habitat responses, or on the relationships between herbivores and predators (Fryxell et al. 1991, Clutton-Brock and Lonergan 1994, Messier 1994, Solberg et al. 1999). Hunting regulations often direct the harvest to particular sex/age categories, depending on the harvest or population goals (Kokko, Lindström, and Ranta 2001). For example, male-only harvest is used in cases where female harvests are expected to decrease the population below the management goal. Adult male and young-of-the-year harvests are often used when populations are at the desired density, and finally all sex/age classes, including adult females, are taken where either the population would grow rapidly in the absence of harvests, or a reduction in density is desired.

In North America, little attention has been paid to the potential selective effects of sport hunting (Harris, Wall, and Allendorf 2002). In parts of Europe, on the contrary, there is a rich tradition of "selective" hunting, sometimes with painstakingly detailed hunting regulations that direct the harvest to particular sex/age classes or even to particular phenotypes. Some of these practices include the selective removal of individuals that appear weak, or with "undesirable" antler or horn shape and size. In some cases, the apparent intent of selective harvests is to decrease intraspecific competition and maintain future recruitment by removing those individuals that are least likely to survive. In other cases, however, the goal of selective harvest is indeed to select, by favoring certain phenotypes over others. There is evidence that European harvest practices can affect the genetic variability of hunted populations, at least for red deer (*Cervus elaphus*) (Hartl et al. 1995, 1991) and foxes (*Vulpes vulpes*) (Frati, Lovari, and Hartl 2000).

In North America, hunting rules are not as detailed as some European regulations, but often go beyond specifying the sex of the animals that can be harvested. For example, a minimum horn size is often set for male pronghorn (*Antilocapra americana*), mountain goat (*Oreamnos americanus*), and mountain sheep (*Ovis* spp.), and a minimum number of antler points for cervids.

In addition to legal requirements, hunters' preferences affect the type of animals they are more likely to harvest. Hunters may avoid shooting females accompanied by young (Solberg et al. 2000). Given a choice, most hunters will take the largest individual, or the one with the largest horns or antlers. Because in many populations of ungulates sport hunting is the principal cause of death for adult animals (Langvatn and Loison 1999), it is reasonable to suppose that nonrandom hunting mortality may have a selective effect. Recent studies of wild ungulates have shown strong heritabilities for morphological traits such as body size, and varying levels of heritability for life history traits, particularly those affecting female fertility (Hewison 1997; Réale, Festa-Bianchet, and Jorgenson 1999; Kruuk et al. 2000). Hunting-induced mortality of nonlactating females may select for increased investment in reproduction by generating an artificial positive correlation between reproductive effort and survival, whereas hunter selection for large-horned males could lead to either a selective advantage for small-horned males or selection for an earlier investment in rutting activities (Heimer, Watson, and Smith 1984).

Trophy hunting is well developed in many parts of the world and is a major economic activity. There is considerable interest in the use of sport hunting as part of a conservation strategy. Trophy hunting of ungulates is particularly appealing from a conservation viewpoint because a very large income can be generated from the harvest of a small number of animals (Lewis and Alpert 1997). Consider for example the markhor (*Capra falconeri*), an endangered species. Like many other ungulates in Asia, it is threatened by poaching and habitat destruction (Shackleton 1997). Trophy hunters will pay several tens of thousands of dollars to kill a mature male. That money could be used for conservation and could show the value of habitat protection to the local population. At the same time, the demographic impact of removing a few mature males is minor. Indeed, although the markhor is listed in Appendix I of the Convention on International Trade of Endangered Species (CITES), a program in Pakistan for limited trophy hunting of this species is supported by the World Conservation Union (IUCN) Caprinae Specialist Group.

There are two questions related to the potential selective effects of trophy hunting. First, what is the effect of increasing the mortality of males with a trait (large horns or antlers) that is favored by sexual selection and is likely correlated with individual reproductive success? Second, if trophy hunting selects for smaller horns or antlers, then it will decrease the availability of large-horned or large-antlered males over the long term. Therefore, what management strategies may ensure that trophy hunting can be sustained, particularly given the direct relationship between the expected trophy size and the amount of money hunters are willing to pay?

There are many possible selective effects of sport hunting upon the hunted species. For example, about half of the adult mortality of snow geese (*Anser caerulescens*) in North America is due to hunting (Gauthier et al. 2001), and there are untested speculations that wild geese have evolved (or learned and then culturally transmitted) behaviors to avoid sport hunters. Sport fishing has been suggested to select for "smarter" fish (Miller 1957), more adept at avoiding anglers' lures. I will consider two specific cases where sport hunting may have a selective effect on large mammals: changes in reproductive strategy caused by high hunting-induced adult mortality, and changes of horn and antler morphology caused by trophy hunting. The evidence for or against a selective effect of sport hunting is limited because this problem has apparently attracted little attention from either wildlife managers or behavioral ecologists (Law 2001). Rather than review all the available evidence, therefore, my goal is to point out that artificial selection through sport hunting can be a serious ecological, economic, and ethical problem, and therefore research is urgently required to determine the extent to which it may occur.

Sport Harvest and Life-History Evolution

For many species of ungulates, hunting, legal or otherwise, is the most common cause of adult mortality. In areas where large predators have been eliminated, hunting and road accidents account for almost all adult mortality (McCorquodale 1999, Ballard et al. 2000). In Europe, outside protected areas, hunting probably accounts for most mortality of adult chamois (*Rupicapra rupicapra* and *R. pyrenaica*), roe deer (*Capreolus capreolus*), wild boar (*Sus scrofa*), moose (*Alces alces*), and red deer. In North America, the same could be said for white-tailed deer (*Odocoileus virginianus*), mule deer (*O. hemionus*), pronghorn antelope, male bighorn (*Ovis canadensis*) and Dall sheep (*O. dalli*), and some populations of moose, wapiti, and black bear (*Ursus americanus*). Modern wildlife management can claim a numerical success: many hunted species are much more abundant now than they have been for several centuries. In these populations, high density coexists with high levels of hunter harvest, a situation made possible by past restraint in harvests, controls over poaching, good habitat, and absence or near-absence of predation on adults. Artificial feeding is also partly responsible for high ungulate densities, particularly in central Europe.

DEMOGRAPHIC EFFECTS OF HUNTING

What are the demographic characteristics of a hunted population, and how do they differ from those of ungulate populations limited by food availability or by predators? There are two major effects of hunting: an age distribution heavily skewed toward younger animals, and a sex ratio biased in favor of females (Squibb 1985, Ginsberg and Milner-Gulland 1994, Laurian et al. 2000).

These effects can be extreme: posthunt sex ratios of less than 5 males per 100 females have been reported for elk (Noyes et al. 1996).

Few studies have measured the survival of marked individuals in sport-hunted populations of ungulates. In a population of red deer in Norway, natural survival of stags from weaning to 5.5 years of age was 56%, but was reduced to 5% by hunting; in the same population and over the same age interval, survival of females was reduced from 59% to 32% (Langvatn and Loison 1999). In a trophy-hunted population of bighorn sheep in Alberta, natural survival of rams from 4 to 8 years was 58%, but actual survival was reduced to 27% by sport hunting. Because that population was partially protected by a wildlife sanctuary where most rams spent most of the hunting season (Festa-Bianchet 1989), it is likely that in other hunted populations survival to 8 years would be even lower. In one population in Alaska, 10 of 23 mature Dall rams were shot within 2 years of marking, an average harvest-induced yearly mortality of about 25% (Heimer, Watson, and Smith 1984). In a population of Norwegian moose, about 15% of adult females were shot each year, in addition to the 2.5% yearly natural mortality (Stubsjøen et al. 2000). Therefore, fewer than 50% of yearling female moose would survive to 5 years in hunted populations, compared to about 90% in unhunted populations. Data on survival of marked individuals from other hunted populations are scarce, but it is reasonable to suspect that in many heavily hunted species, fewer than 5 to 10% of yearling males and perhaps fewer than 15 to 20% of yearling females survive to 5 years. In unhunted populations the corresponding figures would be about 50 to 60% for males and 60 to 70% for females (Loison et al. 1999a, Gaillard et al. 2000). Because almost all studies of marked individuals report that adult survival of ungulates is not density-dependent, natural survival should not be lower in unhunted than in hunted populations (Gaillard et al. 2000).

SPORT HUNTING AS A SELECTIVE PRESSURE FOR REPRODUCTIVE STRATEGY

Sport hunting causes high mortality of prime-aged adults, whereas most natural mortality affects young of the year and senescent individuals (Gaillard, Festa-Bianchet, and Yoccoz 1998; Gaillard et al. 2000). Life-history strategy and demography are linked: early comparative approaches to life-history evolution suggested trade-offs between, for example, age of first reproduction and longevity (Harvey and Zammuto 1985), or litter size and juvenile survival (Promislow and Harvey 1990). Over the long term, however, those trade-offs are inevitable: a species where first reproduction occurs late in life and average life span is short will go extinct and therefore will not be around for biologists to study. If adult mortality is high, either fecundity or juvenile survival must be high, or extinction will follow. Conversely, if adult mortality is low, either fecundity or juvenile survival will decrease because populations cannot increase indefinitely.

If ungulates evolved with low adult mortality, what are the possible consequences of high adult mortality through hunting? The most likely consequence is an increase in reproductive investment by young adults. In ungulates, strong iteroparity and small litter size select for low maternal investment to avoid compromising the female's survival and future chances to reproduce, particularly when combined with high and temporally variable juvenile mortality, much of which is independent of the amount of maternal care (Festa-Bianchet and Jorgenson 1998). Indeed, interspecific comparisons show that the survival of prime-aged females (before senescence) in unhunted populations is high and varies little, regardless of the causes of mortality (disease, predation, starvation, weather) (Gaillard, Festa-Bianchet, and Yoccoz 1998; Gaillard et al. 2000). A female with a 92 to 96% yearly survival probability should not increase her current maternal investment to a point where it may affect her viability, given that her offspring faces a much lower and widely varying probability of surviving to 1 year, and then a yearling survival that is typically lower than adult survival (Gaillard et al. 2000).

In heavily hunted populations, however, female survival is greatly diminished, independently of current reproductive effort. In addition, a dependent offspring may increase survival, as hunters are often reluctant to kill lactating females (Solberg et al. 2000). Hunting regulations for alpine chamois in many jurisdictions prohibit the killing of lactating females. Similar regulations protect members of grizzly (*Ursus arctos*) and black bear family groups in much of North America. In Alberta, there is a high proportion of 2-year-old ewes among the harvest of "nontrophy" bighorn sheep. Two-year-old ewes often do not produce lambs, and hunters may select ewes without lambs (W. D. Wishart, 1982, pers. comm.). In hunted populations, therefore, there could be selection for increased maternal expenditure. In species like chamois and bighorn sheep that are usually hunted in open areas, selection against females without dependent offspring is likely stronger than for forest-dwelling species such as white-tailed or roe deer, where hunters have fewer opportunities to evaluate female reproductive status before they shoot.

When populations are kept below carrying capacity through hunting, female reproductive performance is enhanced: age of primiparity decreases, whereas fecundity, juvenile survival, and litter size usually increase (Swenson 1985; Jorgenson, Festa-Bianchet, and Wishart 1993; Jorgenson et al. 1993; Swihart et al. 1998). Over the short to medium term, these effects can largely be explained by density-dependent mechanisms: observational and experimental studies of ungulates show that female reproduction, particularly age of primiparity, is very sensitive to resource availability (Langvatn et al. 1996). An additional, potentially confusing variable is the modified age distribution, which in hunted populations is typically heavily skewed toward younger and more productive age classes. This latter effect, however, should be weak: reproductive senescence in female ungulates occurs at an age reached by a very small proportion of females even in unhunted populations (Benton, Grant, and Clutton-Brock 1995; Bérubé, Festa-Bianchet, and Jorgenson 1999). Age differences between hunted and unhunted populations, however, are very likely to cause differences in mortality because survival senescence typically sets in several years before reproductive senescence (Benton, Grant, and Clutton-Brock 1995; Loison et al. 1999a). Therefore one may expect greater natural (i.e., unhunted) female survival in hunted than in unhunted populations, simply because in hunted populations there are few if any females older than 8 to 10 years.

In addition to the ecological effects due to lowered intraspecific competition, I suggest that heavy harvest may select for a life-history strategy that is normally disadvantaged in natural populations. Consider a set of genes whose phenotypic expression led to females that invested heavily in early reproduction, leading to early primiparity and an increase in offspring survival at the expense of maternal survival. In a naturally regulated population of ungulates, that genotype would be selected against because longevity is the greatest determinant of lifetime reproductive success for females (Clutton-Brock 1988; Bérubé, Festa-Bianchet, and Jorgenson 1999). If very few females survive more than two to four hunting seasons, however, a reproductive strategy leading to greater reproductive success early in life would be favored even if it had a negative effect on life span. If the average life span including natural and hunting mortality is 5 years, a gene that increased mortality of 6 to 10 years of age would not be selected against. Selection for high maternal investment would be strengthened by hunter preference for females without dependent offspring. This scenario is not dissimilar to what may be expected in other ungulate populations under artificial selection, such as domestic sheep, cows, or goats. Domestic ungulates have a shortened life expectancy compared to their wild counterparts, possibly because of artificial selection for increased reproduction (or milk production) early in life.

We readily accept that many traits of domestic animals are the result of artificial selection, and some life-history traits of wild animals could also be affected by artificial selection. With sport hunting, most adult mortality is human-caused and human predation is not random with respect to reproductive status or morphology. Obviously, a major methodological challenge in studying the selective effects of hunting is to separate the environmental effects due to lowered intraspecific competition and the genetic effects due to selection for a less iteroparous reproductive strategy. A modeling exercise (Benton, Grant, and Clutton-Brock 1995) suggested that the reproductive strategy of red deer hinds that were hunted until a few generations before the study may be suboptimal, possibly because it was shaped by culling that for many generations resulted in a high level of adult mortality. Researchers have recently expressed concern that the life-history strategies of moose in heavily harvested populations in Sweden may be affected by hunting-induced mortality, which may select for high reproductive effort in early life and lead to premature senescence (Ericsson and Wallin 2001, Ericsson et al. 2001).

My review concerns large herbivores (and possibly some bear populations) that are subject to intense sport hunting, but a similar line of reasoning could apply to large carnivores that are the target of both sport hunting and trapping (or even predator control programs): for example studies of wolves (*Canis lupus*) outside protected areas report very high levels of human-caused mortality (Potvin et al. 1992).

The potential selective effects of harvesting have preoccupied some fisheries scientists for a long time (Miller 1957). Heavy fishing pressure may have not only a demographic effect on fish populations but also a selective effect (Kirkpatrick 1993, Policansky 1993, Reznick 1993, Rochet et al. 2000). Fishing disproportionately increases mortality of adult fish, and nets with mesh sizes allowing the escape of some of the smaller individuals further select against large fish (Law 2001). A logical outcome of these selective pressures is an earlier age of maturity, as reported by a number of studies (Rijnsdorp 1993, Rowell 1993, Rochet 1998). It is often problematic to partition environmental and genetic effects: early reproduction could occur in the absence of selection simply because resources may be more abundant in heavily harvested populations (Rochet et al. 2000).

High predation on adult guppies (*Poecilia reticulata*) is associated with earlier maturation, higher reproductive effort, and more and smaller offspring compared to populations where predation is mainly on juveniles. Differences in life-history strategies are heritable. Translocation experiments to areas where predation was mostly on juveniles led to life-history changes in 11 years (30–60 generations) (Reznick, Bryga, and Endler 1990), providing experimental evidence that life-history traits respond quickly to strong selective pressures. Fisheries scientists are interested in the possible evolutionary impacts of fishing upon fish populations that are exploited either commercially or for sport fishing. Because most of these populations are very difficult to study, however, the evidence for genetic changes consists mostly of phenotypic measurements on exploited stocks and controlled experiments in short-lived species that are not exploited (Reznick, Bryga, and Endler 1990;

Reznick 1993). A more direct approach is possible with exploited ungulates, where individual-level information on morphology, life-history, and genotype can be obtained. Different ungulate hunting regimes in adjacent areas offer great potential to compare life-history differences associated with differences in age-specific mortality.

Trophy Hunting and Selective Pressures on Horns and Antlers

Trophy hunting has a competitive component. Complex scoring formulae measure various aspects of an animal's horns, antlers, or skull, and records are kept by a number of organizations. Trophy scores are strongly correlated with size, therefore most trophy hunters seek adult males with large horns or antlers. Trophy hunting is big business: hunters are willing to pay very large sums in the hope of harvesting a "record book" trophy. Guides and outfitters typically advertise the trophy scores of animals shot by their clients, and areas reputed for producing large trophies attract much greater revenues than areas where males have smaller horns or antlers. For example, consider the bids received by the Foundation for North American Wild Sheep during its auctions of special permits for bighorn sheep (Erickson 1988). These permits are offered by some American states and Canadian provinces to the highest bidder, and typically sell for tens to hundreds of thousands of dollars that should then be used for conservation, research, or wildlife management activities. Recent auction results reveal that, although most jurisdictions obtain bids for special permits of between $20,000 and $60,000, those with a reputation for producing large rams (Alberta, Montana, Arizona) regularly receive up to 10 times as much, with bids topping $400,000 (http://www.fnaws.org/page1.html). Some hunters are willing to pay great sums of money to obtain a few extra centimeters of horn, and the availability of large trophy males can play a strong role in the economics of a wildlife management program.

By definition, the trophy hunter selects according to morphological criteria. For most bovids, the criterion is simply horn size; for cervids, the number of tines, antler symmetry and branching pattern can affect a trophy's score. Given that a proportion of the variability in horn and antler size is genetically determined (Fitzsimmon, Buskirk, and Smith 1995; Hartl et al. 1995; Lukefahr and Jacobson 1998; Moorcroft et al. 1996), trophy hunting may create the somewhat paradoxical situation of selecting against the preferred phenotype. It is therefore surprising that wildlife managers, especially in North America, have paid so little attention to the genetic effects of trophy hunting (Harris, Wall, and Allendorf 2002).

ECOLOGICAL VARIABLES AND ARTIFICIAL SELECTION

The strength of artificial selection caused by trophy hunting will depend upon ecological variables and harvest regulations. Obviously, a high level of harvest of trophy-class males should have a stronger selective effect than a low level of harvest. Harvest regulations based on a simple morphological criterion, without a limit on the number of permits issued, are likely to have a stronger selective effect than management regimes that limit the number of males harvested within each age class or morphological grouping. The timing of the hunt in relation to the reproductive cycle will also affect the selective pressure caused by trophy hunting: a pre-rut hunt will have a stronger effect than a post-rut hunt. The pattern of age-specific horn growth may also play a strong role. For example, species like chamois, mountain goat, and roe deer have a relatively rapid horn or antler growth: mountain goats and chamois achieve over 90% of their horn growth by 3 years of age (Côté, Festa-Bianchet, and Smith 1998). In these species, males become desirable trophies at a relatively young age, and therefore large-horned individuals risk being killed before contributing to future generations. The horns of ibex (*Capra ibex*), on the other hand, grow substantially up to about 10 to 12 years, and ibex may reproduce actively for several years before being selected by trophy hunters (Toïgo, Gaillard, and Michallet 1999). Bighorn sheep are somewhat intermediate; the horns of 6-year-old rams are about 90% of the length they will attain by 9 or 10 years (Jorgenson, Festa-Bianchet, and Wishart 1998). The mating system will also affect the strength of artificial selection for small horns or antlers caused by trophy hunters: where alternative mating tactics account for a substantial proportion of paternities (Hogg and Forbes 1997), selection is likely weaker than where paternities are monopolized by a few highly successful males (Apollonio, Festa-Bianchet, and Mari 1989).

If mating success is affected by both weapon size and male age, an intense level of trophy hunting of young males will have a stronger selective effect than in species where only older males are removed by hunting. For example, although precise information on male reproductive success is not available, studies of both chamois and ibex suggest that in unhunted populations most matings are achieved by males 10 years of age and older (Lovari and Cosentino 1986; Toïgo, Gaillard, and Michallet 1999). An ibex male may not achieve "trophy" status until about 10 to 12 years of age, but the horns of a 5-year-old chamois are not much smaller than those of a 10-year-old. If in a trophy-hunted population of ibex most matings are done by 10-year-olds rather than 12-year-olds, there will still be 10 years of time for natural selection to potentially affect pre-mating male survival. In trophy-hunted populations of chamois, on the other hand, most matings may be by males aged 4 to 5 years because few males may survive to older ages, possibly allowing some reproduction by males that normally would not survive to mating age.

The potential selective strength of trophy hunting is illustrated by fallow deer (*Dama dama*), where a single male can mate with 25% of the females during one rut (Apollonio, Festa-Bianchet, and Mari 1989). If the traits that favor male reproductive success were the same as those selected by trophy hunters, a single male shot before the rut could lead to a large difference in the genetic makeup of fawns born the following year.

Male reproductive success in most ungulates appears to be determined mainly by an individual's ability to beat other males. Antler or horn size is, presumably, only one component of fighting ability: body size and condition can also play a role, especially if very large weapons suffer a risk of breakage (Alvarez 1994). Both size and shape of antlers and horns could be modified by selection to preserve their effectiveness as intraspecific weapons but make them less attractive as trophies. For example, in most of the Canadian province of Alberta, hunting regulations specify that only bighorn rams whose horns describe at least four-fifths of a curl can be shot.

A ram with a large body mass and whose horns were massive but did not reach the minimum legal size until 6 or 7 years of age would enjoy greater survival than a ram with fast-growing horns that became "legal" at 4 or 5 years of age (Jorgenson, Festa-Bianchet, and Wishart 1998). In areas with good hunter access, few rams survive more than one hunting season after becoming legal, and in areas with moderate access, about 30 to 40% of legal rams are shot each year (Festa-Bianchet 1986). A ram that survives the hunting season will face little competition during the following rut because many potential competitors will have been shot. It is therefore reasonable to predict that any genetic trait that retards the age at which a ram's horn becomes legal will be strongly selected for. There is considerable interindividual variability in the age at which rams reach legal status, from as early as 3 years in exceptional cases, to never (Jorgenson, Festa-Bianchet, and Wishart 1998). Rams that reach legal status later in life may have greater lifetime reproductive success than those whose horns are legal by 4 or 5 years of age. In addition, recent evidence suggests that horn size plays an important role in male mating success only after about 7 years of age (Coltman et al. 2002). Rams with fast-growing horns therefore risk being shot before their large horns give them a reproductive advantage, compounding the potential selection for small horns.

Similarly, imagine a wapiti or red deer male with large antlers but with only a few tines: such an individual would do well in an area where hunting regulations state a minimum number of tines for harvestable males, or could enjoy greater survival under a trophy hunting regime simply because hunters would "pass him up" in favor of what they may see as a more attractive set of antlers. Trophy hunting favors a "nontrophy" phenotype by increasing its survival relative to the population mean, and by removing potential competitors. The harvesting scheme prevalent in parts of Europe, where "undesirable" horn or antler phenotypes are selectively harvested in addition to trophy-class males, would obviously complicate the situation.

Of course, the preceding scenario does not take into account potential gene flow among populations subject to different hunting regimes, changes in hunting regulation or harvest levels, and the strengths of several competing selective pressures, many of which are likely temporally variable. For example, there could be a net outflow of genes from protected into hunted areas because males who survived the hunting season by staying within protected areas would be in a very good position to compete for estrous females in neighboring populations where most resident males were shot by hunters (Hogg 2000).

In addition to selection for horn or antler morphology, a high level of trophy hunting may select for greater reproductive effort by young males. Over the short term, there may be a demographic effect without evolution of novel mating strategies: if most mature males are removed by hunters, younger males may take over the role of breeders and possibly suffer higher mortality as a result, as suggested by Geist (1971) and Heimer, Watson, and Smith (1984).

Over the long term, selection could favor males with high reproductive effort over their first few years of life, possibly including faster growth, lower fat reserves, and riskier behavior during the rut. A shortened life expectancy would weaken selective pressures for less risky behavior that may increase the chance to survive to breed again. The consequence could be higher nonhunting mortality for young males. Consider the many white-tailed deer, roe deer, chamois, or moose populations that are subject to very high harvest levels: in these populations very few males survive past 2 or 3 years of age. In three management areas in Oregon, over 90% of wapiti males were killed before 4 years of age (Biederbeck, Boulay, and Jackson 2001). High hunting mortality of males could lead to a high selective advantage for those few that survive beyond 4 years (possibly because they have small horns or antlers, or because their behavior decreases their chance of being shot), or strong selection for early reproduction. In either case, sport hunting could lead to evolutionary change.

The Implications for Consumptive Management

Harvest of large mammals through sport hunting can lead to economic and social benefits that can stimulate conservation. It is therefore important that management decisions be based upon the best available information.

It is reasonable to suspect that any selective harvest may have evolutionary consequences by altering selective pressures and gene frequencies compared to naturally regulated populations. There is clearly a need for more information, particularly about the levels and types of hunting that may lead to evolutionary change. Sexual selection and possibly female choice may favor males with large horns or antlers, and partly compensate for the effects of selective hunting. If the hunting mortality is not very high, it may be insufficient to change the genetic makeup of future generations. Immigration from protected areas may reduce the potential for selection for a "short and fast" reproductive strategy among both sexes. Finally, harvest schemes that simply stipulate a minimum size or minimum number of tines required for legal harvest will likely have stronger selective effects than the more complex harvest strategies prevalent in central Europe.

Three potential problems should be considered. First, some current harvest policies may select for unwanted morphological or life-history attributes that may lead to loss of economic and recreational opportunities. This would be the case for selection for small horns or antlers by high levels of trophy hunting, but also for selection of a reproductive strategy favoring high early investment in reproduction, if it increased nonhunting mortality of young adults. Selective hunting may lead to a loss of genetic variability (Hartl et al. 1995), which may negatively affect a population's ability to survive environmental changes over the long term.

Poaching of African elephants (*Loxodonta africana*) for the illegal ivory trade may select for tusklessness (Jachmann, Berry, and Imae 1995). Second, artificial "adaptive" changes in hunted populations may compromise their long-term ability to persist. A cessation of hunting may have unpredictable consequences for a population that has undergone adaptations to a high level of hunting mortality: both evolutionary and demographic effects should be considered when hunting is stopped because of changes in land designation. Artificial selection is not necessarily reversible (Law 2001). Third, there are ethical concerns: should hunting shape evolution? Much of the nonhunting public and many hunters dislike the competitive nature of scoring trophies. The competitive aspect of trophy hunting spurs a negative reaction by many people that accept or even support other forms of sport hunting. As public attitudes change, the conservation of ungulates will increasingly require the support of people with little interest in hunting. I suggest that the best outcome for both hunting and conservation would be a decreased emphasis on trophy scores, and more emphasis on the enjoyment of hunting, independent of the particular attributes (sex, age, horn size) of the animals that are harvested.

Conclusions and Recommendations

The ideas I have put forth in this chapter, if correct, justify changes in several sport hunting practices. If these ideas are incorrect, however, changes in wildlife management would not be required and could have a negative effect. It is therefore important to test these ideas, ideally through long-term studies conducted in cooperation with researchers, wildlife management agencies, and sport hunting groups. Wildlife management agencies can do the required experiments by manipulating hunting regulations. For example, an experimental change in the definition of legal ram was approved in Alberta partly to test the effects of different management schemes on bighorn ram survival and harvest. Changes in regulations, however, require the support of the hunting public. Future research should combine the analysis of genotype frequencies, morphology, and life-history attributes in populations subject to different levels of hunting or to different harvest regimes.

An alternative to experimental manipulation of hunting regulations would be to better exploit available information. There are vast repositories of data on morphology, sex, and age of harvested animals, in computers and file drawers of wildlife management agencies all over Europe and North America. Additional information on morphological measurements (or trophy scores) is available from private organizations and individuals, including records and actual specimens (stuffed heads) from several decades ago. This information could be used to investigate hypotheses about the selective effects of sport hunting, or to form the basis of future research programs. There are several recent examples of how long-term information gleaned from wildlife management agencies can provide very valuable scientific contributions (Loison, Gaillard, and Jullien 1996; Post et al. 1999; Schneider and Wasel 2000).

The diversity of wildlife management schemes in different areas, including different sex/age restrictions, could also be used to test specific hypotheses. The main difficulty will be teasing apart environmental and genetic effects: a high level of harvest that reduces population density will almost certainly lead to a phenotypic response, but it may or may not also select particular genotypes. The most powerful test of these hypotheses will be a long-term study of the survival and reproduction of a large sample of marked individuals. Long-term studies of marked large mammals are rare, and very few have been done in hunted populations (Festa-Bianchet 1989; Jorgenson, Festa-Bianchet, and Wishart 1993; Langvatn and Loison 1999), partly because researchers are reluctant to invest time and money for marking animals that may be shot within a few months or years. As a result, much of the information on the evolutionary ecology of wild ungulates comes from populations that are either unhunted or very lightly hunted (Byers 1997; Clutton-Brock, Rose, and Guinness 1997; Festa-Bianchet, Gaillard, and Jorgenson 1998; Gaillard et al. 1998a), and little is known about the evolutionary effects of sport hunting. Because of the high cost of marking and monitoring programs, and because a long-term study in a hunted population would be unable to consider many questions of theoretical interest, there is a need for government agencies to become involved. The long-term monitoring program of polar bears (*Ursus maritimus*) in Canada (Messier, Taylor, and Ramsay 1992; Derocher and Stirling 1998) is an excellent example of a successful study supported by government agencies.

The effects of gene flow in and out of protected areas is a research subject that holds particular promise and particular urgency, for both its practical and its theoretical interest. The amount of gene flow among areas subject to different harvest regimes will likely decrease the selective pressures brought about by selective hunting. On the other hand, selective hunting may itself affect the rate and direction of gene flow (Hogg 2000). There are complex patchworks of protected and exploited ungulate populations that would lend themselves to a very productive study.

The possibility that life-history strategies of large mammals have been shaped by hunting also has potential applications for our understanding of interspecific differences in behavior and reproductive strategies (Benton, Grant, and Clutton-Brock 1995). Consider two mountain ungulates, the alpine ibex and the bighorn sheep. The former has been protected from hunting in most of its range since early in the twentieth century, and is still protected from legal harvests in both Italy and France. Bighorn sheep, on the other hand, have been and are heavily hunted for trophies in most of their range in North America. Ibex males have a very high survival rate until about 11 to 12 years of age (Girard et al. 1999; Toïgo, Gaillard, and Michallet 1997) and a very gradual pattern of age-specific horn development (Toïgo, Gaillard, and Michallet 1999), whereas bighorn rams have low survival at 3 to 8 years of age (Jorgenson et al. 1997, Loison et al. 1999a), rapid horn growth (Jorgenson, Festa-Bianchet, and Wishart 1998), and subadult adoption of risky but successful alternative mating strategies (Hogg and Forbes 1997). These interspecific differences could be due to a wide range of plausible ecological explanations but may also result from selection for greater reproductive effort at a younger age in bighorn sheep, brought about by high hunting mortality over the last century. If this is the case, then one may predict higher natural mortality rates and faster horn growth of ibex in areas where they are hunted, such as in Switzerland (Giacometti et al. 1997), and higher survival and slower horn growth (but not smaller asymptotic horn size) of bighorn rams in protected areas, such as large national parks. Information on genetic differences, however, would also be required to test this prediction because differences in survival could be due to changes in age ratios and therefore in age-specific rutting behavior (Heimer, Watson, and Smith 1984), and changes in horn growth would be expected simply from differences in population density (Jorgenson, Festa-Bianchet, and Wishart 1998).

A Final Thought: Is Human-Induced Selection a Modern Phenomenon?

The current extinction crisis caused by human activities is unprecedented, but there is evidence that humans have had a strong impact on the species composition of several ecosystems for thousands of years (Kay 1994a, Balmford 1996, Caughley and Gunn 1996), although the exact nature and strength of historic human impacts are unclear and often controversial (Beck 1996, Choquenot and Bowman 1998). Nevertheless, it is reasonable to suspect that changes in density, distribution, and behavior of many species of large mammals have been affected by human hunters for a long time.

Consider the differences in behavior toward humans of brown bears in Europe and North America. European bears are less aggressive, possibly as a result of coevolution with humans, who may have selectively killed aggressive individuals. Similarly, although North American wolves appear unable to survive outside wilderness areas (Mladenoff, Sickley, and Wydeven 1999), in parts of Europe wolves coexist with very high human population densities (Okarma 1993, Meriggi and Lovari 1996). Differences in response to habitat fragmentation and other human activities also appear to vary according to the potential for coevolution of humans and other species, measured by the length of time since recorded human occupancy (Balmford 1996, Martin and Clobert 1996). Hunting by humans has likely affected adult mortality of many large mammal species in much of the world for several centuries, possibly for millennia. If this is the case, then the reverse argument of the one I have presented may have some merit: the "new" selective pressures may be those experienced by ungulates in several European and North American national parks, particularly southern parks without large predators.

We should be concerned about the potential selective effects of sport hunting because they may limit the future ability of populations to adapt to a changing environment, or future opportunities for trophy hunting. There is also an ethical concern that sport hunting may lead to "artificial" selection. If we wish to avoid the evolution of "artificial" phenotypes, however, we must know what is "natural." Establishing what is "natural" for species whose evolution has been shaped by human predation may be very difficult.

Summary

Game management is mostly concerned with what determines the size and sex/age composition of populations of hunted animals. Consequently, principles of population dynamics are most often applied to wildlife management, including considerations of sex- and age-specific survival and reproductive rates. It is often assumed that sport hunting affects population dynamics but is not a selective force. For many game species, however, avoiding getting shot is a major selective force because most mortality is due to human hunters.

The age-specific mortality caused by sport hunting of large mammals is usually very different from natural mortality. Hunters often kill prime-aged individuals, which normally have a very high survival rate. Regulations often specify the sex and the age class of animals to be killed. Hunters may select prey according to sex, age, reproductive status, or morphology. In much of Europe, morphology-based harvests favor certain phenotypes, particularly with regard to antler or horn size. The term *selective hunting* is somewhat foreign to North American managers, but it is often used in Europe. In North America, harvest is directed to certain age classes through morphology-based definitions of what can be killed, particularly with regard to horn size and antler points. Principles of evolutionary theory suggest that "selective" harvesting may indeed "select," but not necessarily with the results that managers or society may seek. Intensive hunting may select for precocious maturity and increased reproductive effort, and trophy hunting may select for small horns or antlers. Long-term management plans must take into account the potential selective pressures of alternative harvest schemes, as is recognized by some fisheries scientists. Because sport hunting is as much a social issue as a biological one, changes in wildlife management require changes in attitudes, particularly in the case of trophy hunting. Relegating the competitive attitude to the past will benefit both hunters and biodiversity.

Acknowledgments

Bill Wishart, Rich Harris, Wayne Heimer, Jon Jorgenson, Val Geist, and Jean-Michel Gaillard helped me develop the ideas presented here, particularly when they vigorously disagreed with my opinions. I thank Marco Apollonio, Steeve Côté, and Sandro Lovari for comments on an earlier draft. My research on ungulate ecology is supported by the Natural Sciences and Engineering Research Council of Canada, the Fonds pour la Formation de Chercheurs et l'aide á la Recherche (Québec) and the Université de Sherbrooke.

Part IV

Genetic Diversity and Individual Differences

All chapters in part IV underline the importance of looking at variability among individuals, either in genotype or in phenotype. The maintenance of genetic diversity within a population is a major preoccupation of conservation biology. Populations with low genetic diversity may be at greater risk of extinction in the face of environmental change, and genetic variability is essential to preserve the potential to evolve. Just how much genetic diversity is required to reduce the risks of extinction, however, is often unclear, and it is only very recently that some studies have provided clear evidence of the deleterious effects of inbreeding and loss of genetic variability in wild populations. From a practical viewpoint, the preservation of genetic diversity must be balanced against other conservation priorities, such as habitat protection or disease prevention.

The next two chapters explore how genetic diversity within a population can be affected by differences in social structure and by variance in individual mating success. First, Dobson and Zinner show the sometimes surprising effects of mating system on effective population size, which is of interest to geneticists because it affects how genetic diversity may be lost over time. Some captive breeding techniques that are commonly used for restocking fish for either commercial or sportfishing lead to variance in reproductive success that is very different from that produced by matings in the wild. Although at first sight any kind of artificial interference with what appears to be "natural" is generally frowned upon, Claus Wedekind points out that in some cases the "natural" system is not the best one for conservation. Animals are not selected to do what's best for their population; they are selected to increase their individual fitness, even if that occurs to the detriment of their population. In some cases, the interests of managers and the interests of the fish can diverge! In other cases, however, artificial insemination techniques can lead to unwanted selection for fish that are adapted to artificial insemination rather than to survival in their natural environment.

One result of genetic variability is phenotypic variability, although individual differences in phenotype can be caused by many environmental and epigenetic effects. As more researchers accumulate long-term databases on individually marked animals, they face the opportunity to look at how persistent individual characteristics affect behavior and reproduction, as well as the problem of how to analyze repeated observations from the same individuals without violating the assumption of nonindependence that is key to statistical analyses. While presenting a primer of recently developed statistical techniques, Brian Steele and John Hogg also show how individual differences can indeed affect our interpretation of results from wildlife studies.

Finally, Peter Arcese examines a series of potential measures of "individual quality" for consistency among years. After analyzing an unusually detailed long-term data set on song sparrows, he concludes that none of the apparent measures of individual quality are very useful to predict population growth. He consequently cautions against the uncritical use of traits that appear correlated with individual fitness in setting priorities for conservation program. His chapter raises a number of important theoretical and practical questions about the differences between individual performance and population growth.

13.

Social Groups, Genetic Structure, and Conservation

F. Stephen Dobson and Bertram Zinner

Many vertebrate species, especially mammals, exhibit social groupings that depend at least to some degree on kinship (Wilson 1975, Trivers 1985). Social groups provide opportunities for a variety of mating patterns, and thus for varying levels of reproductive competition. Mating patterns influence the genetic properties (or "gene dynamics") of populations (Wright 1969). The influences of different mating patterns on gene dynamics have been studied mainly by theoreticians and are controversial (Nunney 1999). These influences are important for two reasons. First, the genetic properties of populations constrain and influence the evolutionary potential of species (Nunney 1999). For example, social behaviors within a population may constrain or promote both cooperation and competition among individuals (Chesser 1998a). At the same time, existing cooperative and competitive social behaviors may influence gene dynamics of a population through their influence on mating patterns (Dobson 1998). Thus gene dynamics and social behaviors probably coevolve, each influencing the properties of the other and their potential for change.

Second, gene dynamics are important in devising effective management plans for species of conservation concern. The genetic properties of populations may influence the flexibility with which species react to environmental changes (Chesser, Rhodes, and Smith 1996; Nunney 1999). In general, it is assumed that populations with greater genetic variation are more viable in the face of dramatic environmental changes. To provide an example of how gene dynamics might inform conservation, Sugg et al. (1996) considered the case of possible translocations of black-tailed prairie dogs (*Cynomys ludovicianus*), given knowledge of the gene dynamics of this species. Although black-tailed prairie dogs are not currently of global conservation concern, a closely related species with very similar biology, the Mexican prairie dog (*Cynomys mexicanus*), is rare and endangered. Sugg et al. (1996) found that translocations of female prairie dogs would break up genetic substructuring of colonies that is caused by social groups of closely related kin, and that this in turn would cause more rapid loss of genetic variation from prairie dog colonies. Low genetic variability could render such colonies more vulnerable to extinction after changes in the environment.

Empirical studies have lagged behind theoretical work on mating patterns and gene dynamics. In some cases, however, it is clear that some social mammalian species have genetic properties that are influenced by their polygynous mating systems and social groups (Schwartz and Armitage 1980, Patton and Feder 1981, Chesser 1983, Pope 1992, Dobson et al. 1998, but see Storz 1999). These studies used Wright's (1965, 1969) *F*-statistics to describe gene dynamics. *F*-statistics describe deviations of a population from random mating: (1) within subpopulations, (2) among subpopulations, and (3) within the population.

The above studies differed from studies of regional gene dynamics under the classical island model of population genetics (reviewed by Slatkin 1987) because they used social groups (or "breeding groups," the lowest level of population structure within which mating is random) as subpopulations, and a colony or an aggregation of families as the population (Fig. 13.1). In general, these studies found that mating among the offspring of a social or family group occurred less frequently than expected if mating was random, and that significant genetic differences occurred among social groups.

Wright (1969, 1978) also devised the concept of effective population size to describe the rate of loss of genetic variation from a population. Effective size is the number of individuals in an ideal population that lose genetic variation at the same rate as the census population. An ideal population is one with equal numbers of randomly mating males and females, and no migration, mutation, or selection. Because effective size is estimated in relation to the actual census population size, it is helpful to compare the two (Nunney 1993, Nunney and Elam 1994). Effective size has been typically estimated at

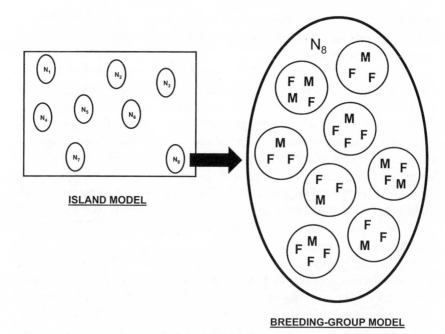

ISLAND MODEL

BREEDING-GROUP MODEL

FIG. 13.1. An illustration of the difference between Wright's classical island model of gene dynamics (left) and population structure at the breeding-group level (right). Because *F*-statistics are hierarchical, these two views of population subdivision are compatible and overlapping. Thus genetic variation can be partitioned within individuals, within breeding groups in subpopulations as on the right, among breeding groups on the right, within subpopulations such as N_8, among the 8 "N" subpopulations on the left, and within the regional population of "islands" on the left. In the original island model, individuals within the subpopulations mated randomly, rather than in breeding groups. M = male, F = female.

between half and the full census population size, indicating that real populations lose genetic variation more rapidly than expected from ideal populations. Over several generations, effective sizes may be lowered even further by historical population declines (Frankham 1996). Recent empirical studies of mammals with social groups, however, indicate that effective size may be nearly as high or even higher than census population size (Long et al. 1998; Pope 1998; Dobson, Smith, and Wang 2000).

In an attempt to model the influence of social groups on gene dynamics, Chesser (1991a,b, Chesser et al. 1993) incorporated the influence of female philopatry and male polygyny on changes in gene dynamics. Matings were assumed to occur only within social or family breeding groups, and models that described gene dynamics of colonies or subpopulations within such groups were called breeding-group models (Sugg et al. 1996, Dobson 1998). Cockerham (1967, 1969, 1973) had shown how Wright's F-statistics could be derived from changes in genetic correlations among individuals in populations, and Chesser (1991a,b) showed how gene correlations, and thus F-statistics, should change over the generations, under conditions of female philopatry and a polygynous mating system. Gene correlations measure such things as the average coefficient of inbreeding and average relatedness within and among breeding groups. With further assumptions, Chesser et al. (1993) showed how effective population size could also be estimated using gene correlations and the F-statistics derived from them. To provide more realistic estimates of F-statistics and effective sizes, Sugg and Chesser (1994) incorporated the influence of multiple paternity, which could result from a female mating with different males during the same or different breeding events.

Not all social birds and mammals are polygynous, and breeding groups in some species contain only two breeders at a time. Apparently monogamous species are quite common in birds (Greenwood 1980, Greenwood and Harvey 1982) and occur less commonly in mammals (Dobson 1982). Of course, many species thought to be monogamous may in fact exhibit extra-pair paternity, or females may change mates from one breeding season to the next (Westneat and Sherman 1997, Goossens et al. 1998). These species may be described as exhibiting social monogamy, but the actual mating system may be somewhat promiscuous or even polygynous (Bouteiller and Perrin 2000). The gene dynamics of socially monogamous species may depend on rates of dispersal away from the natal area (Balloux, Goudet, and Perrin 1998), as well as on the degree of multiple paternity (Parker and Waite 1997).

The breeding-group model makes some important predictions for both conservation and evolutionary biology. The essential insight of the model is that genetic structure may occur within colonies; that is, within what Wright called subpopulations (Sugg et al. 1996). If this finer scale of genetic structure is ignored, aspects of a population's gene dynamics that occur at the level of social breeding groups are also ignored. The first prediction that can be made from the model is that when social groups are based on kinship, there should be genetic differences among the social groups (Dobson 1998). Relatives within a group should be more closely correlated to each other through common descent, than they are to individuals in other groups. A second prediction is that if dispersal evolved to reduce inbreeding (reviewed in Greenwood 1980, Dobson 1982, Thornhill 1993), then the F-statistics that compare the rate of inbreeding to random mating within the breeding groups should be strongly negative (Dobson et al. 1997). Finally, under a pattern of strong female philopatry and harem polygyny by a single male, the rate of loss of genetic diversity should be considerably slowed (Chesser et al. 1993; Chesser, Rhodes, and Smith 1996).

The breeding-group model and its empirical tests indicate that social groups can be extremely different genetically (reviewed by Dobson 1998, but see Storz 1999). The loss of genetic variation from socially structured populations may be much slower than predicted by classical approaches to population genetics (Sugg et al. 1996), although this conclusion has been questioned by Nunney (1999). Thus estimates of gene dynamics and our understanding of the genetic properties of populations could be very different if the breeding-group approach were supported. In fact, the breeding-group model has been supported when tested against two other means of calculating gene dynamics: biochemical data and estimates from pedigrees (Dobson et al. 1997, 1998).

Our purpose in this chapter is threefold. First, we will review the predictions of the breeding-group model. Second, we will examine the empirical evidence that supports the model's usefulness for polygynous species, including an example that assumed both multiple paternity and its absence. And third, we will examine theoretical and empirical tests of the model's usefulness for monogamous species, and briefly consider the possible influence of multiple paternity on the gene dynamics of socially monogamous species.

Model Description

Mathematical details of the breeding-group model appear in Chesser (1991a,b), Chesser et al. (1993), Chesser, Willis, and Mathews (1994), and Sugg and Chesser (1994). Rather than repeat the mathematical details, we will explain the reasoning that forms the basis for operation of the model. Our purpose is to provide a guide to application of the breeding-group model to real situations in nature. Consequently, we will restrict our description to the most familiar gene dynamics, F-statistics and effective population sizes, that are commonly reported in the literature. The model has been extended to autosomal inheritance (Chesser 1998b), as well as to various forms of paternal and maternal inheritance (Chesser and Baker 1996), but we will not discuss these further advances.

Cockerham (1967, 1969, 1973) defined three gene correlations: F, the inbreeding coefficient; θ, the average genetic correlation between individuals in a group (in the absence of inbreeding, this is half the average coefficient of relatedness = r); and α, the average genetic correlation between individuals that are not in a group together. These gene correlations can be used to calculate Wright's (1965, 1969) F-statistics (after Chesser 1991a,b):

$$F_{LS} = \frac{\theta - \alpha}{1 - \alpha}, \qquad F_{IL} = \frac{F - \theta}{1 - \theta}, \qquad F_{IS} = \frac{F - \alpha}{1 - \alpha}. \qquad \text{(Equation 1)}$$

The first F-statistic, F_{LS}, indicates the degree of genetic differentiation among social or family breeding groups (L stands for the "lineage" in a social group and S for "subpopulation"), so if its value is 0.15, the breeding groups are on average 15% different genetically. The second F-statistic, F_{IL}, shows how the rate of inbreeding compares with that expected from random mating among the offspring of a breeding group (where I stands for "inbreeding"). If F_{IL} is positive, then offspring mate with closer kin than expected at random. If negative, mating is with more distant kin, as expected with dispersal out of the breeding group. A value of zero indicates that mating among offspring within the breeding group is about as expected at random. These same interpretations apply to the last F-statistic, F_{IS}, but here the random mating expectation is for the whole colony or subpopulation, rather than the breeding group.

F-statistics are hierarchical, so that the genetic properties at greater levels of spatial scale are easily examined. Thus an F-statistic, F_{ST}, indicates the degree of differentiation among colonies or subpopulations in a larger regional population. At this level, F_{IS}, the deviation of the rate of inbreeding from that expected under random mating of the subpopulation, also occurs and is the same value as the F_{IS} calculated above, assuming that all subpopulations have this same value. Finally, F_{IT} is the deviation of the inbreeding coefficient from that expected if matings within the entire population were random. The notation of F_{ST}, F_{IS}, and F_{IT}, were what Wright (1965, 1969) originally used when he defined F-statistics because he envisioned genetic differentiation and inbreeding in populations that were isolated or semi-isolated by geography, rather than by social grouping (Fig. 13.1).

In addition to the F-statistics, effective population sizes can be estimated from gene correlations, according to a simple formula developed by Wright (1969):

$$N_{eI} = \frac{1}{2\Delta F}, \qquad N_{eV} = \frac{1}{2\Delta\alpha}. \qquad \text{(Equation 2)}$$

N_{eI} is termed the inbreeding effective population size, and N_{eV} is called the variance effective size. At genetic equilibrium, these estimates give the same value. Thus, for practical purposes, it doesn't matter which formula is used to estimate effective size. It may sometimes be easier to use formulas for effective size that are based on the F-statistics (Balloux, Goudet, and Perrin 1998), however, and a formula for doing so was presented in Chesser et al. (1993) and Sugg and Chesser (1994):

$$N_e \cong \frac{4s - 3F_{IS} - 1}{6(F_{LS} - F_{IS})} \qquad \text{(Equation 3)}$$

where s is the number of breeding groups in the population, and the F-statistics can be determined from Equation (1).

Chesser's (1991a,b) basic method was to track the change in gene correlations over time. Gene correlations of offspring can be calculated from the gene correlations of their parents (Cockerham 1967, 1969, 1973):

$$\theta_{i,j} = \frac{1}{4}\left(\theta_{S_i S_j} + \theta_{S_i D_j} + \theta_{S_j D_i} + \theta_{D_i D_j}\right) \qquad \text{(Equation 4)}$$

Here, i and j refer to two different individuals, and S and D refer to their sires and dams, respectively. By averaging across pairs of individuals, this formula can estimate the average genetic correlation between individuals of the offspring generation that are in the same breeding group (viz., θ), and the average gene correlation between individuals that are in different breeding groups (viz., α). Also needed is the average inbreeding coefficient ($F = \theta_{S_i D_i}$), the correlation due to common ancestry between the gametes that form an offspring individual. Given a pedigree, estimates of gene correlations are fairly easy to make, even after only a few generations (Dobson et al. 1997, 1998). Unfortunately, such pedigrees are seldom available for wild populations.

Chesser (1991a) recognized that the change in gene correlations from one generation to the next depends on patterns of mating and thus could be calculated using a transition matrix that represented the way that the mating system influenced gene correlations from the parental to offspring generations. First, he began with the notion of an ideal population that was subdivided into several social or family breeding groups. He assumed that generations do not overlap, to make the accounting of gene correlations simpler. He defined terms for the average number of adult males and f emales in breeding groups, the number of breeding groups, and dispersal among breeding groups. The dispersal term was based on the rate of both male and female movements, and thus could indicate the importance of female philopatry on changes in gene correlations. Because several females may occur in a breeding group, and their reproductive success may vary, he also incorporated a term ϕ_f, the mean probability that two randomly chosen surviving offspring in a breeding group share the same mother (this must be calculated or estimated over the lifetimes of mothers, due to the assumption that generations do not overlap). In the transition between generations, only individuals that potentially breed should be counted, so "surviving offspring" refers to young that both survive to adulthood and mate successfully. ϕ_f varies from 1 for monogamy to lower values as the number of mothers in a breeding group increases.

Next, Chesser (1991b) incorporated the influence of male polygyny. If one male does all the breeding in a group, the offspring of different females are more closely correlated genetically than if mothers mated with different males. For this, he introduced the term ϕ_m, the mean probability that two randomly chosen surviving offspring of different mothers in a breeding group have the same father. Again, this must be calculated or estimated over the lifetimes of mothers. This term is zero when there is only one mother (Sugg and Chesser 1994, Balloux, Goudet, and Perrin 1998), but for multiple mothers in a breeding group, it primarily depends on the number of fathers and their contributions to offspring. If only one male breeds in each generation, the value of ϕ_m is 1, and declines as more males father surviving offspring. The last innovation was to incorporate the influence of multiple paternity, which was not taken into account in earlier modeling. For this, the term ϕ_w was defined as the mean probability that two randomly chosen surviving offspring, produced over the lifetime of a female, have the same father (Sugg and Chesser 1994). The value of ϕ_w is 1 for single paternity, and zero for complete multiple paternity. Notice that multiple paternity can occur in two different ways: from a mother mating with different males in different years, and from multiple paternity within a clutch or litter.

Nunney (1999) questioned the efficacy of the breeding-group model and the realism of its assumptions (nonoverlapping generations and the "local" population regulation that philopatry of females indicates). As an alternative, he offered models for estimating effective population sizes, based on *F*-statistics, that have arbitrary groupings that might be treated as social groups, as well as a variable that represents the degree of local population regulation. In a locally regulated population, group size and group number are relatively constant over time. This alternative breeding-group model requires estimation of variation in reproductive success for males, females, and the designated groups. Dispersal patterns of males and females among groups do not explicitly appear in Nunney's models and thus do not need to be estimated. Because multiple mating precludes knowledge of paternity patterns for many species, however, this method may be difficult to apply broadly. Also, means of estimating the degree of local population regulation are unknown. Applications of empirical data to this model have not yet been made. However, if the above difficulties could be overcome, more accurate estimate of effective population sizes might be produced than with Chesser's models, at least under some conditions of mating and dispersal (Basset, Balloux, and Perrin 2001).

From the preceding description, it is clear that some sort of "group" model could be used to predict the gene dynamics of a monogamous population. Assuming single paternity, Chesser, Willis, and Mathews (1994) did this and found that the breeding-group model reduced to the classical gene dynamics predicted by Wright's (1969) methods, with the important difference that *F*-statistics are calculated at the breeding-group level, no matter how small the breeding group is. Basset, Balloux, and Perrin (2001) showed, however, that Nunney's (1999) group model was much more accurate for monoga-mous species than Chesser et al.'s (1993) breeding-group model. For those bird species that are genetically monogamous, a family group (equivalent to a breeding group) consists only of the paired male and female and their cur-rent offspring. A similar mating and family situation might occur in a few mammalian species as well (but see Balloux, Goudet, and Perrin 1998; Goossens et al. 1998).

Tests of the Model

Two questions need to be answered concerning the breeding-group model. First, is the concept of breeding groups important? That is, do social breed-ing groups have distinct genetic properties? This question can be answered in the affirmative (Dobson 1998). Three studies of rodent species have found that polygynous breeding groups are significantly genetically differentiated from one another (Schwartz and Armitage 1980, Patton and Feder 1981, Dobson et al. 1997, 1998). Three studies of primates, one of red howler mon-keys (*Alouatta seniculus*) and two of humans (*Homo sapiens*), yield similar results (Long 1986, Long et al. 1998, Pope 1992, 1998). The primate studies also found that estimates of effective population sizes were elevated when polygynous social breeding groups were analyzed. Second, does the breeding-group model reflect the same patterns of gene dynamics that can be found empirically from studies of pedigrees or biochemical alleles? This question can be answered by comparing predictions of the model with other empirical results.

The efficacy of the breeding-group model was examined in a case study, and indicated empirical verification of the model under a polygynous mating system, male dispersal, and a high degree of female philopatry. From a long-term study of social breeding groups (called coteries) in a colony of black-tailed prairie dogs, Dobson et al. (1997, 1998) estimated *F*-statistics from three different sources of data: the breeding-group model, pedigrees, and allozyme alleles (Table 13.1). They found that the first prediction, genetic dif-ferences among breeding groups, was strongly supported: all three methods indicated about 17% (viz., $F_{LS} \approx 0.17$) genetic differentiation among the coteries of prairie dogs. Previously, Hoogland (1992) had shown that mating

TABLE 13.1. *F*-statistics from studies of the gene dynamics of mammals

SPECIES	F_{LS}	F_{IL}	F_{IS}	N_e	N_C	RATIO
Black-tailed prairie dog[1]	0.16	–0.18	0.01	95	85	1.12
Black-tailed prairie dog[2]						
breeding-group model[3]	0.16	–0.18	0.00	89	74	1.21
pedigrees	0.19	–0.23	0.00	79	74	1.07
allozyme alleles	0.17	–0.21	–0.01	89	74	1.21
Plateau pika[4]						
single paternity	0.30	–0.37	0.04	61	67	0.91
multiple paternity	0.28	–0.34	0.04	66	67	0.99
White-toothed shrew[5]	0.09	–0.02	0.08	26	44	0.59

[1]Sugg et al. 1996, based on Hoogland 1995. N_e calculated from Equation (3).

[2]Dobson et al. 1997, 1998, unpublished data. N_e calculated from Equation (21) in Sugg and Chesser (1994), of which Equation (3) is an estimate.

[3]The model estimate of N_e based on Equation (2) was 93, yielding a ratio of 1.25.

[4]Dobson, Smith, and Wang 2000. N_e calculated from Equation (2).

[5]Balloux, Goudet, and Perrin 1998; Bouteiller and Perrin 2000. N_e calculated after Nunney (1999).

Definitions of *F*-statistics are given in the text, except for N_C, the census number of breeding adults in the study population.

with respect to kinship was close to random within the colony, with the exception that parents and offspring, and full siblings, seldom bred together. This exception, along with a strong pattern of male-biased dispersal away from the natal area and equally strong female philopatry, was sufficient to fulfill the second prediction: within breeding groups, inbreeding was much less frequent than expected if offspring were mating randomly in their natal coteries ($F_{IL} \approx -0.20$). Again, all three methods of estimating gene dynamics produced very similar results and the same conclusion. These results confirmed similar conclusions by Sugg et al. (1996), who used results from Hoogland's (1995) book on the prairie dogs to estimate *F*-statistics using the breeding-group model to produce similar estimates of fixation indices (see Table 13.1).

Sugg et al. (1996) also used the breeding-group model to estimate effective population size and found that effective size was about 12% greater than the size of the census population (see Table 13.1). Estimates of effective population sizes over a 10-year period during the prairie dog study, using estimates from the breeding-group model, pedigrees, and allozyme data and comparing them to the actual mean census population size, confirmed the above results using the complete data set on the prairie dogs (harmonic means used in all calculations; see Table 13.1). Estimates of effective population sizes averaged about 10 to 20% larger than the census population size. It may seem anomalous that effective size can be greater than the number of individuals in the census population. Effective population size, however, does not reflect an actual number of individuals. Rather, effective size is a measure of the rate at which genetic variation is being lost from the population (equation [2]). If the census population loses genetic variation more slowly than would be expected under an equal number of randomly mating males and females, then effective size can be larger than census size (Chesser et al. 1993). Because the great majority of surviving females settle in the breeding group where they were born, social structure causes slower loss of genetic variation due to the genetic isolation of unrelated females from each other. Different breeding groups may lose genetic alleles, but they will often lose different alleles, thus slowing the loss of genetic variants from the overall colony (Chesser, Rhodes, and Smith 1996).

One reason why Nunney (1999) criticized the results presented by Sugg et al. (1996) was because prairie dogs have overlapping generations, and the breeding-group model assumes nonoverlapping generations. Dobson et al. (2000) applied the breeding-group model to highly social Tibetan plateau pikas (*Ochotona curzoniae*). Plateau pikas have tightly knit family groups, variable mating patterns that average out to a low degree of polygyny, and philopatry and dispersal by both sexes (Dobson, Smith, and Wang 1998). When dispersal among pika families does occur, it is male-biased but restricted to very short distances, so that some close inbreeding is possible. Most individuals live for only about 1 year, and they do not breed until about a year old. Thus they are nearly an "annual" species, and they come close to having nonoverlapping generations. Unfortunately, patterns of paternity were unknown, but some females mate multiply when there are more than two males in a family (Smith and Wang 1991). Thus analyses were run twice: first under the assumption of single paternity, and then assuming complete multiple paternity wherever it was possible.

Results of the breeding-group model indicated that pika families were even more genetically differentiated than coteries of prairie dogs, at about 29% genetic differentiation among families (see Table 13.1; Dobson, Smith, and Wang 2000). Again, inbreeding was much less frequent than expected from random mating of offspring within families (reflected by strongly negative F_{IL} values), despite the limited dispersal pattern. Inbreeding within the pika colony could barely be distinguished from random, as indicated by

F_{IS} values that were close to zero. Effective population size was slightly lower than the number of breeding adults in the population; but classical estimates of effective size that do not take breeding groups into account produced much lower values (at about 42 adults), probably due to the fact that some females were more successful at leaving surviving offspring than others (variance in female reproductive success ≈ 4.20, expected Poisson variance $= 2.00$). Analyses that assumed single and multiple paternity yielded very similar results and the same conclusions.

One socially monogamous species, the greater white-toothed shrew (*Crocidura russula*), has been studied using the breeding-group model. Balloux et al. (1998) used biochemical estimates of gene dynamics and knowledge of the mating system to "back-calculate" the influence of dispersal patterns. F-statistics indicated that "family" groups of shrews were based on male philopatry, producing significant genetic differentiation among families (see Table 13.1). Female dispersal from the natal family was only about 40%, and thus considerable inbreeding occurred within subpopulations of shrews. This study showed the utility of the breeding-group model, even under the case of monogamy, as an exploratory tool. Balloux et al. (1998) concluded that the female-biased pattern of dispersal probably accounted for an extremely high effective population size. The analysis assumed single paternity, and that was likely common (Bouteiller and Perrin 2000). This last study, however, showed that, although they are socially monogamous, the mating system of the shrews is slightly polygynous due to some males mating with the females that live with other males. Multiple paternity of litters was not found, however, so the breeding-group model was probably appropriate for modeling the shrew population. Bouteiller and Perrin (2000) recalculated effective population size for the shrews (using the approach of Nunney 1999), and found that it was about 60% of the census population size.

Basset, Balloux, and Perrin (2001) compared estimates of effective population sizes from Chesser et al.'s (1993) version of the breeding-group model to an alternative model designed by Nunney (1999). A major difference between the models is the timing of estimates of gene dynamics, which occurs before dispersal in Chesser's model and after dispersal in Nunney's model. Effective population sizes were overestimated by the Chesser model under the conditions of monogamous mating and equal dispersal of the sexes. Otherwise the models yielded similar results. This indicates that Nunney's (1999) model, which is more difficult to apply because it requires more parameters (especially variation in male reproductive success and the degree of local population regulation), should be used for monogamous species that exhibit little sex bias in natal dispersal.

In summary, the studies of black-tailed prairie dogs verified the breeding-group model by showing that the model results (estimated from demography, dispersal, and mating patterns) were closely consistent with empirical results from pedigrees (where genetic correlations due to descent can be calculated directly) and allozyme alleles (where F-statistics are estimated indirectly from heterozygosity, and effective sizes are estimated from F-statistics). Although the three sources of data came from the same prairie dogs and were thus biologically interdependent, there was little overlap in the application of variables used to estimate gene dynamics under the three methods. So the different methods were as independent as one might expect. The results of the breeding-group model were strongly supported by the empirical results. The study of somewhat polygynous plateau pikas applied the breeding-group model to a species with little overlap of generations and indicated the importance of family social structure on gene dynamics. The pikas also indicated little influence of multiple paternity on gene dynamics in a slightly polygynous (though with variable mating systems among families) species. Finally, the study of white-toothed shrews indicates the usefulness of the breeding-group approach for estimating gene dynamics of monogamous species. The group model of Nunney (1999) should be applied, however, to obtain more accurate estimates of effective sizes under monogamy or equal dispersal of males and females (Basset, Balloux, and Perrin 2001), and when accurate information about F-statistics; variation in male, female, and group reproductive success; and mode of population regulation are all available.

Multiple Paternity

With the exception of Sugg and Chesser's (1994) breeding-group model, the influence of multiple paternity on gene dynamics is seldom taken into account. Social species may not always exhibit single paternity and thus may not always exhibit the mating system that is apparent from the composition of social groups. For example, multiple male matings with individual females were found in polygynous social groups of prairie dogs and plateau pikas (Smith and Wang 1991, Hoogland 1995). In an apparently different mating system, litters of the socially monogamous alpine marmot (*Marmota marmota*) were found to average only about 70% of offspring sired by the family male (Goossens et al. 1998). To apply genetic models that incorporate social groups to these situations, it is necessary to take multiple paternity into account. Naturally, if a pedigree were available for a population, accurate estimates of gene dynamics could be made. However, species like alpine marmots are long-lived. This has several consequences. First, the estimate of multiple paternity for this species is an underestimate because females may mate with several males over their lifetimes. Second, some degree of multiple mating by males (polygyny) may occur, both within a single breeding season and over the lifetimes of males. Third, the generations overlap (Nunney 1999), though this problem might have more influence on the number of generations that it takes to reach genetic equilibrium, rather than the equilibrium values of gene dynamics (Hill 1979; but see Nunney 1993). In short, the influence of multiple paternity on socially monogamous species needs to be investigated.

There are two ways that multiple paternity might influence the gene dynamics of polygynous species. One is when males that otherwise would not mate gain access through extra-pair fertilizations to females that are otherwise mated to dominant or territorial males. In this case, effective population size likely increases because more individuals participate in breeding, and the variance in male reproductive success of males becomes lower. The other is when all males are mated, but some males gain extra copulations with neighbors. In this case, variance in male reproductive success could become lower or higher (or even remain unchanged), and effective population size would accordingly become higher or lower, respectively. The influence of multiple paternity on F-statistics is unclear. But in general, when more males breed in local social groups, genetic correlations within the breeding groups will be diluted as the number of fathers increases. With the value of θ lower, Equation (1) suggests that genetic differentiation among breeding groups (viz., F_{LS}) should become lower as well.

Socially monogamous species should also have higher effective population sizes if more males father offspring and the variance in male reproductive success decreases. The number of fathers in such populations may exceed the number of breeding females. If the number of fathers decreases, producing de facto polygyny, then effective population size should decrease (Parker and Waite 1997). It is not so clear, however, how multiple paternity would influence effective population size if all breeding males in a population were otherwise mated. With single paternity, the variance in male and female reproductive success would be equal. As multiple mating occurs, differences among males, but not females, in reproductive success might be diluted. If the variance in male reproductive success was lowered via multiple paternity, the effective size of the population should be increased (Nunney 1999). We examined this possibility in a simple model.

Consider a population of n couples (male and female). Denote by X the number of total offspring of a given female during a certain time interval. Assume that a certain proportion of her offspring are not fathered by her partner. Suppose that this proportion is p, on average. Denote by Y the number of total offspring fathered by her partner, a given male, during the same time interval. Note that these may or may not be born to the given female. We are then interested in how the variance of Y is related to p.

We made the following assumptions. Each of the X offspring is fathered by the given male with probability $1 - p$. Each male also has a chance q of being the father of any of the offspring of a number k of other females.

Denote the total number of offspring of these k females by Z. Denote by Y_1 the number of offspring fathered by the given male and born to his partner, the given female, and denote by Y_2 the number of offspring fathered by this male but born to other females. Then the distribution of Y_1 given $X = x$ is binomial with parameters x and $1 - p$ and the distribution of Y_2 given the number of offspring from the other k females, say $Z = z$, is also binomial with parameters z and q.

Since the mean of Y must be the same as the mean of X, one can express q in terms of p using the Law of Total Expectations:

$$E[X] = E[Y] = E[Y_1 + Y_2] = E[Y_1] + E[Y_2] = E[E[Y_1|X]] + E[E[Y_2|Z]]$$
$$= E[(1-p)X] + E[qZ] = (1-p)E[X] + kqE[X],$$

which implies

$$q = \frac{p}{k},$$

Then

$$Var(Y_1) = E[Var(Y_1|X)] + Var(E[Y_1|X])$$
$$= E[p(1-p)X] + Var((1-p)X)$$
$$= p(1-p)E[X] + (1-p)^2 Var(X)$$

and

$$Var(Y_2) = E[Var(Y_2|Z)] + Var(E[Y_2|Z])$$
$$= E[q(1-q)Z] + Var(qZ)$$
$$= kq(1-q)E[X] + kq^2 Var(X)$$
$$= p\left(1 - \frac{p}{k}\right)E[X] + \frac{p^2}{k} Var(X).$$

Since Y_1 and Y_2 are independent, $Var(Y) = Var(Y_1 + Y_2) = Var(Y_1) + Var(Y_2)$ and therefore the previous two results yield

$$Var(Y) = E[X] + (Var(X) - E[X]) \left(\frac{p^2}{k} + (1-p)^2 \right).$$

Since $0 \leq \frac{p^2}{k} + (1-p)^2 \leq 1$, it follows from this equation that $Var(Y)$ must always be contained in the interval determined by the values of $E(X)$ and $Var(X)$, regardless of the values of p and k. In particular,

$$\left| Var(Y) - Var(X) \right| \leq \left| Var(X) - E[X] \right|.$$

So far we did not specify the distribution of X. Usually X is assumed to be Poisson distributed. In this case $Var(X) = E[X]$ and the equation above implies that $Var(Y) = Var(X)$ regardless of the values of p and k. When the expected value of X is approximately equal to the variance of X, then the equation above implies that the $Var(Y)$ must also be approximately equal to the $Var(X)$. For example, suppose the distribution of X would be a truncated Poisson distribution with parameters m and λ, where m is the maximum number of surviving offspring over a female's lifetime (thus truncating the Poisson distribution) and λ is the mean of the untruncated Poisson distribution:

$$P(X = x) = \begin{cases} \dfrac{\lambda^x}{x!} e^{-\lambda}, & x = 0, \ldots, m-1 \\[3mm] 1 - \displaystyle\sum_{j=0}^{m-1} \dfrac{\lambda^j}{j!} e^{-\lambda}, & x = m \end{cases}$$

Then $Var(Y) \approx Var(X)$ regardless of the values of p and k, provided that m is not too small. For instance, if $m = 6$ and $\lambda = 2$ one calculates that $Var(X) = 1.943$ and $E[X] = 1.994$, accurate to three digits. Therefore $1.943 \leq Var(Y) \leq 1.994$ for any p and k. As m becomes smaller, $Var(X)$ declines relative to $E[X]$.

This simple model suggests that the variance in male reproductive success, and therefore effective population size, may change only slightly for multiple paternity when the mating system is socially monogamous. The effect of this slight change, under the model conditions, is a slight decrease in the rate of loss of genetic variation from the population, and thus a slightly larger effective size. Although preference by females for particular males for extra-pair matings might be expected to cause a shift from strict monogamy to some degree of promiscuity or even polygyny (Parker and Waite 1997), multiple paternity in the absence of such preferences likely causes little change in effective size of populations unless otherwise unmated males gain matings.

Conclusions and Recommendations

Chesser's breeding-group model (Chesser 1991a,b, Chesser et al. 1993, Sugg and Chesser 1994) uses behavioral ecology data to calculate the gene dynamics of kin-based social groups. Because of this, the model could be applied to studies of polygynous and promiscuous species, where information about demography, dispersal, and mating patterns has been studied. Breeding groups appear to have important implications for gene dynamics and thus for genetic management of endangered species. Family breeding groups can be extremely different from one another genetically. Dispersal generally reduces the rate of inbreeding, and rates of loss of genetic variation may be slowed in populations that have polygynous social breeding groups. Under any mating system, an increased number of fathers should lead to decreases in genetic correlations within breeding groups, decreased genetic differentiation among breeding groups, and increased effective size. In socially monogamous species where all males are mated, however, and other things being equal, effective size should remain little changed when multiple paternity occurs.

Study of the sensitivity of model estimates to the different model parameters is needed for both the Chesser and the Nunney approaches. Also, population genetic models generally assume equilibrium populations, but real populations rarely exhibit demographic or genetic equilibrium. In particular, populations of conservation concern may be out of genetic equilibrium due to population decline (with attendant genetic sampling effects associated with small population size), or due to population increases because of management. The influences of deviation from genetic equilibria on both genetic models and the gene dynamics of wild populations are largely unknown and need to be investigated. Applications of Nunney's (1999) models to organisms in the wild should prove useful, both for verifying the models and for finding ways to estimate the models' parameters. Such models are needed for estimating the gene dynamics of monogamous species and those with equal male and female dispersal patterns (Basset, Balloux, and Perrin 2001). The strength of Chesser's breeding-group model is that gene dynamics can be estimated from the sorts of behavioral and demographic data that are often collected in field studies of behavioral ecology. The utility of this breeding-group model, however, appears limited to polygynous mating systems with sex-biased dispersal.

Summary

Maintenance of genetic diversity is a concern for conservation biologists. Genetic diversity may be lost through inbreeding and genetic drift, but the rate of loss depends on mating patterns. In social species, such as some birds and many mammals, matings often occur within kin-based groups. Breeding-group models of gene dynamics describe the partitioning and loss of genetic variation in such social groups, primarily through estimates of *F*-statistics and effective population sizes. Our purpose was to review evidence on the utility of breeding-group genetic models. When male polygyny and female philopatry are coupled with tight social structuring of populations, considerable genetic differences among breeding groups occur. In addition, the loss of genetic variation may be considerably slowed by the presence of sociogenetic structuring within breeding groups. Under monogamous mating systems, breeding-group models must be chosen with caution because some produce biased estimates of gene dynamics. Under polygynous mating systems, multiple paternity may slow the rate of loss of genetic variation, especially if cuckolding males have no other mating options. The same is likely under "socially" monogamous mating systems (apparent monogamy, but actual promiscuity). We show, however, that when all socially monogamous males are paired, multiple paternity should not greatly influence effective population size. The most practical aspect of breeding-group models (termed demographic models by Nunney and Campbell 1993) is that they can be used to estimate gene dynamics from the sorts of data that behavioral ecologists frequently collect in their research: mating and dispersal patterns and population demography.

Acknowledgments

F.S. Dobson owes a great debt of thanks to Ron Chesser for instruction on fundamentals of population genetics, as well as the details of breeding-group models. Ron's patience with almost constant questioning extended over the four summers that F.S. Dobson spent at the Savannah River Ecology Laboratory as a visiting faculty fellow (1993, 1994, 1996) and as an Oak Ridge Institute for Science and Education research fellow (1995). Thanks also to other mentors in this process of edification: D.W. Foltz, J. Goudet, J. Long, L. Nunney, N. Perrin, D. Sugg, and M.C. Wooten. We thank M. Apollonio and M. Festa-Bianchet for the invitation to share this work with students and colleagues at the International School of Ethology, Erice, Sicily. M. Festa-Bianchet and M. Winterrowd provided thorough and helpful reviews of the manuscript. F.S. Dobson thanks the Oak Ridge Institute for Science and Education and the Savannah River Ecology Laboratory (University of Georgia) for funding, and M. Smith, the director of SREL, for his encouragement.

14.

Pathogen-Driven Sexual Selection for "Good Genes" versus Genetic Variability in Small Populations

Claus Wedekind

Even in cases where we know little about a particular species or population, it is safe to assume that mating is not normally at random (Andersson 1994), and that nonrandom mating and sexual selection influence the genetics of the next generation. Therefore, mate choice and breeding systems are important topics for the management of free or captive populations—especially in small or declining populations. An important management concern is whether free mate choice should be encouraged or prevented. The best management practice will depend on population size, mating system, and the mate preferences that would be expressed if individuals were given the opportunity to choose.

In this chapter I briefly outline some principles about supportive breeding and breeding in captivity, and explain some terminology (see also Lacy 1995). I summarize, from an evolutionary point of view, why we should expect nonrandom mating in nature, and what factors are most likely to influence mate choice. I concentrate on the genetic aspects of mate choice, rather than on choice based on body condition, territory quality, mating gifts, and other mate preferences that may be roughly classified as "preferences for good fathers." Many supportive measures, especially in captive populations, reduce the effect of different paternal investments, whereas genetic aspects become increasingly important with decreasing population size.

There are a number of so-called good genes hypotheses for sexual selection. A common prediction of these hypotheses is that mate choice increases the survival prospects of the average offspring. However, it would be too simplistic to conclude from this that managers should generally support free mate choice. Mate choice often leads to variance in reproductive success: some males are more successful than others. Anything that increases the variance in reproductive success also increases inbreeding and thereby the rate by which a small population loses genetic variation and rare alleles, the rate of decrease in average heterozygosity, and the loss of any heterozygote advantage. Minimizing inbreeding on the one hand and allowing for potential "good genes" effects of mate choice on the other hand is an optimization problem. This optimization problem cannot easily be solved, especially when we do not know how important genetic aspects of mate choice are in a particular population, and to which extent good genes effects from free mating can potentially outweigh the loss of genetic variability due to increased inbreeding. I discuss this in the context of management recommendations.

Free or suppressed mate choice is relevant not only to the genetics future generations but also to life-history decisions of the parents. Life-history decisions, such as the amount of resources a mother is willing to invest in a particular offspring, often evolved under conditions different from those faced by small or declining populations, especially captive ones. In some cases, maternal investment could be manipulated to optimize breeding success, and here I suggest some preliminary ideas about the methods for such manipulation.

N_e, Supportive Breeding, and Reproduction in Captivity

The more a population decreases in number, the more likely it is that the mean fitness of its members, and thereby the population's probability of long-term persistence, will decrease due to genetic problems (Lande 1998b). These problems include (1) an increase in inbreeding and hence an increase in inbreeding depression (Hedrick and Kalinowski 2000), (2) a general loss of genetic variability and therefore a reduced potential for adaptation to changing environments, and (3) the risk of fixation of deleterious mutations due to random drift. Therefore, an obvious aim in conservation is to stop the decrease in number of individuals and even to achieve population growth.

The genetic problems just outlined are only indirectly dependent on population size (i.e., the census size), but are directly related to the genetically effective population size, N_e (e.g. Hartl 1988). N_e is therefore an important parameter in population genetics and conservation biology (Caughley and Gunn 1996). N_e is the size of an ideal population that would lose genetic variability at the same rate as the real population. Hence, N_e corrects for a number of factors, such as age-related differences in reproductive rates, unequal family size, or unequal numbers of males and females. These corrections are necessary because any skew in sex ratio increases the variance in individual reproductive success and leads to increased inbreeding and further loss of genetic variability.

Supportive breeding, the practice of supporting weak, wild populations by releasing captive-bred individuals, is a technique used in the conservation of many species. Its first aim is obviously to maintain or increase population size, or to provide more harvest opportunities for species that are sport harvested. Supportive breeding, however, could have harmful long-term effects, and its potential impact on the genetics of a population must be given serious consideration (Hedrick and Miller 1994, Lande 1998b).

A first important rule is to avoid mixing individuals from different populations to reduce the risk of maladaptive hybridization and outbreeding depression (Hindar and Balstad 1994). Interpopulation mixing may only be desirable in very exceptional cases, such as for populations suffering from severe inbreeding depression (Madsen et al. 1999).

A second potential genetic problem of supportive breeding (especially of captive populations, as will be explained here) is the potential effect of artificial selection on the offspring that leads to a certain genetic adaptation to captivity. Adaptation to captivity is likely to reduce reproductive fitness of individuals reintroduced to the wild (Bryant and Reed 1999, Frankham 1999). Fleming and Gross (1993) compared hatchery-born and wild Coho salmon (*Oncorhynchus kisutch*) in direct competition and found that males and females from hatcheries had significantly reduced breeding success.

A third important genetic risk or disadvantage of supportive breeding is the fact that dividing a given population into a wild breeding and a captive breeding segment could have a negative impact on N_e. According to Ryman and Laikre (1991), splitting a population into wild and captive breeders can increase inbreeding in the total population because supportive breeding increases the variance in reproductive success by giving a reproductive advantage to a few randomly chosen individuals (Ryman, Jorde, and Laikre 1995; Nomura 1999; Ryman, Jorde, and Laikre 1999; Wang and Caballero 1999). Recent empirical studies provide support for this "Ryman–Laikre effect" (Tessier, Bernatchez, and Wright 1997). An increase in the inbreeding coefficient can eventually lead to population extinction (Saccheri et al. 1998).

The fourth potential genetic problem of supportive breeding is the elimination of mate choice. In nature, mating is not normally random with respect to genetic characters (Andersson 1994, Møller and Alatalo 1999, Wedekind 2002a). Circumventing mate choice may be significant with respect to offspring genetics and viability and may also decrease the amount of parental investment an offspring will obtain.

Some species have been or still are maintained under semi-natural conditions in captivity for several generations because their natural habitat is largely destroyed or the wild population is severely endangered. Hunter (1996) lists a number of species that would probably be extinct today without ex situ conservation. The list includes such charismatic species as the California condor (*Gymnogyps californianus*), the wisent (*Bison bonasus*), the Przewalki's horse (*Equus caballus przewalski*), the black-footed ferret (*Mustela nigripes*), and the Arabian oryx (*Oryx leucoryx*), but also contains less known species such as several cichlids of Lake Victoria (Kaufman 1992) or several viviparous tree snails (see also the list at www.earthwitness.com). In such cases, the mating system and natural mate choice may remain intact. However, some mating systems may be very damaging to the genetics and demographics of a population over the long term, even leading to extinction of captive populations. It may therefore be of interest to manipulate mate choice and breeding behavior to prevent the captive population from generating high inbreeding coefficients.

A recent study on the effective population size of a captive-bred fish (Fiumera, Parker, and Fuerst 2000) provides an alarming example. The piscivore *Prognathochromis perrieri* is a highly endangered Lake Victoria cichlid (Kaufman 1992). As part of a species survival plan, this fish is bred in captive populations spread over several institutions. Fiumera, Parker, and Fuerst (2000) collected data from subpopulations spanning five captive generations. Using microsatellite DNA markers, they found that N_e ranged from 2.5 to 7.7 individuals per subpopulation, which was far smaller than the census sizes of 32 to 243 individuals and suggested that very few individuals contributed to reproduction. Through inbreeding, approximately 19% of the initial alleles were lost within the first four generations of captive breeding. The loss of heterozygosity at each generation was 6 to 12%. A first (but late) measure against this loss of genetic variability was the removal of dominant breeders after they had reproduced to encourage reproduction by other individuals.

Why sex at all? And why are pathogens important for the maintaince of sex? Before I discuss different forms of mate choice, I will briefly summarize the prevailing theory as to why sexual mixing of genes is such a successful evolutionary program. The evolution of sex is somewhat difficult to understand because sex involves a number of significant evolutionary disadvantages

(Michod and Levin 1987, Stearns 1987). The major disadvantage has been termed the cost of meiosis: a female that reproduces sexually is only 50% related to her offspring, whereas an asexual female transmits 100% of her genes to each of her daughters. Hence, gene transmission (and population growth in terms of absolute numbers) is about twice as efficient in asexuals as it is in sexuals. If asexuals had the same survival probability as sexuals, a mutation causing a female to produce only asexual daughters would, when introduced into a sexually reproducing population, rapidly increase in frequency and outcompete sexuals in numbers within just a few generations (Williams 1975, Maynard Smith 1978). Why doesn't this happen? What are the evolutionary advantages of sex? or what are the disadvantages of asexual reproduction?

"Muller's ratchet" (Muller 1932) is one major disadvantage of asexuality: it predicts that slightly deleterious mutations accumulate in asexuals from generation to generation until the genome no longer codes for a viable organism, and the population goes extinct (Andersson and Hughes 1996). Thus, at first glance one may think that sex must be so successful because recombination and natural selection can efficiently remove deleterious mutations. Asexual reproduction is so much more efficient than sexual reproduction that, all else being equal, asexuals would need only a few generations to outcompete sexual conspecifics, probably long before the effects of Muller's ratchet become significant (Kondrashov 1993).

A second set of hypotheses suggests that sex enables the spread, or even the creation of, advantageous traits. These hypotheses require that the direction of selection be continuously changing, therefore the main source of fitness reduction must be short-term environmental changes. This condition is especially likely in the coevolution of hosts and pathogens because host–pathogen systems are more deeply interdependent than predator–prey or competitor systems. Host resistance genes that are advantageous today will become disadvantageous in the near future if pathogens evolve to overcome them. Therefore, hosts must continuously change gene combinations, and sex is an efficient means to do so (Hamilton, Axelrod, and Tanese 1990). Genetic heterogeneity within a sexually produced clutch may increase the chances that the clutch contains an optimal genotype ("lottery model"; Williams 1975, Kondrashov 1993), and it may decrease the risk of competition between relatives (the "elbow-room model"; Maynard Smith 1978, Kondrashov 1993). The possibility that sex reduces the risk of transmission of pathogens between relatives because of their reduced genetic similarity (Baer and Schmid-Hempel 1999) can be seen as a variant of Williams's lottery model.

To summarize (see also Howard and Lively 1994), the coevolutionary conflict between pathogens and hosts selects for sexual mixing of genes as a diversity-generating mechanism, which allows both parties, hosts and pathogens, to survive Muller's ratchet.

Inbreeding Avoidance

If sex evolved as a diversity-generating mechanism, it is not surprising that there is mate choice and that mate choice often takes the degree of kinship between two individuals into account. Many population models, however, assume random mating, even with respect to kinship. Many breeding programs, such as those in fish hatcheries, do not account for potential inbreeding avoidance in the wild. Examples where there is no evidence for kin recognition and inbreeding avoidance span a broad taxonomic range, from invertebrates (Baur and Baur 1997, Peters and Michiels 1996) to vertebrates (Keane, Creel, and Waser 1996; Keller and Arcese 1998). However, there are probably more examples of species where inbreeding avoidance is known to occur, often through sex-biased dispersal (Pusey and Wolf 1996). Active avoidance of kin as mates has been demonstrated in a number of species, often relying on recognition of familiarity (Penn and Potts 1998, Clarke and Faulkes 1999). In some species, odors that reveal information about highly polymorphic loci like the major histocompatibility complex (MHC) seem to play a crucial role in kin recognition and inbreeding avoidance (review in Brown and Eklund 1994, Penn and Potts 1999).

"Good Genes" Models of Sexual Selection

Apart from inbreeding avoidance, there are a number of more sophisticated forms of sexual selection based on general phenotypic appearance or on sexual signals such as odors or secondary sexual ornaments. The literature usually groups criteria for mate choice into three classes (Andersson 1994): (1) direct benefits, such as parental care or nuptial gifts; (2) "Fisherian-traits," which are attractive to members of the other sex but do not reveal anything else (Fisher 1930); and (3) good genes (Zahavi 1975, Hamilton and Zuk 1982, Grafen 1990, Wedekind 1994a, Johnstone 1995). The third class of criteria is of special interest here, because good genes are expected to increase offspring survival. Good genes in the context of sexual selection are mainly alleles at loci that are important in the coevolution between hosts and their pathogens because continuously changing selection pressures in this coevolutionary process maintain enough additive genetic variability for sexual selection to act on. Mate choice for good genes may therefore be important in determining virulence in natural host–pathogen systems. Hence, not only sex itself but also some forms of sexual selection could be strongly influenced by the coevolutionary dynamics of host–pathogen systems.

Møller and Alatalo (1999) concluded that sexual selection for good genes is widespread across taxa, but its effect on offspring survival varies. Their meta-analysis found that male sexual characters chosen by females on average accounted for 1.5% of the variance in offspring viability, but they stressed that many of the studies included in their analysis may only partly estimate the full fitness consequences of mate choice for offspring survival. The effects were generally stronger for studies where the target of selection had been identified than for those with an unknown target of selection. Indeed, a recent experimental study demonstrated that the good genes effect of mate choice can be very strong: optimal mate choice in a whitefish would reduce pathogen-correlated egg mortality by 67% as compared to random mating (Wedekind, Müller, and Spicher 2001). Differences in male breeding ornamentation accounted for 32% of the variance in offspring mortality.

Levels of Selection

Mate choice for criteria that reveal good genes is only one possible level of pathogen-driven sexual selection. Other possible levels may include selection of sperm within the female reproductive tract, selective fertilization, or selective support of the embryo or the offspring (Wedekind 1994a). All these levels could potentially be connected to host–pathogen coevolution, or at least help to prevent inbreeding.

Preferences for sperm of genetically dissimilar types have been observed in vertebrates and invertebrates (reviewed in Eberhard 1996, Birkhead 2000). In many plants, growth of the pollen tube is often affected by the stigma and depends on the combination of male and female alleles on the self-incompatibility locus (Jordan et al. 2000). In all these examples, the connection between cryptic female choice and the offspring's immunocompetence is unclear. In mice, however, a series of experiments revealed that gamete fusion is not random with respect to the sperm's and the egg's MHC-type, and with respect to current epidemics of hepatitis virus (Wedekind et al. 1996, Rülicke et al. 1998). In these studies, mice of two inbred strains that were bred to differ only in MHC but otherwise had an identical genetic background, and F1's of both strains, were paired or used for in vitro fertilization experiments, and the MHC of the resulting blastocysts was analyzed by polymerase chain reaction. Infected mice produced more heterozygous blastocysts than sham-infected mice. The difference was not a result of selection by the pathogen but was due to nonrandom fertilization of the oocytes. The physiology behind such nonrandom fertilization is, however, unknown.

Different Types of Mate Preference versus N_e

An important aspect of mate preferences is whether they are uniform, and all members of one sex have the same preference, or variable, and mate preferences vary among individuals. As I will outline in the following, whether preferences are uniform or variable influences the variance in reproductive success among individuals. Variance in reproductive success in turn influences effective population size N_e (Hartl 1998), and with it the long-term viability of small and endangered populations.

Hamilton and Zuk (1982) suggested that individuals in good health and vigor are preferred as mates because they are likely to possess heritable resistance to predominant pathogens. By preferring healthy partners one may thereby produce robust, vigorous, and resistant progeny, better adapted and therefore less susceptible to local pathogens (review in Grahn, Langesfors, and vonSchantz 1998; Westneat and Birkhead 1998; Møller, Christe, and Lux 1999). There is much empirical support for this hypothesis (Møller, Christe, and Lux 1999). The mechanisms Hamilton and Zuk (1982) and later von Schantz et al. (1999) suggested, however, lead to populations where all individuals of one sex have the same mate preference, so that members of the opposite sex can be ranked in a universally valid order of attractiveness, and less attractive individuals will only mate if the more attractive ones are not available. A universal ranking would increase variance in reproductive success (Petrie and Lipsitch 1994) and decrease N_e. The effect often exists even in species thought to be monogamous because females sometimes solicit extra-pair copulations with more attractive or more viable males than their social partner (Kempenaers et al. 1992; Hasselquist, Bensch, and von Schantz 1996; Petrie, Doums, and Møller 1998).

Inbreeding avoidance can be seen as a simple form of variable preferences because members of the opposite sex cannot be ranked in a universally valid order of attractiveness according only to inbreeding. Consequently, in populations where inbreeding avoidance is the only criterion for mate choice, the variance in reproductive success is expected to be lower than where mate choice is based upon universal criteria. The difference between N_e and the actual population size, N_a, would also be lower.

Inbreeding avoidance often leads to offspring with increased degrees of heterozygosity, especially on important loci such as the MHC. Those loci often affect body or urine odors used in kin recognition (Brown and Eklund 1994, Penn and Potts 1999). In some systems, however, mate preferences may specifically aim at reaching heterozygosity on the MHC and other important loci (Brown 1997, Wedekind and Füri 1997). Although there is a conceptual difference between such a mating system and inbreeding avoidance, in practice it is difficult to discriminate between these two types of mating preferences. The mechanisms that lead to heterozygosity on specific loci could have evolved to avoid inbreeding, and their effect on N_e may be about the same as the effect of inbreeding avoidance: in both cases the deviation from N_a would be small.

Some good genes models predict variable mate preferences and therefore differ from the original Hamilton and Zuk (1982) model, because an offspring's level of resistance would depend on the genetic contribution of both parents. At loci important for the host–pathogen interaction (e.g., immunogenes), certain combinations of alleles may be more beneficial than others. If individuals choose their mates to obtain such beneficial allele combinations, their preferred mate should vary according to their own genotype. Individuals with different resistance genes should then show different preferences, and there would be no universally valid order of sexual attractiveness with respect to signals that reveal heritable disease resistance or immunogenes (Wedekind 1994b). Wedekind and Füri (1997) specifically searched for evidence for such preferences but did not find any. Experimental evidence for such a mating preference was, however, provided by Rülicke et al. (1998), who showed that gamete fusion depends on the MHC and on the presence or absence of mouse hepatitis virus. In the absence of further examples for such mate preferences, their relevance for conservation programs remains unclear. Nevertheless, this form of good genes sexual selection would lead to variable mate preferences. Hence, the variance in reproductive success would be low, and inbreeding would not increase much if one would allow free mate choice in small or captive populations.

Fisherian models of sexual selection, in which preferences exist for traits that are sexually attractive but do not reveal anything else, typically assume that females in a population have similar levels of preferences for male ornamental traits. These models therefore predict an increased variance in individual reproductive success and thereby a decrease in N_e.

Many of the preceding models of sexual selection predict rather uniform mate preferences, whereas a number of empirical studies provide evidence for genetic variability in female mate preference (review in Bakker and Pomiankowski 1995, Jennions and Petrie 1997). In real systems, different forms of sexual selection may act at the same time and interfere with each other. It is therefore difficult to make clear predictions in a given population about the effects of natural mate preferences on the variance in reproductive success and on N_e. Moreover, the competition of members of the same sex for access to mating partners often interferes strongly with free mate choice. The combination of inter- and intrasexual selection leads to the mating system of a population and should have a strong impact on the life-history decisions of parents.

Parental Investment

Mate choice has often been shown to influence not only the genetics of offspring but also different kinds of parental life-history decisions, including offspring sex ratio, the timing of reproduction, and the amount of parental investment. Each one of these parental life-history decisions could potentially be manipulated.

Evolutionary theory predicts that parents should invest in each offspring according to the potential fitness return of that offspring (Fisher 1930). If, for example, the relative reproductive value of sons and daughters differs for different females or different males, sex allocation theory predicts that females should adjust the sex ratio of their offspring according to their own condition or according to their mate's attractiveness (Trivers and Willard 1973). If sex ratio adjustments led to an uneven population sex ratio, they would become important for conservation. Inbreeding and loss of genetic variation increase with any deviation from an even sex ratio because $N_e = 4N_mN_f/(N_m + N_f)$ (Hartl 1988). If the number of available eggs limits population growth, however, there are female-biased sex ratios that lead, in the longer term, to increased N_e and to population sizes that are safe against the Allee effect. Because there are a number of potential invasive and noninvasive methods to influence sex ratio, sex ratios could be optimized with respect to N_e and the long-term survival of a population (Wedekind 2002b).

Many birds and mammals with attractive mates increase their investment into their current offspring, probably decreasing their future reproductive potential. Parental investment conditional on mate quality is predicted from life-history theory (Williams 1966) and was first demonstrated in experiments and field studies that showed that females increase their investment into the current brood when mated with a preferred male (Burley 1982, Delope and Møller 1993, Petrie and Williams 1993). Increased parental effort may lower one's own survival and future reproductive potential (Saino et al. 1999).

Recent studies on birds have identified the mechanisms of these life-history decisions. Some female birds lay more eggs (Petrie and Williams 1993) or larger eggs (Cunningham and Russell 2000) after copulating with preferred males. In the latter case the females produced offspring of better body condition when paired with preferred males. Gil et al. (1999) found that females deposit higher amounts of testosterone and 5 alpha-dihydrotestosterone in their eggs when mated to attractive males. In kestrel, maternal hormones influence offspring survival (Sockman and Schwabl 1999), and in canaries, chick social rank is positively correlated with concentration of yolk testosterone in the eggs from which they hatched, suggesting that the development of aggressive behavior of offspring may be modified by maternal testosterone (Schwabl 1993; Schwabl, Mock, and Gieg 1997).

The preceding examples are mostly from birds, but there is evidence that the effect exists also in other taxa. The tapeworm *Schistocephalus solidus*, for example, produced large eggs if given the opportunity to outbreed, but relatively small ones if forced to reproduce by selfing (Wedekind, Strahm, and Schärer 1998). There is even evidence that a somewhat comparable effect exists in humans: in some populations, the degree of MHC-similarity influences mate preferences (Ober et al. 1997, Wedekind and Füri 1997), and in one study, baby birth weight and weight of the placenta were negatively correlated to the degree of similarity of MHC-alleles between the parents (Reznikoff Etievant et al. 1991).

In species where male characteristics influence decisions about maternal investment, it may be possible to exploit the rules used by the females to determine the attractiveness of a given male. It is rather unlikely that the decision rule about whether a given male is perceived as attractive or not is entirely genetically fixed (Real 1991). In most species such a decision rule may be adaptable to a certain degree to a sampling template given by the population (Milinski and Bakker 1992). If, for example, females have a general preference for males with a red belly, then a male with a medium red belly might be perceived as very attractive in a population of (real or dummy) dull males, but as unattractive in a population of bright red males. To make the female invest much into the offspring of a medium red male in response to his attractiveness, it may be useful to avoid exposing her to bright red males before mating. Moreover, instead of changing the template a female uses to judge the relative attractiveness of a male, it may sometimes be possible to alter the attractiveness of the male directly, for example by cutting or attaching feather ornaments. However, it will often be necessary to test whether such an option for noninvasive manipulation exists in a given species.

Conclusions and Recommendations

Traditional methods in breeding programs normally attempt to avoid inbreeding (Montgomery et al. 1997, Bryant et al. 1999, Frankham 1999, Frankham et al. 2000), but breeding programs could potentially be further optimized with respect to genetics. Apart from the general potential problems of supportive breeding programs that have been outlined here, details about the methods used in supportive breeding can also be crucial with respect to N_e. Imagine, for example, the situation in a fish hatchery where only the sperm of a few males was available to fertilize the eggs of many females. The skewed effective sex ratio would decrease N_e, which would decrease further if variance in male reproductive success was artificially increased by using different amounts of sperm from the few males available, or by using their milt sequentially, striping the males' milt directly one after the other into the container that already holds all the eggs, and thereby giving the first males a reproductive advantage.

If we assume that all males are of equal genetic quality to all females, and if our only concern was a high genetic variation in the offspring, probably the best method for supportive breeding in fish hatcheries would be the following (analogous methods could be used in other breeding programs): (1) Catch a number of adults that result in a N_e of at least 50 (a rule of thumb, suggested by Ryman, Jorde, and Laikre 1999). (2) Catch an equal amount of males and females, or at least avoid using far fewer males than females because $N_e = 4N_mN_f/(N_m + N_f)$; that is, using only six males to fertilize the eggs of 20 females leads to an effective population size of less than 20! If, however, more males than females are available, they should all be used, regardless of the skewed sex ratio, because N_e in the captive population increases with every male used. (3) Pool and carefully mix equal amounts of sperm of all the males (using appropriate methods sperm can normally be stored for some hours), and use this mix to fertilize the eggs, either together (Billard 1985) or each female's batch separately. When fertilizing each batch separately, potential negative effects of swollen or broken eggs on the fertilization of healthy eggs would be avoided. Alternatively, combining the gametes of one male and one female could minimize potential negative effects of sperm interactions and restrict transmission of pathogens (Crim and Glebe 1990).

There are several possible methods to incorporate the fact that males normally vary in genetic quality and that female mate choice may take this variability into account.

1. Allowing free mate choice in monogamous or nearly monogamous species. Variance in reproductive success is lower in monogamous than in polygamous breeding systems. Therefore, N_e is not much affected by mate choice. Free mate choice may be beneficial for a number of reasons: inbreeding avoidance and hence reduced inbreeding depression, good genes

effects of sexual selection, and, possibly, increased parental investment of females (or males) with attractive mates.

2. Allowing for some degree of mate choice but keeping N_e as close as possible to the census population size, by avoiding high reproductive variance among the males and females of a population. Females should be presented to different males so that mate choice can happen, but if a few individuals become highly successful at the cost of the reproduction of others, the manager should remove them for part of the breeding season. In cases where presenting different males is not possible, mate preference tests of the type that have been used in behavioral research (Andersson 1994) may be considered. If, for example, odors are important in sexual communication, it may be possible to provide odor samples of different males and assess female reaction before bringing male and female together. Such behavioral tests could also be performed in advance of artificial insemination and other reproductive technologies often used in captive breeding programs (Gibbons, Durrant, and Demarest 1995; Dobson and Lyles 2000). Mate preference tests may increase the success of assisted reproductive technologies by reducing the risk of miscarriage (Wedekind 1994a).

3. Allowing free mate choice in nonmonogamous populations, disregarding the possibility that it may lead to high reproductive variance and hence lower N_e. The negative effect of increased inbreeding would then need to be compensated by the good genes effects of sexual selection. Comparing good genes effects in sexual selection with the effects of increased inbreeding coefficients is a challenge that requires good data and realistic models for different types of populations (see also, for example, Lacy 2000). It may therefore be too early to suggest a rule of thumb for conservation managers. Probably, the larger the population, the more likely that good genes effects will outweigh increased inbreeding coefficients caused by high reproductive variance. Although our knowledge of good genes effects on sexual selection is limited, it is probably reasonable to suggest that for small and medium-sized populations, free mate choice without any efforts to keep N_e close to the census size may be detrimental (Fiumera, Parker, and Fuerst 2000).

Summary

Sexual mixing of genes has two main evolutionary advantages; namely, that recombination followed by selection results in the efficient removal of deleterious mutations, and that it creates genetic diversity, which is important in evolutionary arms races, especially in host–pathogen coevolution. It may therefore not be surprising that mating in nature is often not random with respect to genetics, and that it may often be linked to host–pathogen coevolution. Although nonrandom mate choice may affect the persistence of small populations, many population models and conservation breeding programs seem to ignore mate choice. Different kinds of sexual selection can have different consequences on the effective population size, N_e, and thereby affect the loss of genetic variability and heterozygosity over time. In some cases, supportive breeding programs may benefit from providing mate choice opportunities, which may sometimes promote offspring health and enable host populations to react to coevolving pathogens. In other cases, however, especially in small populations with high variance in reproductive success, natural breeding systems and free mate choice could have severe negative effects on the long-term survival chances of a population. In such cases, free mating should not be fully allowed, and natural breeding systems need to be manipulated to avoid an extensive reduction of N_e.

Acknowledgments

I thank Marco Festa-Bianchet and Marco Apollonio for organizing the meeting in Erice and for inviting me there. I also thank Nick Colegrave, Tom Little, Margaret Mackinnon, Andrew Read, Ana Rivero, Mirjam Walker, and especially Marco Festa-Bianchet, Morris Gosling, and the anonymous reviewer for discussion and/or comments on the manuscript. The Swiss National Science Foundation provided support through an IHP-fellowship.

15.

Measuring Individual Quality in Conservation and Behavior

Brian M. Steele and John T. Hogg

Individual variation in phenotype is ubiquitous across species and traits. When associated with traits related to individual fitness, such variation can, in principle, lead to population-level dynamics different from those predicted on the assumption that individuals do not vary, or in the absence of information on the full extent and nature of individual differences (Huston, DeAngelis, and Post 1988; Łomnicki 1988). Nonetheless, the practical value of considering individual variation in conservation programs has not been fully determined. This is partly because individual-based population models are a relatively recent addition to conservation's toolbox (Judson 1994). Another reason is that appropriate methods for the analysis of individual variation in natural populations have not been generally available. An appropriate method is one that partitions individual variation into its sources. This, in turn, requires statistical models that, although developed, have not been widely adopted in conservation and behavior. In this chapter, we present an introduction to the application of these models (referred to herein as generalized linear mixed models) for the analysis of individual variation in phenotype.

Individual variation has two fundamental components. To see this, consider that individuals vary in phenotype for many reasons. These include developmental stage, genetic architecture, environment, and the effects of purely stochastic processes (Williams 1992). These sources of variation differ, among other things, in the persistence of their phenotypic effects. Now consider a behavior or life-history event that is repeated at intervals over an individual's lifetime. At one extreme, genotype can affect every repetition of the trait in a consistent direction. At the opposite extreme, certain environmental influences might affect single repetitions in directions that are independent of the individual's performance in any past or future repetition. Other causes of individual variation (e.g., repairable injury) could influence more than one but not all repetitions in a consistent manner. "Individual quality" can be thought of as a measure of the extent to which individual differences in a specific phenotypic trait are due to genetic and environmental causes that have persistent versus transient phenotypic effects.

Under this definition, individual quality is similar to Kokko et al.'s (1999) notion of "propensity," and variation in individual quality is similar to Nunney's (1996) concept of fixed variation. Fixed variation, in Nunney's lexicon, originates from differences among individuals that are maintained throughout their lifetime, whereas random variation originates from non-age-related events that generate transient effects on the expression of a trait. To avoid confusion with the statistical meanings of fixed and random effects, we use *persistent* and *transient* rather than *fixed* and *random*, respectively, when referring to these two components of individual differences. We emphasize that these terms refer to the duration of the phenotypic effect rather than the duration of the condition or event giving rise to the effect. These durations can be very different. For example, an environmental cause of individual variation may operate only briefly yet have lifelong consequences for the affected individuals (e.g., food deprivation in early development).

The distinction between persistent and transient individual differences is relevant to both behavioral ecology and conservation science. For example, the extent to which males differ consistently and heritably in some aspect of performance is important to the "good genes" class of theories for the evolution of female mate choice (Andersson 1994). In conservation settings, persistent individual differences are under scrutiny for their effects on the likelihood of population extinction. Goss-Custard and Sutherland (1997) suggested that individual differences in resource acquisition provide a demographic buffer to extinction by directing mortality onto lower-quality individuals when resources decline. Similarly, Conner and White (1999) found that populations experienced lower rates of extinction when persistent individual heterogeneity in mean fecundity and survival was added to a demographic model. The population genetic effects of persistent individual differences may have opposite consequences for population viability. By increasing individual variation in lifetime reproductive success, persistent differences may lower effective population sizes and accelerate the rate at which deleterious mutations are fixed by genetic drift (Nunney 1996). Transient effects on individual reproductive success, in contrast, contribute relatively little to variation in lifetime reproductive success because they tend to average out across individuals over lifetimes (Nunney 1996).

In both disciplines, the concept of individual quality is typically used as a theoretical or heuristic construct. The question of how to measure individual quality in field settings has received much less attention. This problem is similar to one faced by animal breeders wishing to identify, for example, the best sires for a particular breeding program and for which formal statistical methods have been developed. Of particular interest here are regression methods based on the generalized linear mixed model. Although mixed model analysis was first introduced over 40 years ago, conceptual subtleties and computational difficulties have until recently largely limited its application to clinical trials and applied quantitative genetics.

Our objective in this chapter is to develop a generalized linear mixed model approach to the measurement of individual quality in conservation and behavior. We begin with a description of statistical concepts and procedures. This section is fairly technical and may prove challenging for some readers. Therefore, we follow with three mixed-model analyses of longitudinal field data from two natural populations of Rocky Mountain bighorn sheep (*Ovis canadensis*). These examples are intended to illustrate the main statistical points in a less abstract way. Finally, we consider the practical interpretation of the "individual effects" predicted by these models and comment on the potential scope of application of generalized linear mixed models in conservation and behavioral ecology.

Statistical Models

We will use *generalized linear mixed model* to refer to regression models that allow for both fixed and random explanatory factors and are appropriate for the analysis of response variables having a variety of distributions including normal, Bernoulli, and Poisson. We use *linear mixed model* to indicate the special case of the generalized linear mixed model in which the response variable is normal in distribution. Because the linear mixed model has been more thoroughly developed than the generalized linear mixed model, we start with it.

Although the following presentation is for the most part general, we have framed it in terms of the measurement of individual quality. Thus we assume that repeated measurements on some fitness-related trait are available for a set of recognizable individuals sampled at random from a population and that individual identity is included in the analysis as an explanatory (random) factor.

The Linear Mixed Model

We begin our discussion of the linear mixed model with a review of the ordinary linear model. The ordinary linear model specifies that the mean, or expected value of the response variable, is a function of one or more of the measured explanatory variables. It is often forgotten that one of the assumptions of the linear model is that the levels, or values, of each explanatory variable are fixed (i.e., chosen) by the researcher. When explanatory variables have fixed levels, such as in a designed experiment, they are referred to as fixed factors. Explanatory variables with levels that are not fixed in this sense are referred to as random factors because the observed levels are the result of a random process such as sampling.

To see the importance of the distinction between fixed and random factors, it is useful to concentrate on the origins of their levels. The levels of a fixed factor are finite in number, and in the control of the researcher, whereas the levels of a random factor are not under the control of the researcher and usually determined by a random mechanism such as sampling. For example, gender is a fixed factor with two levels, male and female. In a designed experiment, the researcher can ensure that equal numbers of male and female animals belong to each treatment group. In contrast, we generally cannot select individuals that are, say, better than the average individual by a particular amount. Instead, the quality of individuals is out of our control. If we think of each individual as a source of variation in the response variable, then individuals correspond to levels of a random factor, which we call identity. The identity factor accounts for persistent differences among individuals, and each level of identity is the difference between one particular individual and the average level of all sampled individuals. If a data set of this type is analyzed using the ordinary linear model by treating identity as a fixed factor, then any conclusions of a statistical nature (e.g., hypothesis tests) apply only to sampled individuals (observed levels). Hence, there is no opportunity for conducting formal statistical inference about the larger population, and the study must be considered anecdotal in nature. This limitation on the scope of inference can be remedied by using linear mixed model methods. Inferences drawn from a linear mixed model analysis using identity as a random factor apply to the sampled population of individuals.

The levels of a random factor are not repeatable. If a study is observational in nature and a factor is random, then we cannot exactly replicate the levels appearing in the sample by sampling again from the population. Often it is said that there are infinitely many levels of a random factor, and that the observed levels are a random sample from this population of levels. It is more practical, though, to say that a factor is random if there exists a population of levels from which the levels have been sampled, and the desired scope of inference is the sampled population.

With these concepts in mind, suppose that a demographic variable such as fecundity has been measured on each of q_1 animals over the course of q_2 years. A useful linear mixed model that identifies animal identity and years as random factors and age as a fixed factor is

$$Y_{ij} = \alpha_0 + \alpha_1 x_{ij} + b_{1,i} + b_{2,j} + \varepsilon_{ij}, \qquad \text{(Equation 1)}$$

where Y_{ij} is an observation on the demographic variable obtained from the ith animal for the jth year of the study, $i = 1, \ldots, q_1, j = 1, \ldots q_2$, and x_{ij} is the age of the ith individual in year j. The parameters α_0 and α_1 are constant and age effect coefficients. Persistent differences between individuals are accounted for through the individual random effects $b_{1,i}$, $i = 1, \ldots, q_1$. We assume the individual effects to be independent and normally distributed with mean 0 and variance γ_1. Similarly, systematic differences between years are accounted for through the year random effects $b_{2,j}$, which are assumed independent and normally distributed with mean 0 and variance γ_2. Residual variation is accounted for by the random errors ε_{ij}, also independent and normal with mean 0, but with variance σ^2. The parameters γ_1, γ_2, and σ^2 are referred to as variance components. Because our interest lies in the population, it is critical to treat identity as a random factor. Years are also treated as random so that statistical inference is not limited to the observed years but to a longer span of time. The relative importance of individuals and years in explaining variation in the demographic rate variable can be assessed by comparing estimates of the variance parameters γ_1 and γ_2. Equation (1) may be extended to allow $p - 1$ fixed factors by writing

$$Y_{ij} = \alpha_0 + \alpha_1 x_{ij,1} + \ldots + \alpha_{p-1} x_{ij,p-1} + b_{1,i} + b_{2,j} + \varepsilon_{ij}.$$

The assumptions of independence and normality of the random effects and errors for this model are the same as for Equation (1). The extension to more than two random factors is straightforward, though the notation is troublesome. See Searle, Casella, and McCulloch (1992) for details.

When a linear mixed model is adopted, there is an implicit assumption that some observations are *not* independent. Specifically, two observations are dependent if they are modeled as a function of the same random effect. Equation (1), for example, implies that observations made on the same individual in different years (Y_{ij} and $Y_{i,j+1}$) are correlated because both observations are modeled as a function of the same random effect $b_{1,i}$. Similarly, observations made on different individuals in the same year are correlated because year is a random factor. This correlation structure may be overly simplistic because the correlation between any pair of observations on an individual is assumed to be the same, regardless of how far apart in time the observations were made. However, the model is tractable computationally and conceptually, and a substantial improvement on the usual linear model. Diggle, Liang, and Zeger (1994) discuss a variety of models with more sophisticated correlation structures.

Parameter Estimation

The parameters of the linear mixed model are the fixed effects parameters $\alpha_0, \ldots, \alpha_{p-1}$, and the variance components $\sigma^2, \gamma_1, \ldots, \gamma_r$, where r is the number of random factors. For convenience, these parameters are collected in the parameter vector θ. The linear mixed model poses some computational difficulties for estimating θ, primarily because observations are not assumed to be independent. There are a variety of methods for parameter estimation, though the most commonly used method is maximum likelihood. Maximum likelihood estimators are popular because they are approximately normal in distribution, nearly unbiased, and have small standard errors (compared to other estimators) when the sample size is large. Moreover, methods of computing maximum likelihood parameter estimates and standard error estimates are tractable (though not necessarily simple). In brief, the maximum likelihood estimate $\hat{\theta}$ is the value of θ that is most consistent with the observed data in the sense that the likelihood of observing the actual data values is at its maximum *for a given probability model* (here normal but possibly Bernoulli, Poisson, etc. in the generalized linear mixed model). In contrast, ordinary least squares estimates are those estimates that minimize the sum of the squared differences between the observed data and the fitted data. Searle, Casella, and McCulloch (1992) and Longford (1993) discuss estimation methods for the linear mixed model.

Prediction of the Random Effects

In the statistical literature, the random effect b_i is said to be predicted rather than estimated. The term *estimation* is usually reserved for parameters, whereas b_i is a random variable. The term *prediction* is used because this task is essentially the same as predicting a future observation, say, the future price of a commodity based on past pricing. The usual prediction method is best linear unbiased prediction (Searle, Casella, and McCulloch 1992). When the b_i's represent individual effects, then the best linear unbiased predictor of the individual effect b_i is a weighted average of the linear mixed model residuals associated with the ith individual. The weights are determined in such a way that, if the model parameters were known, then the sum of the mean squared prediction errors for the b_i's is minimized.

Hypothesis Testing and Confidence Intervals

The problem of inference for the linear mixed model is relatively straightforward for the fixed effects parameters $\alpha_0, \ldots, \alpha_{p-1}$. Arguably, the most important inferential procedure is determining the significance of an individual explanatory variable, say, x_k. This is accomplished by testing whether α_k is 0 or not. Specifically, we test $H_0 : \alpha_k = 0$ versus $H_1 : \alpha_k \neq 0$ using the test statistic $Z = \hat{\alpha}_k/\hat{\sigma}(\hat{\alpha}_k)$ where $\hat{\sigma}(\hat{\alpha}_k)$ is an estimate of the standard error $\sigma(\hat{\alpha}_k)$. When the sample size is large, Z is approximately standard normal in distribution. Consequently, p-values and confidence intervals (Verbeke and Molenberghs 1997) for fixed effect parameters are readily obtained with statistical packages such as S-PLUS or SAS.

The significance of a random factor is assessed by testing whether the associated variance component is zero. Suppose that γ is the variance component of interest. A test of the hypothesis $H_0 : \gamma = 0$ against the alternative $H_1 : \gamma > 0$ is used to determine if the random factor is a significant source of variation in the response variable. If H_0 is rejected in favor of H_1, then it is concluded that the random factor associated with γ accounts for some of the variation of the response variable. In the ordinary linear model analysis, an analysis of variance would compare the regression sums of squares with and without the factor in the model via an F-statistic. In a linear mixed model analysis, the likelihood ratio statistic $l = 2\log[L(\hat{\theta}_2;y)/L(\hat{\theta}_1;y)]$ compares the likelihood $L(\hat{\theta}_1;y)$ of the model without γ and with parameter estimate $\hat{\theta}_1$ to the likelihood $L(\hat{\theta}_2;y)$ of the model that includes γ and with parameter estimate $\hat{\theta}_2$. The distribution of the likelihood ratio statistic is somewhat unusual (see Verbeke and Molenberghs 1997). For example, if the linear mixed model contains a single random factor with variance component γ, then the distribution of l is approximated by a mixture of two chi-square distributions having degrees of freedom 0 and 1, respectively. An approximate p-value for this test is p-value $\approx \frac{1}{2}P(\chi_1^2 > l)$, if $l > 0$, and p-value ≈ 1, if $l = 0$, where χ_1^2 denotes a chi-square random variable with 1 degree of freedom. Alternatively, Pinheiro and Bates (2000) suggest a conservative approximation given by p-value $\approx P(\chi_1^2 > l)$. Khuri, Mathew, and Sinha (1998) discuss exact tests for specific linear mixed models.

The Generalized Linear Mixed Model

A limitation of the linear mixed model is that the random errors, or residuals, are assumed to be independent and normally distributed with a common variance σ^2. Often, the residuals fail to meet these assumptions, and consequently, some or all of the parameter estimates and test statistics may be biased. A plot of the residuals against the fitted values is usually sufficient to detect failures of model assumptions. Draper and Smith (1998), Bryk and Raudenbush (1992), and Ramsey and Schafer (1997) discuss model checking and transformations. Certain types of data are not amenable to transformation. Two important cases are presence/absence data and counts that are dominated by small values (e.g., 50% or more of the counts are < 5). In the case of small counts, the distribution of the residuals from the fitted model is likely to be right-skewed, and not even approximately normal. A transformation yielding approximately normal residuals usually cannot be found when the plurality of observations is zero, and consequently, inferences derived from a linear mixed model analysis will be suspect. An example of such a response variable might be the number of offspring sired in a given year by individual males in a highly polygynous mating system.

Here we provide a brief discussion of models appropriate for response variables that are not normal. These models extend both the generalized linear and the linear mixed models, and hence, are named generalized linear mixed models. The generalized linear model (Dobson 1989, McCullagh and Nelder 1992, Fahrmeir and Tutz 1994) is used for a variety of response variables, but most importantly, for those that are binomial, multinomial, or Poisson in distribution. The generalized linear model specifies that some transformation of the mean, or expected value, of the response variable, say $\mu = E(Y)$, is a linear function of fixed explanatory variables $x_1, \ldots, x_{p-1}$. We write the model as $k(\mu) = \alpha_0 + \alpha_1 x_1 + \ldots + \alpha_{p-1} x_{p-1}$. The function k is called the link function, and the right-hand side of the model is called the linear predictor η, i.e., $\eta = \alpha_0 + \alpha_1 x_1 + \ldots + \alpha_{p-1} x_{p-1}$. In the case of logistic regression, Y is assumed to be a Bernoulli random variable taking on the value of 1 with probability π, and 0 with probability $1 - \pi$. The expected value of Y is $\mu = \pi$, and k is known as the logistic function; thus $k(\mu) = \log[\mu/(1-\mu)] = \log[\pi/(1-\pi)]$. The probability of success can be recovered by computing $\pi = e^\eta/[1+e^\eta]$. Parameter estimation and hypothesis testing are accomplished by maximum likelihood methods, and there is a large literature on practical aspects of generalized linear models. Most statistical analysis packages produce parameter estimates for the more common generalized linear models such as logistic and log-linear models.

The extension of the generalized linear model to the generalized linear mixed model is simple in principle: introduce random effects into the linear predictor. For example, we may model reproductive success of the ith female in year j by defining Y_{ij} to be 1 if at least one offspring survives 1 year, and 0 if not. The probability that $Y_{ij} = 1$ is π_{ij}, and we assume that Y_{ij} is a Bernoulli random variable with expected value $E(Y_{ij}) = \mu_{ij} = \pi_{ij}$. Moreover, we assume that there are persistent differences among individual females with respect to the likelihood of offspring survival. In this case, we assume that differences may be accounted for by independent random effects $b_i \sim N(0,\gamma)$, $i = 1, \ldots, q$. A generalized linear mixed model of the response variable states that if b_i were known and included in the linear predictor, then Y_{ij} has a Bernoulli distribution, and the parameter π_{ij} is related to η_{ij} through the logistic link function. Mathematically, we write $Y_{ij} \mid b_i \sim$ Bernoulli (π_{ij}) where $\log[\pi_{ij}/(1 - \pi_{ij})] = \eta_{ij}$ and $\eta_{ij} = \alpha_0 + \alpha_1 x_{1,ij} + \ldots + \alpha_{p-1} x_{p-1,ij} + b_i$. The notation $Y_{ij} \mid b_i \sim$ Bernoulli (π_{ij}) is shorthand for stating that Y_{ij} is conditionally Bernoulli in distribution given the b_i's.

The application of the generalized linear mixed model is, however, complicated by the fact that, in the case of nonnormal response variables, there are no general and exact methods of computing maximum likelihood estimates of model parameters. Breslow and Clayton (1993), McCulloch (1997), and Steele (1996) provide approximate estimation methods. We use the approach proposed by Steele (1996) in our examples of the generalized linear mixed model (below). Although these methods often produce relatively unbiased estimates of parameters, standard errors, and confidence intervals for the fixed model coefficients $\alpha_0, \ldots, \alpha_{p-1}$, tests of the variance components are not well understood and hence not currently available. The mixed model analysis of binary data presents some additional difficulties. The method of McCulloch (1997) produces unbiased maximum likelihood parameter estimates, yet the algorithm is both difficult to program and not widely available. Earlier methods, such as those proposed by Breslow and Clayton (1993) and Steele (1996) are somewhat simpler but also not widely available. Both algorithms are known to yield biased parameter estimates when applied to binary response variables (Shun 1997). Consequently, the binary response, generalized linear mixed model is at this time largely a theoretical model, though one holding substantial promise for future empirical application.

In the generalized linear model setting, a test of the hypothesis $H_0 : \alpha_i = 0$ against the alternative $H_1 : \alpha_i \neq 0$, where α_i is a fixed effect parameter, is best carried out using a likelihood ratio statistic (see McCullagh and Nelder

[1992] for details). As discussed previously, the likelihood ratio statistic *l* compares the likelihood of the model without α_i to the likelihood of the model that includes α_i. The likelihood ratio test statistic is well approximated by a χ_1^2 random variable when the sample size is large. The test may be extended to simultaneously test the significance of more than one parameter; if so, then the asymptotic distribution of *l* is chi-square where the degrees of freedom are the difference in numbers of parameters between the full and reduced models. In the generalized linear mixed model setting, we informally use this test statistic to assess the importance of a fixed effect. We use an informal treatment because the generalized linear mixed model parameter estimates are usually not exactly equal to the true maximum likelihood estimates, and, consequently, the large-sample approximation of the distribution of *l* by the chi-square distribution may not be very accurate.

Computation Aids

There are good procedures for linear mixed model analysis in several widely available statistical packages. We mention two because there are also detailed books dedicated to linear mixed model analysis using these packages. Specifically, the book by Verbeke and Molenberghs (1997) discusses linear mixed model analysis in SAS, and the book by Pinheiro and Bates (2000) discusses the use of S and S-PLUS for linear mixed model analysis. Computational support for mixed model analysis of nonnormal response variables is not well developed at this time. Because S-PLUS is widely used for research in the statistical community, some authors have made programs for computing generalized linear mixed model estimates available to other S-PLUS users. The SAS GLIMMIX macro for fitting generalized linear mixed models is also widely available. Finally, a set of GAUSS programs for linear and generalized linear mixed model estimation can be obtained from the GAUSS archives at www.american.edu/academic.depts/cas/econ/gaussres/regress/GLMM/GLMM.HTM.

The Generalized Linear Mixed Model and Individual Quality

Field studies in conservation and behavior frequently focus of necessity on small populations. Even when the population of interest is quite large, practical constraints may limit the number of individuals that can be recognized and studied in detail. Repeated measurements on individuals in such situations are possible and may be unavoidable. These considerations alone suggest wide applicability of the generalized linear mixed model in both conservation and behavior. In many applications, the generalized linear mixed model may be helpful as a way of removing the effects of statistical dependence among repeated measurements and allowing evaluation of the effect of an ecological or other factor of interest. In contrast, when applying the generalized linear mixed model to estimate individual quality, one is primarily interested in statistics associated with the variables representing individual identity. For this application, repeated measurements are necessary assets rather than a potential complication.

We have chosen life-history traits relevant to individual-based, demographic models as a context for illustrating generalized linear mixed model analyses of individual quality. In this setting, it is not so much the variance component associated with individual identity that is of interest, but the distribution of individual random effects. This is because these demographic models introduce variation in demographic rates by incrementing some base rate by an amount appropriate to each individual's circumstance (e.g., see Harris and Allendorf 1989). In the case of differences due to identity, this increment is assigned at birth and drawn from some specified distribution of increments (Conner and White 1999). The generalized linear mixed model is well suited for providing a data-based estimate of this distribution; the (persistent) effects of identity are predicted as individual-specific deviations from a population mean once the independent effects of other variables are accounted for. Although the conceptual relevance of the individual effects is particularly clear in this context, the generalized linear mixed model has not, to our knowledge, been previously applied to problems of this kind.

Equation (1) is a suitable framework for assessing most, if not all, of the four sources of process variation affecting demographic performance (individuality, environmental change in time and space, and stochasticity) (Akçakaya, Burgman, and Ginzburg 1999). This model explicitly addresses (1) individual variation due to developmental stage (age) and persistent phenotypic differences (identity), (2) temporal variation in the environment (year), and (3) variation due to demographic stochasticity. The latter is part of the unexplained variation represented by the error term. Spatial variation in the environment is not explicitly addressed. However, to the extent that there is population-wide competition for better sites, and individuals vary consistently in competitive ability, spatial variation in demographic performance may be reflected in the individual effects.

A model such as Equation (1) does not provide insight into the causes of persistent individual differences. It merely provides a basis for allocating variation in the response variable to persistent differences versus all other terms in the model, including residual, or unexplained, variation. When identifying specific causes of persistent variation is of interest and the relevant field measurements are available, it might seem straightforward to test a plausible explanatory factor (inbreeding coefficient, territory quality, social dominance, etc.) by adding it to the regression and testing for significance. However, if an added variable can change value during an individual's lifetime (e.g., territory quality), it may test significant because it is a source of persistent individual differences or because it is a source of transient individual differences. These cases might be distinguished by examining changes in variance components. The variance component for identity should decline relative to the basic model in the first case, whereas the variance component for the error term should decline in the second. The important point in the present context is that, when an added variable is a source of differences in individual quality, individual effects predicted by the more elaborate model are no longer measures of individual quality. Rather, they are measures of residual individual quality. Similar comments apply in the case of attempts to identify the specific causes of random year effects.

Response Variables

In a demographic context, the ideal response variables for which to obtain estimates of individual quality are fertility, fecundity, and survivorship. Because many species of conservation concern have extended periods of parental care, it may often be appropriate to also consider the survivorship of dependent young as a function of the individual quality of one or both parents rather than that of the offspring. In sufficiently long-lived species, fertility, fecundity, and offspring survival are subject in principle to repeated measurement as required for the estimation of individual effects in a generalized linear mixed model framework.

Modeling individual heterogeneity in survival probabilities poses fundamental difficulties associated with the fact that individuals may meet (survive) many environmental challenges but can die only once. Consequently, there is no opportunity to collect more than one observation of failure on any individual, and no possibility of using generalized linear mixed model analysis to estimate the proportion of variation in survival attributable to individual quality. Conner and White (1999) suggested that one might circumvent this problem by identifying traits that correlate with survival and then sampling individual adjustments of survival probabilities from a distribution modeled on that of the surrogate trait. They further suggested that age-specific size may often be an appropriate surrogate trait and developed an example based on the result from Bartmann, White, and Carpenter (1992) that over-winter survival of mule deer (*Odocoileus hemionus*) fawns was correlated with fawn weight at the beginning of winter.

The problem with this method is that it does not utilize repeated measures and hence cannot establish the extent to which individual variation in the surrogate trait reflects persistent versus transient individual differences. In the mule deer example, individuals were weighed once as fawns. Variation in fall fawn weight was then some unknown mix of persistent and transient sources of individual variation. It follows that this method also cannot establish or estimate persistent individual heterogeneity in survival probability. A generalized linear mixed model version of Conner and White's suggestion is to estimate individual effects for the surrogate trait and then sample individual-specific adjustments of survival probability from a distribution modeled on that of the *individual effects* estimated for the surrogate trait.

Examples

We will illustrate the application and interpretation of the generalized linear mixed model in a demographic context with three data sets from a field study of bighorn sheep. In each example, we develop a model based upon Equation (1) and tailored to the response variable of interest. Estimates of the parameter vector, random effects, and likelihoods were, in all cases, obtained using GAUSS programs written by B. Steele.

Normal Response Variable: Breeding Date

We begin with the relatively straightforward case of traits with the normal distribution. Traits reflecting overall physiological condition (quality) should often approximately follow a normal distribution because they summarize success across many independent, fitness-related activities. For our example, we have chosen annual date of first breeding by bighorn females. Females of many mammalian species are known to breed earlier when in better nutritional condition (Mitchell and Lincoln 1973, Frisch 1984, Byers and Hogg 1995).

A linear mixed model analysis of individual breeding date predicts timing of breeding for the ith animal in year j, given random effects $b_{1,i}$ and $b_{2,j}$ accounting for differences among individuals and among years. The conditional mean is denoted by $\mu_{ij} = E(Y_{ij} \mid b_{1,i}, b_{2,j})$. Our model of μ_{ij} involves parameters $\alpha_0, \alpha_1, \alpha_2$ and variables $x_{1,ij}, x_{2,ij}$ observed on the ith animal in year j. Specifically, the model is

$$\mu_{ij} = \alpha_0 + \alpha_1 x_{1,ij} + \alpha_2 x_{2,ij} + b_{1,i} + b_{2,j}, \qquad \text{(Model 1)}$$

where $x_{1,ij}$ is a dummy variable identifying whether the ith animal in year j is a member of the 1-year-old age class, and, similarly, $x_{2,ij}$ is a dummy variable identifying membership in the 3-year or older age class. Females in our study population typically had a first estrus at 1 year of age. Thus, these age classes, and the reference 2-year-old age class, generally indicate first-time breeders (age 1 year), second-time breeders (age 2 years), and veteran breeders (age $\geq$ 3 years). The random effects are assumed to be realizations from independent normal distributions with means 0 and standard deviations σ_1 and σ_2.

Copulation may occur over a period of 2 or more days in bighorn. Therefore, we used the day of ovulation to represent breeding date, assumed that ovulation occurred on the last day of behavioral estrus (Hogg, Hass, and Jenni 1992), and measured ovulation date categorically as the number of days from November 1 to the ewe's final day of estrus. We obtained a total of 464 measurements on 118 ewes from the Sheep River (Alberta) population in the period from 1989 to 2000. The average individual was measured for ovulation date 3.9 times. However, one and two observations were most frequent (Fig. 15.1).

A summary of regression results is given in Table 15.1. The interpretation of the fixed effect coefficients is straightforward. They are the advance (if negative) or delay (if positive) in ovulation date, measured in days, relative to the reference age category (here 2 years). There is no comparable coefficient summarizing the effect of the two random factors because the effect of each individual and year on ovulation date is assumed to be different; that is, the random effects have a distribution rather than a value. The distribution of the individual-specific coefficients (effects) for the 118 ewes

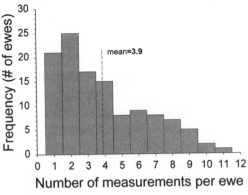

FIG. 15.1. Frequency distribution of the number of years in which individual bighorn ewes were measured for date of first ovulation. A total of 464 such measurements were obtained on 118 different ewes.

TABLE 15.1. Summary of a linear mixed model analysis of ovulation date as a function of ewe identity, ewe age, and year.

	COEFFICIENT	SD[1]	SE[1]	TEST STATISTIC[2]	df	p-VALUE
Fixed effects						
Ewe age						
1 year	9.30	—	1.48	6.27	—	< 0.001
≥ 3 years	−2.53	—	0.57	−4.41	—	< 0.001
Intercept	32.55	—	0.90	—	—	—
Random effects						
Ewe identity	—	3.33	0.30	121.11	1	<< 0.001
Year	—	2.31	0.53	81.32	1	<< 0.001
Residual	—	3.45	0.13	–	—	—

[1]Estimated standard deviations and errors.
[2]Z for fixed effects and *I* for random effects.

in this sample is presented in Figure 15.2. Individual and year-specific effects may be interpreted similarly to the fixed effect coefficients. They are estimates of the individual and year-specific advance or delay in ovulation date, again measured in units of days. Figure 15.2 shows that the predicted random effects for identity are symmetrically distributed about zero. This indicates that our use of the normal as a model for the random effects distribution was appropriate.

A simple way to evaluate the overall (versus year or individual-specific) contribution of the random effects to total variation in the response variable is to compare standard deviations (see Table 15.1). For example, we can say that, roughly speaking, the average (absolute) effect of identity on ovulation date (3.3 days) exceeded that for year (2.3 days) and was comparable to the average (absolute) residual effect (3.4 days). Similarly, the average (absolute) effect of identity (3.3 days) exceeded the fixed effect attributable to development from age 2 years to the age class 3 years and older (2.5 days) but was one-third of that associated with development from 1 to 2 years of age (9.3 days).

In Table 15.1, we also present the results of the approximate test of significance for random factors. Recall that this test involves calculating log likelihoods for the model with and without the random factor of interest (here identity then year), calculating an (approximate) chi-square test statistic equal to twice the absolute difference in these likelihoods, and then dividing the associated probability by 2. Taken together, the results shown in Table

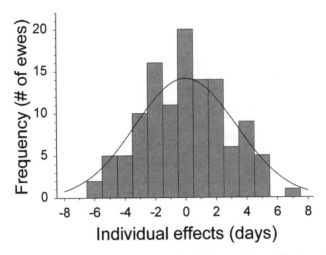

FIG. 15.2. Frequency distribution of individual effects on date of first ovulation for 118 bighorn ewes as predicted by linear mixed model regression. Histogram bins include the midpoint ± 0.5 days. The superimposed curve gives expected frequencies under the assumption that the individual effects are normally distributed with mean zero and standard deviation equal to the observed value (3.33).

15.1 provide strong evidence for large persistent individual differences in breeding date in bighorn. Excepting early development, persistent individual differences were the primary source of explained variation in ovulation date in this model, in this population, and over this time period. This is not an obvious outcome. For example, the year effects reflect the cumulative impact of temporal changes in the environment, and it would be reasonable to expect relatively large environmental influences in the case of a condition-dependent trait like breeding date.

Poisson Response Variable: Annual Male Reproductive Success

For polygynous mating systems, it may often make sense to model male reproductive success as a Poisson random variable, particularly when the expected number of offspring per male tends to be small (e.g., five or fewer). Both the number of offspring and the Poisson response variable are nonnegative counts. Moreover, although normalizing transformations, such as the square root or natural logarithm (Ott and Longnecker 2000), are often successful with count data, they will not be effective if a substantial number of the counts are zero as would be expected with a high degree of polygyny.

A Poisson generalized linear mixed model for male reproductive success models the expected number of successful matings for the ith animal in year j, given random effects $b_{1,i}$ and $b_{2,j}$ accounting for differences among individuals and among years, respectively. The conditional mean is denoted by $\mu_{ij} = E(Y_{ij} \mid b_{1,i}, b_{2,j})$. It is assumed that μ_{ij} is related to the linear predictor η_{ij} through the natural logarithm of μ_{ij}; that is, $\eta_{ij} = \log(\mu_{ij})$. The model of η_{ij} involves parameters α_0, α_1 and α_2, and variables x_{ij} and x_{ij}^2 observed on the ith animal in year j. Here, x_{ij} is ram age in years and x_{ij}^2 is ram age squared. This model allows the effect of age to follow a quadratic function, including one in which maximum reproductive success occurs at less than maximum age. Mathematically, the model is

$$\eta_{ij} = \alpha_0 + \alpha_1 x_{ij} + \alpha_2 x_{ij}^2 + b_{1,i} + b_{2,j}. \qquad \text{(Model 2)}$$

Finally, the random effects are assumed to be independent observations from normal distributions with mean 0 and standard deviations σ_1 and σ_2. Coltman et al. (2002) used a similar regression equation to analyze annual male reproductive success but did so in a linear mixed model framework. Parameter estimates for a Poisson model may be computed using any one of several methods (Breslow and Clayton 1993, Lee and Nelder 1996, Lin and Breslow 1996, and Steele 1996). We used Steele's (1996) method.

To illustrate the application of Model (2) in estimating persistent individual differences in male reproductive success, we will use a data set composed of paternity assignments for 83 bighorn lambs conceived during nine breeding seasons (1988–1996) on the National Bison Range (Montana). Fathers for these lambs were identified by genetic exclusion using multilocus microsatellite genotypes (Hogg and Forbes 1997). Estimates of annual reproductive success were obtained for 43 rams measured in an average of 4.3 different years (range 1–9 years). When totals for all 9 years are considered, it is clear that there was large individual variation in success (Fig. 15.3). The two most successful rams fathered 21 lambs in this period, whereas almost half of the remaining 41 rams fathered none. However, some, and potentially all, of these differences could be attributable to the fact that rams were often measured at different ages and for different numbers of years. Thus, whereas the pattern in Figure 15.3 leaves open the possibility of large differences in individual quality regarding male competitive ability, it does not demonstrate or estimate such differences.

The results of a generalized linear mixed model analysis of these data using Model (2) are presented in Table 15.2 and Fig. 15.4. We can interpret the fixed and random effects on both the natural logarithm and the original count scales. The interpretation on the natural logarithm scale is the same as when a linear mixed model is used, except that the random effects are modifying the natural logarithm of the expected number of paternities rather

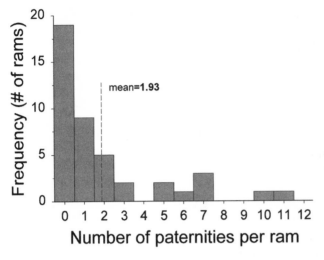

FIG. 15.3. Frequency distribution of the number of offspring conceived by 43 bighorn rams during a 9-year period (1988–1996).

than the expected number of paternities. The entries in Table 15.2 are on the natural logarithm scale. If it is desirable to examine the random effects on the original count scale, though, we compute the antilogarithm of the predicted random effects. To see the implications of this, consider that taking the antilogarithms of the left- and right-hand side of Model (2) and factoring the right-hand side yields $\mu_{ij} = \exp(\alpha_0 + \alpha_1 x_{1,ij} + \alpha_2 x_{2,ij} + b_{2,j})\exp(b_{1,i})$.

TABLE 15.2. Summary of a generalized linear mixed model analysis of annual male reproductive success as a function of ram identity, ram age, and year. The response variable was assumed to be Poisson, given random effects for identity and year.

	COEFFICIENT	SD[1]	SE[1]	TEST STATISTIC[2]	df	p-VALUE
Fixed effects						
Ram age				15.83	2	—
Age	0.57	—	0.19	—	—	—
Age²	−0.03	—	0.01	—	—	—
Intercept	−2.98	—	0.58	—	—	—
Random effects						
Ram identity	—	0.72	0.19	—	—	—
Year	—	0.03	0.07	—	—	—

[1]Estimated standard deviations and errors.

[2] *t* for fixed effects; not available for random effects.

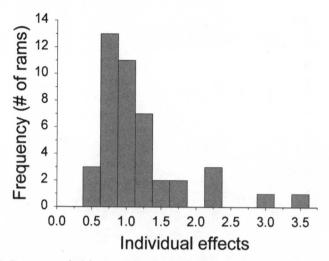

FIG. 15.4. Frequency distribution of individual effects on annual male reproductive success for 43 bighorn rams as predicted by generalized linear mixed model regression. Individual effects in this figure modify the expected number of paternities (see text). Thus a value of 1.0 indicates no effect of individual identity on reproductive success. Histogram bins include the midpoint ± 0.125 units.

This shows that $\exp(b_{1,i})$ is a multiplicative term expressing the extent to which male *i* differs from a typical male. Male *i* is typical with respect to reproductive success if $b_{1,i} = 0$, because $\exp(0) = 1$. Similarly, if $b_{1,i}$ is substantially larger than 1, then male *i* is regarded as more successful at reproducing than the typical male by a factor equal to its individual effect.

In our data set, annual male reproductive success was a nonlinear function of age (see Table 15.2). Maximum success occurred at 9 or 10 years of age and declined thereafter. The year random factor functioned primarily to account for annual variation in the number of lambs assigned to fathers. Because this variation was small, year effects were negligible (see Table 15.2). In contrast, individual effects varied substantially; values on the original count scale ranged from 0.51 to 3.48 (see Fig. 15.4). Thus the highest-quality male was predicted to be 3.48 times as successful as a typical male at each age (and seven times as successful as the lowest-quality individual). A formal test of significance is not available in the Poisson case. However, this level of variation in predicted individual effects is consistent with a conclusion that biologically significant differences in quality were present in this sample of rams, particularly when one considers that the multiplicative individual effects apply at each age and can therefore sum to large absolute differences in lifetime reproductive success. For example, we can estimate the difference in

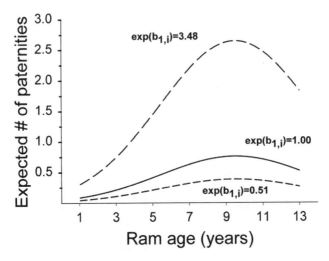

FIG. 15.5. Expected annual male reproductive success as a function of ram age for the typical [$\exp(b_{1,i}) = 1.00$], highest-quality [$\exp(b_{1,i}) = 3.48$], and lowest-quality [$\exp(b_{1,i}) = 0.51$] rams in the National Bison Range data set (Fig. 15.4). Expected success was calculated as $\mu_{ij} = \exp(-2.98 + 0.57x_{1,ij} - 0.03x_{2,ij} + b_{2,i})\exp(b_{1,i})$ (see text for derivation). Year effects ($b_{2,i}$) were set to zero and fixed effect coefficients are from Table 15.2. The scale of the y-axis reflects the average annual number of lambs assigned paternity (9.2; SD = 1.4).

expected lifetime reproductive success for the highest- versus lowest-quality individual in this data set as the sum of the difference in annual reproductive success at each integer age (Fig. 15.5). Although the maximum difference in expected annual reproductive success was only two offspring, the difference in expected lifetime reproductive success was 16 on the assumption both individuals lived 13 years and total lamb production stayed at approximately 10 per annum (see Fig. 15.5). In the context of individual-based demographic models, multiplicative individual effects are functionally equivalent to the additive individual effects obtained in the normal case and can be readily used to increment average probabilities of paternity.

Binary Response Variable: Annual Female Reproductive Success

In this last example, we illustrate the application of the generalized linear mixed model to binary demographic data, specifically annual reproductive success among bighorn ewes. We measured female success as the number of offspring alive in the fall of year 1 (0.5 years postpartum) when lambs are typically weaned and, alternatively, in the fall of year 2 (1.5 years postpartum) when reproductive maturity is typically reached. Bighorn litters are almost invariably singletons so these numbers are either 0 or 1 in practice. Female infertility was not distinguished from offspring mortality in these data. However, the former is uncommon relative to the latter in this population (Festa-Bianchet 1988b; Bérubé, Festa-Bianchet, and Jorgenson 1999), particularly at those ages (4–15 years) constituting the bulk of our sample. Thus, although reproductive success in this case is technically a combination of three demographic rates (fertility, fecundity, and offspring survival), differences in success among females were largely due to differences in offspring survival.

The generalized linear mixed model for annual female reproductive success models the probability of success (offspring counted alive) for the ith animal in year j, given random effects $b_{1,i}$ and $b_{2,j}$ accounting for differences among individuals and among years. The conditional probability is denoted by $\pi_{ij} = P(Y_{ij} = 1 \mid b_{1,i}, b_{2,j})$. It is assumed that π_{ij} is related to the linear predictor η_{ij} through the log-odds of π_{ij}; that is, $\eta_{ij} = \log[\pi_{ij}/(1-\pi_{ij})]$. Similar to the Poisson model of male reproductive success, η_{ij} involves parameters α_0, α_1, α_2 and variables $x_{1,ij}$, $x_{2,ij}$ observed on the ith animal in year j. Specifically, the model is

$$\eta_{ij} = \alpha_0 + \alpha_1 x_{1,ij} + \alpha_2 x_{2,ij} + b_{1,i} + b_{2,j}, \qquad \text{(Model 3)}$$

where $x_{1,ij}$ is a dummy variable identifying whether the ith animal in year j is a member of the 2-year-old age class, and, similarly, $x_{2,ij}$ is a dummy variable identifying membership in the 4-year and older age class. These age classes, and the reference 3-year-old age class, generally reflect first-time mothers, second-time mothers, and veteran mothers. The random effects are assumed to be realizations from independent normal distributions with mean 0 and standard deviations σ_1 and σ_2.

It is difficult to obtain unbiased parameter estimates for generalized linear mixed models involving binary response variables. We used Steele's (1996) method for purposes of illustration. The reader should recognize that parameter estimates obtained with this and other approximate methods should not be used for formal inference and that, without unbiased parameter estimates, formal statistical analysis of the predicted random effects is unwarranted.

The data set consisted of 574 (0.5 years) and 563 (1.5 years) measurements of reproductive success obtained from 1989 to 2000 for 127 different ewes in the Sheep River (Alberta) population. The average number of repeat

TABLE 15.3. Summary of a generalized linear mixed model analysis of annual female reproductive success (measured 0.5 and 1.5 years postpartum) as a function of maternal age, maternal identity, and year. The response variable was assumed to be Bernoulli given random effects for identity and year.

	0.5 YEARS POSTPARTUM			1.5 YEARS POSTPARTUM		
	Coefficient	SD[1]	SE[1]	Coefficient	SD[1]	SE[1]
FIXED EFFECTS						
Maternal age						
two years	−1.65	—	0.45	−1.68	—	0.66
≥ four years	0.77	—	0.29	0.74	—	0.37
Intercept	−0.45	—	0.29	−1.70	—	0.42
RANDOM EFFECTS						
Maternal identity	—	0.50	0.17	—	0.78	0.17
Year	—	0.35	0.13	—	0.67	0.20

[1]Estimated standard deviations and errors.

measures per female was 4.5 (range = 1–12) at 0.5 years postpartum and 4.4 (range = 1–12) at 1.5 years postpartum. The analysis of these data using Model (3) is summarized in Table 15.3. The interpretation of the fixed and random effects in this case can be based on the log-odds model of μ_{ij}, so that $b_{1,i}$, say, expresses the difference between the ith animal and the population mean with respect to the log-odds of survival. The entries in Table 15.3 are on this scale. A simple transformation from log-odds to the probability of success scale is not available, though the following loose interpretation may be helpful. Suppose that the fixed effect portion of the linear model is 0; that is, $0 = \alpha_0 + \alpha_1 x_{1,ij} + \alpha_2 x_{2,ij} + \alpha_3 x_{3,ij}$, and $b_{2,j} = 0$. Then, $\eta_{ij} = b_{1,i}$, which implies that $\pi_{ij} = [1+\exp(-b_{1,i})]^{-1}$. Note that if $b_{1,i} = 0$, then $\pi_{ij} = \frac{1}{2}$, and the difference $\frac{1}{2} -[1+\exp(-b_{1,i})]^{-1}$ roughly expresses the quality of the ith female relative to the typical female in terms of the probability of reproductive success.

Cumulative relative frequencies for each set of (transformed) individual effects predicted by Model 3 are given in Figure 15.6. The general shape and horizontal location of these distributions show that the individual effects on reproductive success at 0.5 and 1.5 years were, in each case, approximately normally distributed with mean zero. Their shapes relative to each other indicate that differences in individual quality among females were markedly greater when reproductive success was measured at 1.5 versus 0.5 years (see

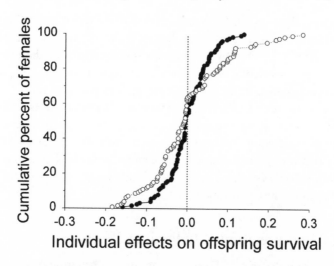

FIG. 15.6. Cumulative relative frequency of individual effects on annual female reproductive success as predicted by generalized linear mixed model regression for
success measured at 0.5 (black circles) and 1.5 (white circles) years postpartum (*n* =
127 females). The (approximate) individual effects are expressed in terms of increments in the probability of success relative to the typical female (see text).

also Table 15.3). Roughly 25% of the individual effects on success at 1.5 years
were greater than 0.10 or less than −0.10, whereas only 6% of the individual
effects on success at 0.5 years were this extreme (see Fig. 15.6). The same
pattern was evident in the predicted year effects; between-year variation in
recruitment to 1.5 years was twice that to 0.5 years (see Table 15.3). Larger
predicted individual effects at 1.5 years are not easily explained as simply
more time for the effects of differences in maternal quality to accumulate.
Offspring were under direct maternal care in the first time period but not the
second, and a substantial majority of offspring mortality in this sample
(68%) occurred during the first time period.

It is more plausible that the increase signals a qualitative change, associated with the onset of winter, in the interaction between maternal differences
and the environment. Climatic conditions are generally benign between
birth and weaning, whereas even the largest and most vigorous lambs are
relatively small and vulnerable to predation for much of this time. It makes
sense that offspring mortality under these conditions would be more subject
to chance (e.g., encounters with predators) and relatively independent of
maternal characteristics or annual variation in seasonal climate. On the other
hand, lamb survival during the first winter in this population is known to be

strongly dependent on fall body mass (Festa-Bianchet et al. 1997), perhaps because larger (fatter) lambs are better buffered against temperature extremes and less affected by poor forage quality. Thus, to the extent that mothers differ consistently with respect to the rate at which they provision lambs during gestation and prior to weaning, or in heritable traits that affect the early growth rate of offspring, one would expect the effects of maternal quality to be more telling and evident during winter (especially severe ones) even though maternal investment has ceased and fewer lambs may die. The details of this interpretation aside, the general point is that differences in individual quality, even for a fixed set of individuals, may not be equally apparent in different environments.

Practical Interpretation of Individual Effects

With these examples in mind, three points should be emphasized regarding the practical interpretation of individual effects that are intended as estimates of individual quality. First, as with any sample-based estimate, variation in the predicted individual effects reflects both sampling and process variation. Sampling variation is a function of sample size at two levels: the number of measurements per individual and the number of individuals for which repeated measurements are available. It is not possible to generalize about minimum sample sizes for a generalized linear mixed analysis of individual differences. Much will depend on the trait of interest. Relatively few repeat measures (e.g., 2–3) may be adequate for simple traits or traits that are relatively insensitive to external influences. For complex traits, or traits subject to many environmental influences, good estimates of mean within-individual performance may require substantially more repeated measurements. Similarly, we can speculate that fewer repeat measurements will be required for organisms with simple life histories (e.g., a small number of growth stages) than for those with more complex patterns of growth and development. Finally, when forced to choose between obtaining more repeat measurements and sampling more individuals, it may be helpful to consider environmental heterogeneity. Other things being equal, high spatial heterogeneity should favor larger (and more broadly distributed) samples of individuals, whereas high temporal heterogeneity may indicate a priority for larger numbers of repeat measurements (see point three following here).

Second, single, or relatively few, measurements will be common even in very long-term studies. Small individual sample sizes arise in longitudinal studies for several reasons. Some individuals may survive only one or a few time periods, whereas others will be old at the beginning of the study or born near its end. There also may be data missing. The generalized linear mixed model predicts an individual effect for all individuals in the sample, even those measured a single time. Despite the fact that such predictions will usually be improved by more repeat measurements, it is not necessarily a good idea to exclude individuals measured fewer than some threshold number of times. To the extent that low-quality individuals survive for markedly fewer repetitions of the focal trait, such filtering may differentially remove low-quality individuals and lead to underestimation of the variance in individual quality. This type of bias may be unavoidably present if low-quality individuals often fail to survive for even one measurement of the trait. Note that individual effects predicted for relatively short-lived individuals, even when based on measurements over that individual's entire life span, should still be thought of as estimates of a quality parameter that would be better revealed by more years of data.

Finally, the distribution of individual quality is not likely to be a species- or even population-specific trait. Individual variation in performance is apt to be highly conditional on social and ecological environments. In particular, we should expect that this variation will be greater in more competitive social situations and under more severe ecological conditions (as we suggested for maternal quality in bighorn). More demanding environments may be required to expose and amplify relatively small phenotypic differences (Keller et al. 1994). This consideration may limit the generality of a given set of estimates for demographic models. However, it suggests another potential use for generalized linear mixed model analysis of fitness-related traits in conservation—that of identifying stressed populations and monitoring populations for deteriorating conditions.

Conclusions and Recommendations

The generalized linear mixed model should be considered whenever repeated measurements on recognizable individuals constitute a substantial fraction of a sample of measurements. Mixed model analysis is mandatory when, in addition, study objectives include (1) drawing conclusions about trait-specific individual effects for the population versus sample of individuals, and (2) obtaining formal estimates of the fundamental components of individual variation (e.g., individual quality).

When the latter objectives apply, field biologists, contrary perhaps to their intuition, should covet and actively seek repeated measurements on marked or otherwise recognizable individuals. The inevitability of incomplete individual records in longitudinal field studies guarantees that the average number of measurements per individual will be lower, perhaps much lower, than the number of sampling intervals (e.g., years) over which measurements have been obtained. Thus individual sample sizes should be monitored carefully. Generally speaking, more repeat observations will be helpful.

Once repeated measurement data are in hand, it is well worth exploring normalizing transformations (if such are indicated) to stay within a linear mixed model framework. The advantages of a linear mixed model approach include well-established methods for analysis and interpretation, widely available software, and straightforward significance tests for random factors. If, however, such transformations do not work, one should not be overly concerned about the limitations of the generalized linear mixed model regarding hypothesis testing. It may still be possible to make judgments about the biological significance of a given variance in individual effects by, for example, extrapolating the lifetime or population-level consequences of the predicted effects. In addition, some insight into biological significance may often be gained by examining the extent to which fixed factor parameters (or other variance components) change when the identity random factor is dropped. Finally, if a given set of individual effects can be shown to be informative in some subsequent analysis, then we may regard this as informal evidence of biological significance. For example, a negative correlation between individual effects on annual male reproductive success and inbreeding coefficient would suggest that the predicted individual effects reflect real (genetic) differences among males.

We encourage investigators to make full use of the predicted individual effects. The further analysis of these random variables should often be rewarding. Important potential applications in conservation and behavior include (1) investigation of the genetic or ecological sources of variation in individual quality and (2) evaluations of the population-level consequences of persistent versus transient individual differences.

Summary

The generalized linear mixed model provides a powerful framework in which to formally evaluate the nature and magnitude of individual differences. We have emphasized an application in demographic models of population persistence. However, the study of individual differences and adaptive strategy in individual behavior is a central activity of behavioral ecology. Behavioral ecologists may find many conceptually distinct applications for such methods in empirical studies of behavioral strategy and, given growing interest in the population-level effects of individuality in conservation, important new applications at the interface of conservation and behavior.

The generalized linear mixed model can be applied for one or more of three reasons. First, it may be necessary to account for lack of independence arising from repeated measurements on individuals in order to evaluate the effect of an ecological variable of interest. Second, it may be of interest to evaluate variance components for the contribution of persistent individual differences to variation in the response variable. Finally, an investigator may desire predictions of an individual quality random variable for use (like any other individual-specific measurement) in some further analysis. The latter may be the most far-reaching and unappreciated use of the generalized linear mixed model in conservation and behavior. The ways in which measurements of individual quality might be used are as various and open-ended as those for any other fundamental individual characteristic.

The power of the generalized linear mixed model comes at a cost. Repeated measurements are a requirement. This implies some system of individual recognition and the ability to relocate study individuals reliably and repeatedly. Such control may be difficult for many long-lived, wide-ranging species of conservation concern. Even when these methodological requirements can be met, there remains the challenge of sustaining studies of natural populations long enough to accumulate sufficient numbers of repeat measures. The extra effort required to surmount these obstacles should often be repaid. Although repeated measurements on known individuals are often regarded in the behavioral sciences as a hindrance to statistical inquiry (e.g., Machlis, Dodd, and Fentress 1985), the opposite is true. When analyzed in a generalized linear mixed model framework, such data improve statistical power (Diggle, Liang, and Zeger 1994) and enable a more thorough description of the nature and consequences of individual variation in natural populations.

16.

Individual Quality, Environment, and Conservation

Peter Arcese

Conservation biologists often try to recover small populations by ameliorating the causes of their decline. A first step in this process is to create ranked lists of factors known to affect population growth. Second, biologists assess which factors are amenable to management by weighing their effect on population growth against the likelihood of influencing them successfully (Boyce 1992, Caughley 1994, Beissinger and Westphal 1998, Holthausen et al. 1999, Morris et al. 1999, Noon et al. 1999). Thus, despite the fact that population growth rate in vertebrates is often most sensitive to variation in adult survival (Caswell 1978), it will sometimes be more practical to manage less influential factors, including the effects of inbreeding (Westemeier et al. 1998) or predators on offspring survival (Caro and Laurenson 1994).

Recently, it has also been suggested that detailed estimates of individual differences in phenotypic or genotypic traits that affect fitness might also be employed to model populations more precisely and manage them more successfully (Clemmons and Buchholz 1997, Connor and White 1999, Cam and Monnat 2000, Steele and Hogg, this volume). It is also discussed casually that managers might use traits of animals indicative of high fitness to select individuals for captive breeding or translocation. In conservation, however, managers often engage in triage to allocate scarce financial and human resources (Sinclair and Arcese 1995b) and generally pursue research on new approaches only when persuasive examples exist. Because applying ideas about individual variation in fitness to conservation will require substantial effort and data (cf. Connor and White 1999, Cam and Monnat 2000), it will be essential to understand the causes of individual variation in fitness and their overall influence on population growth.

To explore this topic further, in this chapter I first review some ideas in population ecology and heterogeneity in individual quality. I then describe individual variation in annual breeding performance over 25 years in a small, unstable, and inbred population of song sparrows (*Melospiza melodia*) resident on Mandarte Island, British Columbia (Arcese et al. 1992, Keller et al. 1994). In general, I ask what value might exist in monitoring or managing one or more phenotypic traits identified as being predictive of "individual quality," which I define here as the tendency for individual animals to perform better or worse than the population mean rate on average. I begin by reviewing some ideas about extrinsic versus intrinsic influences on individuals and populations because the former are often the focus of species recovery plans, whereas the latter include factors more often linked to heterogeneity in individual quality and behavior.

Extrinsic versus Intrinsic Influences on Individuals and Populations

Much of conservation biology focuses on the dynamics of populations and the various environmental and other extrinsic effects that limit population growth (Caughley 1994, Newton 1998). Caughley (1994) argued that the extrinsic influences of habitat loss, fire, weather, and introduced predators, competitors, and disease often have overwhelming effects on populations relative to those occurring as a consequence of intrinsic differences between individuals or, cumulatively, between populations. In contrast to extrinsic factors, intrinsic factors that influence fitness might include traits related to genotype, inbreeding, or heterozygosity, as well as phenotypic traits that vary as a consequence of maternal or developmental effects (Stearns 1992, Schlichting and Pigliucci 1998, Cam and Monnat 2000, Santos 2001).

Recall, however, that both extrinsic and intrinsic factors can act simultaneously on individuals and populations. Extrinsic influences on individual fitness occur, for example, when a parasite reduces the reproductive output of its host (Smith 1981a). This extrinsic effect will impact population fitness if it affects enough individuals to reduce the mean rate of reproduction compared to that in populations without parasites (Arcese, Smith, and Hatch 1996). Similarly, in the case of intrinsic effects, we can contrast the fitness of inbred individuals to outbred ones in the same population (Keller 1998) and also compare the fitness of populations with different mean rates of inbreeding (Gilpin and Soulé 1986). Overall, this distinction between individual and population effects is crucial to many arguments about the relevance of behavioral or other individual-based approaches to conservation. Even though we can often demonstrate substantial impacts of a particular effect on individual fitness, conservation teams will be compelled to consider ameliorating such effects only when it is also shown that their magnitude and frequency are sufficient to influence population dynamics overall (Arcese, Keller, and Cary 1997; Beissinger 1997).

Caughley's view of the predominance of extrinsic influences on conserved populations persists today. This is because very few studies have demonstrated strong links between heterogeneity in the intrinsic quality of individual animals and variation in demographic rates at the level of populations. In addition, many case studies in conservation demonstrate an overwhelming influence of extrinsic factors on population growth, especially via habitat loss (Noon and McKelvey 1996), predator introduction (Reichel, Wiles, and Glass 1992), exploitation by humans (Brashares, Arcese, and Sam 2001), and disease (Atkinson et al. 1995). Moreover, techniques for estimating individual quality and incorporating it into predictive models of populations are recently introduced to conservation and in various states of development (Conner and White 1999, Kendall and Gordon 2002, Steele and Hogg, this volume). Thus, until recently, it has generally seemed practical to assume that management will be most effective when it focuses on the amelioration of detrimental extrinsic effects, while acknowledging that a focus on individuals may be appropriate in remnant or captive populations (Caughley 1994, Beissinger and Westphal 1998, Morris et al. 1999, Noon et al. 1999). However, this view may be changing.

In particular, several studies now suggest that the dichotomy of extrinsic versus intrinsic influences on populations is too simplistic overall, and that the potential interplay of individual heterogeneity and environment is more influential on population trajectory than assumed previously. For example, in some populations, individual differences in fitness vary more or less synchronously in time as a consequence of feedback between the state of the population or environment and the quality of individual animals or their young (Cam and Monnat 2000, Coulson et al. 2001, Reid et al. in press). Summed over individuals, these shared differences in fitness have the potential to cause cohort or other group-specific differences in vital rates that may

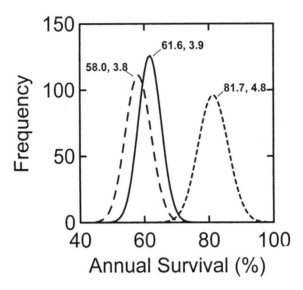

FIG. 16.1. Normal approximations of the distribution of annual survival rates in "native" (long dash) and "F_1 hybrid" (short dash) male song sparrows (cf. Marr, Keller, and Arcese 2002:136), and for the weighted mean of the pooled estimates (solid line; based on 500 random draws given indicated mean, SD). Natives had no immigrant parents or grandparents; F_1s produced by immigrant-native matings. Distributions represent a case wherein two classes of individuals exist in one population and survive at different rates; pooled distribution depicts the mean, SD, that might be reported by researchers and modeled by managers in the absence of data on individual heterogeneity in fitness.

in turn influence the mean expected rate of population growth (Fig. 16.1). In the Mandarte Island song sparrow population, for example, small population size and coincident inbreeding can reduce individual fitness (Keller 1998; Marr, Keller, and Arcese 2002; see Fig. 16.1) and, presumably, mean expected population size. Successful immigrants to the island, however, raised hybrid offspring that survived better than natives and constituted a substantial fraction of the smallest populations (Marr, Keller, and Arcese 2002). It is plausible, therefore, that immigration acts to increase the population's growth rate most when it is small by raising the fraction of hybrid offspring and population fitness overall. Demonstrating that such mechanisms operate in nature argues strongly for the management of immigration in fragmented systems of populations and, perhaps, also for monitoring the average quality of individual breeders.

Covariance in fitness components among classes of individuals can also lead to structural dependencies that affect the precision and accuracy of predictive models and the stability of populations (Engen, Bakke, and Islam 1998; Connor and White 1999; Cam and Monnat 2000; Fox and Kendall 2002; see Fig. 16.1). Among Soay sheep (*Ovis aries*), for example, weather and population density each affect population stability directly but also interact with population age structure, sex ratio, and individual fitness as a consequence of additive genetic variation in body size, parasite resistance, and feeding apparatus (Coltman et al. 1999a,b, Smith et al. 1999, Milner et al. 2000). Forchhammer et al. (2001) showed further that density-dependent and climatic conditions experienced by individuals before birth as a consequence of maternal effects, and after birth as a consequence of direct environmental influences on development, varied by cohort and resulted in persistent differences in mean fitness among cohorts. Coulson et al. (2001) suggested that differences among cohorts are sufficient to influence population growth and, in particular, to facilitate declines. Chitty (1967, 1999) also described relationships between population size and fecundity, aggressive behavior, and body size in voles (*Microtus* spp.), some of which are indirect consequences of changes in the biotic environment (Norrdahl and Korpimaki 2002).

Individual by environment interactions may also destabilize populations when inbreeding depression is expressed as a nonlinear, additive mortality during periods of severe environmental stress (Keller et al. 1994, Keller and Waller 2002; Fig. 16.2). In island-living song sparrows and large-billed ground finches (*Geospiza magnirostris*), for example, Keller et al. (1994, 2002) showed that selection was concentrated on inbred individuals during population crashes that coincided with periods of cold and drought, respectively. If selection purges populations of deleterious alleles responsible for inbreeding depression, it may also ameliorate population-level effects of inbreeding on population growth (Keller and Waller 2002). However, even rare instances of immigration may be sufficient to reestablish deleterious alleles that facilitate inbreeding depression (Keller et al. 2001, Keller and Waller 2002). Thus, where immigration occurs and inbreeding depression is expressed only under the most stressful conditions, it becomes plausible that the average level of inbreeding might rise in the period between stressors, and destabilize populations overall.

Taken together, these examples suggest that predictive models of conserved populations might capture more accurately their dynamics and predict better the influence of particular vital rates on population growth by taking into account the interplay of intrinsic and extrinsic influences via their links to individual quality (Engen, Bakke, and Islam 1998; Connor and White 1999; Coulson et al. 2001; Kendall and Gordon 2002). Such models might include parameter estimates for subpopulations of individuals that

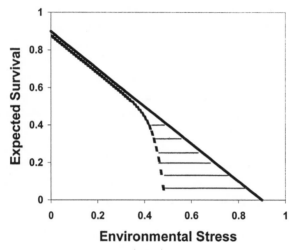

FIG. 16.2. Hypothetical relationships between the expected survival of two classes of individuals that vary intrinsically in quality, in relation to "environmental stress." Solid line depicts a negative linear relationship wherein 90% of high-quality individuals will survive the most benign years but 0% survive very stressful conditions. Dashed line depicts the case wherein impaired individuals survive about normally in benign to moderately stressful years, but experience catastrophic mortality in more stressful years. Horizontal lines emphasize the difference in expected survival of each class in stressful years. Note that the effects of stress on the size of a population composed of these two classes of individuals will depend on the fraction in each class at the time of a stressful event, and that buildups in the fraction of impaired individuals in the population between stressful events will destabilize the population overall.

vary intrinsically in quality as a consequence of genetic, maternal, or environmental effects (Connor and White 1999, Cam and Monnat 2000; see Figs. 16.1 and 16.2). Ideally, these models should improve the efficiency of management and triage exercises by reducing uncertainty about the response of populations to alternate management proposals. Managers might also use traits indicative of individual quality to identify candidates for captive breeding, reintroduction, or translocation likely to contribute positively to population growth.

Many questions remain about the utility of focusing on heterogeneity in individual quality in conservation. In particular, we need to know (1) what are reliable indexes of individual quality, (2) which traits have large and consistent influences on fitness, (3) whether interactions between individual quality and extrinsic factors are common features of populations, and (4) whether such interactions are often sufficient to affect population growth. I focus on the first two questions in the remainder of this chapter because identifying reliable and influential indexes of individual quality will be a first step to incorporating the concept into conservation practice.

Currencies and Repeatability of Individual Quality

References to individual quality nearly all consider measurable traits that are known or assumed to predict individual fitness (Grafen 1988, Hochachka 1993, Cam and Monnat 2000). However, a brief survey of recent literature shows that no universally applied index exists (see Table 16.1). In contrast, researchers typically adopt as candidate indexes various traits linked to survival or annual or lifetime reproductive success by statistical correlation, including the ability to defend territory (Arcese 1987), breed early (Hochachka 1993, Catry et al. 1999), or attain high social status (Dufour and Weatherhead 1998), or to maintain high body condition (Faivre et al. 2001), bilateral symmetry (Bowyer et al. 2001), immunocompetence (Hasselquist, Wasson, and Winkler 2001), or low parasite loads (Engen and Folstad 1999; see Table 16.1). This partial list is a testament to the potential of individual variation in quality to become a unifying theme in population, behavioral, and conservation ecology.

However, the diversity of candidate indexes listed here and in Table 16.1 also raises challenges to applying ideas about individual quality to conservation. These challenges arise, first, because much exploratory work will be needed in all but the most thoroughly studied species to identify and validate reliable and influential indexes of individual quality. In many cases, this work will be prohibitively expensive, time consuming, or logistically impractical. As a consequence, proposals to pursue such work will have little weight in the allocation of scarce conservation resources except where substantial data and candidate indexes already exist.

A second challenge to applying ideas about variation in individual quality to conservation involves the inconsistent definitions in the literature of quality as a fixed versus ephemeral trait of animals. Cam and Monnat (2000) point out that several different views on individual quality exist in the literature, with different implications for population dynamics and structure. Thus, whereas positive relationships between survival and reproduction are often accepted as evidence of heterogeneity in individual quality within populations (Smith 1981b, Stearns 1992), there is no consensus on whether quality is an ephemeral, static, or dynamic trait of individual animals (Cam and Monnat 2000). For example, body condition, typically expressed as mass relative to structural size, is often related positively to other direct correlates of

TABLE 16.1. Traits considered as potential indexes of individual quality in animals, with their stated or implied cause of variation.

TAXA	INDEX OF INDIVIDUAL QUALITY	ORIGIN OF EFFECT	SOURCE
BIRDS			
Bucephala clangula	plumage ornaments	—	Ruusila et al. 2001
Sterna hirundo	condition, fate, ARS	—	Becker et al. 2001
Sterna hirundo	age, condition	Intrinsic	Wendeln et al. 2000
Pagodroma nivea	body size	Intrinsic	Barbraud et al. 2000
Rissa dactyla	reproductive state	Intrinsic, repeatable	Cam and Monnat 2000
Puffinus tenuirostris	breeding frequency	Intrinsic, repeatable	Bradley et al. 2000
Catharacta skua	breeding date	—	Catry et al. 1999
Diomedea chrysostoma	ARS	Intrinsic, repeatable	Cobley et al. 1998
Tyto alba	plumage ornaments	Intrinsic, repeatable	Roulin et al. 1998
Meleagris gallopavo	spur length asymmetry	Intrinsic, repeatable(?)	Badyaev et al. 1998
Dendrocopos minor	clutch size, lay date, ARS	—	Wiktander et al. 2001
Tachycineta bicolor	immunocompetence	—	Hasselquist et al. 2001
Hirundo rustica	tail feather asymmetry	Intrinsic, repeatable	Shykoff and Møller 1999
Manorina melanocephala	mobbing behavior	—	Arnold 2000
Pica pica	tail length	—	Blanco and Puente 2002
Turdus merula	condition, bill color, lay date	—	Faivre et al. 2001
Parus caeruleus	size, mite load, lay date, survival	Intrinsic, repeatable	Leech et al. 2001
Parus caeruleus	fecundity, parasitism, immunocompetence	Intrinsic/extrinsic	Merila and Andersson 1999
Parus major	plumage ornaments, carotenoids	—	Horak et al. 2001
Alauda arvensis	aerial display/song	—	Hedenstrom and Alerstam 1996
Ficedula albicollis	feather wear	—	Merila and Hemborg 2000

Sturnus unicolor	resource holding potential, androgens	Intrinsic	Veiga et al. 2001
Sturnis vulgaris	body mass, clutch size, lay date	Intrinsic/extrinsic	Christians et al. 2001
Passer domesticus	plumage ornaments	Intrinsic, repeatable	Gonzalez et al. 2001
Passer domesticus	plumage ornaments, androgens	—	Buchanan et al. 2001
Melospiza melodia	territory tenure	Intrinsic, repeatable	Arcese 1989
Melospiza melodia	repertoire size	Intrinsic, repeatable	Heibert et al. 1989
Melospiza melodia	lay date, clutch size, condition, ARS	Intrinsic, some repeatable	Hochachka 1993
Melospiza melodia	inbreeding	Intrinsic, repeatable	Keller et al. 1994
Melospiza melodia	heritage (immigrant, resident, hybrid)	Intrinsic, repeatable	Marr et al. 2002
MAMMALS			
Microtus agrestis	body condition, shape	Extrinsic via population size	Norrdahl and Korpimaki 2002
M. rossiaemeridionalis	body condition, shape	Extrinsic via population size	Norrdahl and Korpimaki 2002
Clethrionomys glareolus	body condition, shape	Extrinsic via population size	Norrdahl and Korpimaki 2002
Peromyscus maniculatus	breeding performance	Intrinsic/extrinsic	McAdam and Millar 1999
Alces alces	horn shape asymmetry	—	Bowyer et al. 2001
Rangifer tarandus	horn size/shape asymmetry	—	Markusson and Folstad 1997
Capreolus capreolus	horn size asymmetry	Intrinsic/extrinsic, repeatable	Pelabon and van Breukelen 1998
Dama dama	social dominance	—	Pelabon and Joly 2000
Oreamnos americanus	horn length asymmetry	Intrinsic, repeatable	Côté and Festa-Bianchet (2001d)
Ourebia ourebi	horn shape asymmetry	Intrinsic, repeatable	Arcese 1994
Ovis aries	parasite load	Intrinsic, repeatable	Smith et al. 1999
Ovis canadensis	body mass	Extrinsic, repeatable	Bérubé et al. 1999

(Continues)

TABLE 16.1. Continued

TAXA	INDEX OF INDIVIDUAL QUALITY	ORIGIN OF EFFECT	SOURCE
INSECTS, FISH, REPTILES AND CRUSTACEANS			
Ageneotettix deorum			
(Acrididae)	predator avoidance	Intrinsic/extrinsic	Oedekoven and Joern 2000
Gadus morhua	parasitism, leukocyte, sperm count	—	Engen and Folstad 1999
Gasterosteus aculeatus	lateral plate assymetry	Intrinsic/extrinsic, repeatable	Reimchen and Nosil (2001)
Anolis cristatellus	pushup display, condition, endurance	—	Leal 1999
Carcinus maenas	fighting performance	—	Sneddon and Swaddle 1999
Drosophila buzzatii	wing-length asymmetry	Intrinsic, repeatable	Santos 2001

Intrinsic effects are those established in the individual phenotype via developmental or genetic effects. Extrinsic effects are those acting on individuals randomly with respect to phenotype (see text); or occasionally interacting with phenotype under particular conditions. Repeatability was either a condition of the authors, implied given the trait or timeframe examined, or uncertain (—). Taxonomic emphasis reflects primarily frequency in literature. ARS: annual reproductive success.

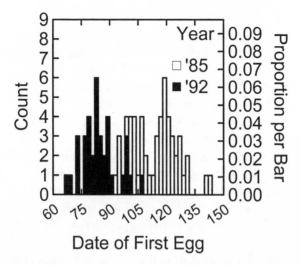

FIG. 16.3. Distribution of laying dates of first clutches by female song sparrows in a cool (1985: mean Julian date = 113, range = 93–143, *n* = 56, degree days of warming = 2719) and a warm year (1992: mean Julian date = 84, range = 66–106, *n* = 36, degree days of warming = 2074, see Methods). Identifying environmental influences on factors such as breeding date will improve models of populations wherein fitness depends on breeding date. But to apply ideas about individual heterogeneity in quality we must also know that identifiable classes of individuals lie, on average, in a similar relative position within distributions sequentially; in this case, high-quality females should appear consistently in the left tail of distributions in successive years because early breeding is related to high ARS (see text; Tables 16.2 and 16.3).

fitness such as survival or reproductive success (Sæther 1990). As a consequence, body condition is sometimes adopted as an index of individual quality (Cam and Monnat 2000; see Table 16.1). However, many fitness correlates, including body condition, also vary temporally as a function of season, environmental condition, population density, habitat quality, or food abundance (Drent and Daan 1980, Boutin 1990, Both 1998, Newton 1998). Thus the mean of condition will vary in time and space and be higher in favorable time periods or study areas as a consequence of the shared response of individual animals to environmental variation (Fig. 16.3). Some fraction of the total variance in body condition observed among individuals will be attributable solely to environmental influence. In addition, however, body condition may also vary as a consequence of maternal or developmental effects with more or less permanent influence on fitness (Norrdahl and Korpimaki 2002) or be under genetic control and reduced, for example, in inbred lines (Wright 1978). In this case, we expect individuals of higher and lower quality to lie nearer the upper and lower tails of the distribution of body condition, respectively, when observed across time and space. Overall, it will be essential to distinguish between ephemeral versus repeatable candidate indexes of individuals' quality to apply them in conservation.

For example, it is axiomatic that as traits measured instantaneously are shown to be predictive of fitness over longer intervals, they will also become more reliable indexes of fitness. Thus high body condition may predict high survival over a period of days in a small bird or mammal; but if condition is random with respect to phenotype, it is unlikely to predict fitness over longer intervals. That is, it is unlikely to be a repeatable trait that can be used to predict individual fitness or, cumulatively, population fitness in the future. Hence, to apply ideas about individual quality to conservation, we also need to ask what particular traits can be used to identify individuals that differ repeatably in quality; that is, those individuals who lie consistently in the upper and lower tails of the fitness distribution through time (see Fig. 16.3; see also Łomnicki 1978).

For some genetic traits (inbreeding, heterozygosity) repeatability will equal 1, subject to the limits of sampling error. For phenotypic traits, however, repeatability will vary depending on a variety of factors, including their heritability and developmental stability, and any interactions of genotype, environment, and life history that affect their expression in time (Stearns 1992, Lynch and Walsh 1998, Schlichting and Pigliucci 1998). If phenotype varies as a function of environment and age, for example, an index of individual quality would require description over the lifetime of individuals, hindering their application in conservation. Candidate indexes shown to be repeatable predictors of fitness over periods of 1 to 2 years may prove unreliable over longer intervals (Catry et al. 1999, Cam and Monnat 2000).

Finally, there are also theoretical reasons to suspect that many repeatable traits will be only weakly related to fitness on average and, therefore, will not affect population growth. The heritability of traits is, on average, related negatively to their influence on fitness (Gustafsson 1986, Merila and Gustafsson 1996). If few indexes of individual quality exist that also account for a substantial fraction of variation in fitness, it will be difficult to demonstrate that variation in the population mean of individual quality will affect population growth or the precision of models which incorporate such indexes.

To summarize to this point, because potential indexes of individual quality are so diverse and because there are so few data demonstrating the magnitude of their influence on population fitness, it is difficult for conservationists to apply ideas about heterogeneity in individual quality. If developing such indexes is desirable for managing species of concern, the majority of which lack longitudinal data on individual animals, detailed studies and subsequent "fishing expeditions" to identify useful indexes might be undertaken, but with high cost and uncertain result. Moreover, even with potential indexes identified, further research will be needed to justify their application by showing that variation in the population mean of individual quality influences population trend or model precision. So far, such demonstrations are rare. Finally, repeatability is required of indexes of individual quality if we are to apply them reliably. Traits predictive of fitness in one year but not others are unlikely to result from genetic effects or other fixed components of phenotype inherited in the broad sense as maternal effects or via developmental interactions with the environment. Traits with low repeatabilities will be poor predictors of performance across years, and of little value as indexes of individual quality.

Given these concerns, it seems prudent to explore them further in species wherein substantial data already exist, and to focus in particular on whether reliable indexes of individual heterogeneity in quality exist and are of sufficient magnitude to affect populations. To begin this exploration, I now describe further the specific goals of my current analysis, and my study site, methods, and statistical approach to identifying reliable indexes of individual quality in the song sparrow.

Individual Quality and Reproductive Output in Song Sparrows

My first goal was to identify traits indicative of individual quality via their correlation with annual reproductive success (ARS) using data from a 25-year study of the semi-isolated song sparrow population that resides year-round on Mandarte Island, British Columbia, Canada. I chose seasonal as opposed to lifetime reproductive success (LRS) as a dependent variable because many studies yield estimates of the former but few estimate the latter. Moreover, early work on components of variation in LRS in the song sparrow found relatively little evidence of repeatable individual variation in LRS (Smith 1988; Hochachka, Smith, and Arcese 1989) but some evidence of repeatability in components of ARS (Hochachka 1993). Finally, perhaps the best empirical analysis of individual heterogeneity in quality to date employed ARS as a dependent variable (Cam and Monnat 2000). My hope is that by also focusing on ARS I will facilitate the transferability of my results to species of concern, wherein data on LRS are typically unavailable. Although estimates of LRS may be desirable (Connor and White 1999) or even necessary (McNamara and Houston 1996) for resolving some questions about individual heterogeneity in quality, the realities of research on rare species make short-term estimates of fitness more attainable in practice.

With correlates of ARS identified, I next rank extrinsic and intrinsic effects in approximate order of influence, and then calculate the repeatability of intrinsic traits identified as being predictive of ARS. I conclude my analysis by speculating briefly on how variation in the average quality of individuals in populations might be expected to influence population growth, and how interactions between the environment and individual quality might affect stability and persistence in this population.

Song sparrows are a short-lived, socially monogamous, open-nesting passerine with several traits and enemies in common with many rare and declining species of passerines, including being a popular host of the brood-parasitic brown-headed cowbird (*Molothrus ater*) and prey to a wide range of other vertebrates (Arcese et al. 2002). The Mandarte Island study population (4–72 breeding females; median = 47) is ideal for my purposes here because about 97% of birds there are hatched locally and of known age. A pedigree of social pairings allows the estimation of inbreeding coefficients of most birds (Keller 1998). Finally, the length of the study and intermittent colonization of the island by cowbirds, an important nest predator and brood parasite (Arcese, Smith, and Hatch 1996), has led to a wide range of environmental and ecological conditions over which birds of various age and inbreeding have been studied. Although many individual effects of density, environment, cowbirds, inbreeding, and age are described elsewhere for this population (Nol and Smith 1985; Keller et al. 1994; Arcese et al. 1992; Arcese, Smith, and Hatch 1996; Keller 1998; Marr, Keller, and Arcese 2002), the relative influence of these factors on ARS has not been examined. Readers are referred to these other papers for details of methods not given here.

My main data set comprised 922 female-years of data collected from 1975 to 1999, based on the onset and fate of each nesting effort by all females. Most (56%) females that breed in this population do so for only one season (Smith 1988) and no female contributed more than seven observations (0.08%) to the overall data set. Thus there was no need to control statistically for the influence of individual females on the statistical significance of results. However, some circumstances did limit which data I included in particular analyses. For example, breeding was not monitored in 1980, adults were of unknown age when the study began, and, due to a gap in the pedigree in 1980, inbreeding coefficients were calculated only for birds hatched after 1981. For all years except 1980, I estimated the total number of independent young raised by each female and the date of first laying by most females. I knew the age of most females and their mates in most cases but pooled birds ≥ 5 years of age to maintain suitable sample sizes.

Extrinsic and Intrinsic Effects in the Song Sparrow

I followed Newton (1998) by including as extrinsic factors variation in climate, population density, and the impact of natural enemies, and extended slightly his definition of intrinsic factors as demographic rates to include phenotypic and genotypic attributes of individual birds. I recognize, however, that interactions may exist between the intrinsic traits of individuals and extrinsic factors, such that inbred, infirm, aged, or other classes of individuals may perform badly when extrinsic factors are severe, but perform about as well as high-quality individuals under benign conditions (Keller et al. 1994, Sinclair and Arcese 1995b, Chitty 1999).

Because of the potential for interaction between intrinsic and extrinsic effects, I attempted to separate their relative influence on fitness in my analyses. For example, consider the mean initiation date of breeding each year, which varies as a function of "spring warming" (cumulative degree-days of warming from January through April) and influences ARS via its effect on season length and the number of broods raised (Arcese et al. 1992, unpublished results). In warm years early bud break in plants and ecdysis in arthropods are mechanistic responses to the environment that act as extrinsic influences on birds via their effects on the availability of food and nesting cover; warm temperatures also act directly on birds via the effect of temperature on the accumulation of breeding reserves (Newton 1998). Around each annual mean, however, there also exists substantial variation in individual response to variation in the environment (see Fig. 16.3). This individual response, in the absence of sampling error, represents the intrinsic component of variation between birds in their response to variation in the environment. Yet, before we can attribute these individual responses to heterogeneity in quality, we must demonstrate further that these residuals about the mean population response are repeatable overall (see following).

Statistical Methods and Terms

I used general linear models (GLM; SYSTAT 1992) to identify extrinsic and intrinsic correlates of individual variation in ARS (the total number of young raised to independence from parental care annually by a female and her mate[s]). I chose variable combinations for entry into stepwise or forced models based on earlier work but only retained models wherein all variables were judged to be statistically significant ($p \leq 0.05$). Some statistically significant effects, however, will be simply a consequence of the number of models explored; I did not correct for this effect in the exploratory models and results presented here. I provided p-values for regression coefficients based on the *t*-statistic and for terms in analysis of variance (ANOVA) (e.g., age) based on the *F*-distribution, and all *p*-values for the models reported were ≤ 0.0001. Overall, my goal was to add terms and interactions to minimize the residual sums of squares and then rank variables by their effect on ARS.

The "intrinsic traits" considered included (1) territory size (residuals of $\log_{10}$ defended area of shrub in late April regressed on total adult breeders, see following text); (2) female age (1 to ≥ 5 yrs); (3) male age (age of female's mate at the onset of breeding; 1 to ≥ 5 yrs); (4) breeding date (residuals date of first egg annually regressed on degree-days of warming, see introduction and following text for rationale); (5) Wright's coefficient of inbreeding (*f* coefficient; after Keller 1998, Marr, Keller, and Arcese 2002); (6) fecundity as the residuals of the ANOVA of eggs laid annually versus year; and (7) efficiency as the transformed ($\text{Log}_{10} [X + 1]$) residuals of the ANOVA of the fraction of eggs laid that survived to independence versus year.

Note that fecundity and efficiency, as defined here, are each related closely to ARS and in their untransformed state should sum to equal ARS. It may be useful to think of fecundity as equaling the potential ARS of individual females in the absence of hatching failure, nest depredation, starvation, abandonment, or other factors likely to compromise the fitness of newly laid eggs. By contrast, efficiency represents the inherent viability of eggs plus the female's alacrity in raising that initial reproductive investment to when offspring are independent of direct parental investment. It should therefore be clear, for example, that females could potentially maximize ARS by laying many eggs of poor quality and caring for them crudely, by laying fewer eggs and caring for them fastidiously, or by pursuing some combination of these tactics. I also standardized fecundity and efficiency by year to estimate the contribution of individual females as opposed to years in subsequent analyses (see e.g., Hochachka 1993).

The extrinsic traits I considered included: (1) warming as the annual degree-days of warming summed from January to April (U.S. National Oceanographic and Atmospheric Administration Meteorological Station at Olga, WA); (2) intensity of nest parasitism as the total number of eggs laid by cowbirds; and (3) population density as the total number of territorial adults present in late April each year. I used the number of cowbird eggs as an index of the extrinsic effect of enemies because cowbirds are the main cause of nest failure and because nest failure is a key factor affecting ARS (Arcese, Smith, and Hatch 1996; Arcese and Smith 1999). Overall, the number of cowbird eggs laid annually was related closely to nest failure ($r = 0.90$, $n_{\text{yrs}} = 24$), but failure was low and about constant in the 7 years when cowbirds were absent (mean percentage of nests failed annually $\pm$ SD = 18.16 $\pm$ 3.98) versus when they were present (mean ± SD = 34.42 ± 13.66). Cowbirds colonized the study region in the mid-1960s.

Although I categorized variables as intrinsic or extrinsic, some are closely related. Median annual breeding date varied widely across years (Julian day 83–126) and was earlier in warmer years ($r = 0.76$, $n = 24$ yrs). However, there was also marked variation in breeding date among individuals within years (see Fig. 16.3). Thus, to reduce the correlation between individual breeding date and spring warming in statistical models, I regressed each female's annual first laying date on annual degree-days of warming, and then used the residuals or the regression as the intrinsic response of individuals to the extrinsic effect of warming (henceforth breeding date). In addition, because territory size and density were negatively related ($r = -0.58$, $n = 898$), I regressed $\log_{10}$ territory size on density to obtain for each individual a measure of the residual variation in territory size in each year. Other correlations between model variables were modest ($r \pm 0.40$).

I calculated the repeatabilities of traits following Lessels and Boag (1987) using data from 61 females that survived to at least 4 years of age. This method uses ANOVA to partition variance in the trait of interest (the dependent variable) among and within study subjects. Repeatability is defined as the intraclass correlation coefficient based on the variance components derived by one-way ANOVA (Lessels and Boag 1987:116). Note, however, that because repeatability is defined mathematically as:

$$r = (V_G + V_{Eg}) / V_P \qquad \text{(Equation 1)}$$

where V_G is the genotypic variance, V_E is the general environmental variance, and V_P is the phenotypic variance (Falconer 1981), the standardization of some traits (e.g., breeding date, territory size) for environmental variation will inflate estimates of repeatability based on these standardized variables (see also Hochachka 1993). I used standardization here to control statistically for annual variation in the population mean of the dependent variable, assuming that the remainder variance after standardization should represent primarily correlated errors due to variation among females plus random error.

Extrinsic Effects on Annual Reproductive Success

Population density, parasitism, and warming were each related negatively to ARS and accounted together for about 16% of variance in ARS across years and individuals (Table 16.2). Tested individually, density, parasitism, and warming accounted for only 11, 10, and 1% of variance in ARS, respectively. Cowbirds were absent in 7 years and laid fewer eggs and spent fewer days on

TABLE 16.2. Extrinsic effects on annual reproductive success estimated in the Mandarte Island song sparrow population over 25 years

EFFECT	COEFFICIENT	STANDARD ERROR	STANDARD COEFFICIENT
Constant	11.7590	1.3906	0.000
Warming	−0.0027	0.0006	−0.262
Parasitism	−0.2560	0.0526	−2.040
Density	−0.0188	0.0028	−0.234
Warming × Parasitism	0.0001	0.00002	1.913

Degree-days of warming, parasitism, and density, and an interaction term, accounted for about 16% of variation in annual reproductive success (ARS) ($r^2 = 0.16$, $F_{4,917} = 43.02$). Parasitism may depress ARS more in cool years because breeding commenced later, and more nests were exposed to cowbirds. Statistics include the estimated regression coefficient, its standard error, and the standardized coefficient. All coefficients had p-values < 0.001.

the island when density was low (Smith and Arcese 1994). Thus the similar explanatory power of parasitism and density in simple regression, and lack of additivity in the GLM, was due partly to a positive correlation between them.

It is also likely, however, that density and parasitism each had independent negative effects on ARS. At high density, the breeding habitat was more finely divided among birds, making reductions in available food and nest sites automatic. Parasites also reduce ARS by removing eggs from nests, in addition to causing nests to fail (Arcese and Smith 1999). Thus it is noteworthy that sparrows with supplemental food nested earlier, had more eggs, suffered less brood parasitism and nest failure, and raised about four times more young than controls (Arcese and Smith 1988). Overall, these results indicate that high density and parasitism each impinged on the reproductive potential of individual birds by making them less able to feed themselves or their young, and/or less able to defend their nests against cowbirds. Consequently, individual heterogeneity in intrinsic traits related to the acquisition, utilization, or defense of food could drive individual variation in ARS.

Intrinsic Effects on Annual Reproductive Success

I explored the effects of age, inbreeding, territory size, breeding date, fecundity, and efficiency by examining the significance of each variable added to the preceding extrinsic model (Table 16.3). Although I was able to fit each variable to improve the extrinsic model, I found that age, inbreeding, territory size, breeding date, and fecundity each had little influence as compared

TABLE 16.3. Effect of intrinsic variables on annual reproductive success measured among song sparrows

EFFECT	MODEL				
	COEFFICIENT	STANDARD ERROR	r^2	F	df
Inbreeding	−3.6720	1.2827	0.19	31.32	5,692
Breeding date	−0.0366	0.0051	0.20	40.25	5,787
Territory size	1.0130	0.2885	0.18	38.75	5,892
Eggs	0.1568	0.0214	0.20	46.86	5,915
Efficiency	13.6409	0.3714	0.64	330.44	5,912
	F	df			
Female age	11.28	4,828	0.22	95.68	8,828
Male age	3.71	4,818	0.19	107.96	8,818

Intrinsic variables were added sequentially to the extrinsic model of Table 16.1. Standardized coefficients were not comparable across models and thus not listed. Female and male age were added as categorical effects. All coefficients had p-values < 0.001 except inbreeding ($p = 0.004$) and male age ($p = 0.005$).

to efficiency (see Table 16.3). Efficiency accounted for 48% of total variation in ARS among years and females in simple regression. Efficient females were those that hatched a high fraction of eggs laid and lost few eggs or young to predators, because starvation, abandonment, and other incidental causes of nest failure and egg loss were rare (Arcese and Smith 1988, Hatch 1996). By contrast, females with high fecundity were primarily those with many re-laying attempts annually, which were associated with high rates of nest failure and the laying of replacement clutches. Females with the highest average ARS laid intermediate numbers of eggs.

Interactions among Intrinsic Effects

Various qualities of individual females are likely to contribute to their ability to raise young efficiently. Keller (1998) noted that low ARS in inbred females was related to poor hatching success as a result of infertility or arrested development. However, I found only weak predictors of efficiency in fecundity (partial $r^2 = 0.044$, $p < 0.001$), female age (partial $r^2 = 0.032$, $p < 0.001$) and inbreeding (partial $r^2 = 0.014$, $p < 0.001$), with about 9% of variation accounted for overall ($r^2 = 0.09$, $F_{6,682} = 52.46$, $p < 0.0001$).

Female age was the next best explanatory variable added to the extrinsic model (see Table 16.3). Age was shown earlier by Nol and Smith (1985) to be related to an initial increase and then plateau in ARS. Thus I expected that age might also affect other intrinsic variables related to ARS. In particular, breeding date was related weakly to age ($r^2 = 0.02$, $F_{4,718} = 4.12$, $p = 0.002$), but the fraction of variation accounted for was increased by the inclusion of inbreeding ($r^2 = 0.07$, $F_{5,597} = 29.80$, $p < 0.001$) and then territory size into the model ($r^2 = 0.09$, $F_{6,585} = 43.03$, $p < 0.001$). This final model suggested that breeding was earliest in relatively outbred 2-year-olds on larger than average territories. Although territory size was also related weakly to the age of the female's mate ($r^2 = 0.03$, $F_{4,804} = 5.67$, $p < 0.001$), other intrinsic effects were unrelated to territory size, and male age was unrelated to breeding date. Thus territory size influenced ARS mainly via its positive effect on breeding date. Finally, inbreeding was also related weakly to breeding date ($r^2 = 0.03$, $F_{1,608} = 21.83$, $p < 0.001$) and efficiency (see earlier), but not to territory size or fecundity.

Interactions among Intrinsic and Extrinsic Effects

The extrinsic effects of nest parasitism and its interaction with warming had the strongest influence on ARS (Table 16.4). Years of high parasitism were associated with poor ARS, but these effects were ameliorated in warm years because more nesting attempts were completed before cowbirds arrived. Density and warming each had less influence than efficiency or fecundity, but more influence than breeding date. These seven variables accounted for 76% of variation in ARS (see Table 16.4).

Although inbreeding was excluded from the final model, it was related negatively to efficiency and parasitism and positively to breeding date (see Table 16.4). Thus inbred females generally bred later, were less efficient, and lived in years with fewer cowbirds. Overall, however, I found no evidence of strong statistical interactions between the intrinsic traits of individuals and extrinsic effects on ARS. In particular, the effect of inbreeding did not depend strongly on other extrinsic effects modeled statistically here. Earlier analyses of survival in this population not only suggest that the magnitude of inbreeding depression was increased during periods of extreme environmental stress (Keller et al. 1994) but also acknowledge that post hoc analyses of observational data must be interpreted cautiously due to the lack of experimental control. A recent experimental study also failed to find a positive association of inbreeding depression and environmental stress in *Drosophila* (Fowler and Whitlock 2002).

TABLE 16.4. Coefficients of the best–fitting statistical model of annual reproductive success for song sparrows that bred on Mandarte Island from 1975 to 1999.

EFFECT	COEFFICIENT	STANDARD ERROR	STANDARD COEFFICIENT
Constant	10.9196	0.8103	0.0000
Parasitism	–0.2066	0.0315	–1.6556
Efficiency	14.2627	0.3442	0.7522
Fecundity	0.2311	0.0146	0.3079
Density	–0.0190	0.0017	–0.2331
Warming	–0.0023	0.0003	–0.2199
Breeding date	–0.0116	0.0031	–0.0754
Warming × Parasitism	0.00007	0.00001	1.5296

$r^2 = 0.76$, $F_{7,785} = 345.25$, $p \leq 0.0001$; all coefficients < 0.0002. Inbreeding was correlated weakly to the variables, in the order listed above: r = 0.06, –0.17, 0.01, 0.19, 0.00, –0.11.

Repeatability of Traits and Candidate Indexes of Quality

I calculated the repeatabilities of each intrinsic trait predictive of ARS in my final model, which included efficiency, fecundity, and breeding date. Of these traits, however, only fecundity and breeding date were repeatable (Table 16.5). This finding presents a predicament in my search for reliable indicators of individual quality because efficiency was the only intrinsic trait with marked influence on ARS. Fecundity and breeding date each accounted for only about 4% of variation in ARS with extrinsic effects accounted for (see Tables 16.2 and 16.3). Thus, whereas breeding date and fecundity each represent statistically defensible indexes of quality, neither has an effect size to recommend it strongly as a candidate for inclusion in a population model, or as the focus of management to improve the performance of wild or captive populations. Overall, therefore, despite "explaining" 76% of variation in ARS over 25 years (see Table 16.4), I did not identify an index of individual quality likely to have a strong influence on population growth.

The fact that efficiency was not repeatable suggests further that factors affecting egg and offspring survival acted randomly with respect to female phenotype. Indeed, it was shown elsewhere that a key factor affecting variation in reproductive output on Mandarte Island was the rate of total nest failure, which itself depends on the intensity of nest parasitism by cowbirds (Arcese, Smith, and Hatch 1996; Arcese and Smith 1999). Hochachka (1993) estimated the repeatability of nest failure at 0.01. Given that the rate of nest failure experienced by individual females was not repeatable, it is unsurprising that efficiency was also unrepeatable.

TABLE 16.5. Repeatabilities of three intrinsic traits predictive of annual reproductive success for 61 female song sparrows that survived to at least 4 years of age, 1975–1999 (see Statistical Methods and Terms).

TRAIT	REPEATABILITY	F	df	P
Efficiency	0.01	1.03	60, 183	> 0.4
Fecundity	0.26	2.38	60, 183	< 0.0001
Breeding date	0.27	2.27	60, 154	< 0.0001

My inability to find an influential index of individual quality is somewhat surprising given earlier work on this population showing marked inbreeding depression in several fitness-related traits (Keller 1998; Marr, Keller, and Arcese 2002). Keller (1998) estimated that the expected LRS of eggs from full-sib matings ($f = 0.25$) is about 45% less than for eggs of outbred matings. LRS is affected primarily by longevity in song sparrows (Smith 1988; Hochachka, Smith, and Arcese 1989). This suggests that a closer focus on LRS and longevity as dependent variables in my current analysis might have yielded more positive results if small individual differences measured annually accumulate over an individual's life span. Alternatively, because the modal life span of breeding female song sparrows on Mandarte Island is 1 year (Smith 1988), and the fraction of the population consisting of highly inbred individuals is usually small (Keller et al. 1994), the cumulative effects of individual quality on LRS and population fitness as related to inbreeding may also be small relative to the components of variance attributable to extrinsic causes.

A further explanation for my lack of success may be that small, short-lived species like song sparrows are simply more susceptible to extrinsic factors than larger, long-lived species (Stearns 1992, Newton 1998). Some of the best examples of phenotypic traits that affect ARS come from studies of long-lived species (Cam and Monnat 2000; Festa-Bianchet, Jorgenson, and Réale 2000; Forchhammer et al. 2001; Reid et al. in press). Finally, traits not examined here, such as those related to morphology (Schluter and Smith 1986), social status (Arcese and Smith 1985), or territory defense (Arcese 1987) might qualify as indexes of quality. Data for these traits are sparse and unevenly distributed by year.

Individual Quality and Song Sparrow Conservation

Key threats to rare and endangered subspecies of the song sparrow in the United States and Mexico include the classical ones of habitat loss and predator introduction (Marshall and Dedrick 1994, Arcese et al. 2002), fitting

Caughley's (1994) declining population paradigm. Once populations are brought to a small size, however, we might expect intrinsic traits of individuals related to genotype or phenotype to become more influential. Low fertility, symptomatic of inbreeding depression, has been observed in subspecies (Sogge and van Riper 1988) and inbred individuals isolated on islands (Keller 1998). Marked variation in ARS and survival by age is also noted among birds on Mandarte Island (Nol and Smith 1985, unpublished results). Including these variables in predictive population models of focal populations may improve decisions about competing management options. Formal modeling will be required to test this idea further and, specifically, to determine if the potential increase in understanding warrants the time and resources required to monitor the age, inbreeding level, and/or fertility of individuals in managed populations. Given my current results, however, it is probably true that indexes of individual quality have their best potential for application in the management of captive or wild populations where extrinsic effects have been ameliorated.

Conclusions and Recommendations

I began by asking what value might exist in monitoring or managing one or more traits predictive of individual quality. The potential reasons for doing so include predicting more accurately the response of populations to alternative management options and identifying better those animals likely to contribute most to the growth of captive and wild populations. My results suggest that reliable indexes of individual quality in the song sparrow have relatively little influence on individual fitness as compared to the extrinsic effects of weather, habitat, and enemies such as brood parasites and nest predators. Overall, this supports Caughley's (1994) suggestion that a focus on extrinsic effects will prove most practical in the majority of cases where managers intervene to conserve populations and species. Confirming these suggestions, however, requires further work to partition the components of variation in population growth among intrinsic and extrinsic effects, and to test further for potential interactions between intrinsic and extrinsic effects that might destabilize populations overall (see Fig. 16.2).

In the wider context of conservation applications it also remains possible that the small, repeatable differences in quality identified here, such as those related to fecundity, breeding date, and inbreeding (see Tables 16.3 and 16.5) could be shown to account for a larger fraction of variation in fitness and population growth in longer-lived species. In some ungulates, for example, it is clear that maternal effects and environmental conditions early in life affect survival and LRS (Festa-Bianchet, Jorgenson, and Réale 2000; Forchhammer et al. 2001), and that individual differences in fitness can influence population dynamics when cohorts vary in the mean of influential vital rates (Coulson et al. 2001). Thus further work to estimate the cumulative effect of individual variation in quality on population growth, stability, or persistence in long-lived species seems warranted, especially where long-term studies have already accumulated detailed data amenable to exploratory analysis. Indeed, the further development of ideas about individual quality to conservation is most suited to common and intensively studied species, many of which are of little specific interest in conservation, and is unsuited to rare species wherein detailed data on individual fitness are not practically collected. Thus further progress in this area probably relies on researchers that focus on intensively studied populations and have a proclivity for integrating classical approaches to evolutionary, behavioral, and population ecology to improve conservation practice.

Whether managers ever focus on indexes of individual quality depends on their demonstrated influence on populations and their ease of observation in live animals. Perhaps the ideal condition is where reliable external badges of individual quality are described and validated as influential indexes of fitness. One interesting and controversial potential index involves bilateral symmetry (Palmer 2000, Santos 2001). In artiodactyls, for example, horn length and shape symmetries have been related to harem size, social status, and annual and lifetime reproductive success (see Table 16.1). Overall, however, the strength and direction of relationships vary markedly (Kimball, Ligon, and MerolaZwartjes 1997; Dufour and Weatherhead 1998; Palmer 2000; Santos 2001; references in Table 16.1), and such indexes remain elusive despite their obvious appeal. Even when candidate indexes of individual quality are identified by their correlation to fitness (see Table 16.1), much work will be necessary to show further that the magnitude of their effect on individuals, and their pervasiveness within populations, justifies their potential application in population management. So far, convincing demonstrations that offer clear lessons for management have not appeared.

Summary

Conservation biologists and managers often try to identify factors influential of population growth, and then devise tactics to ameliorate deleterious effects to reverse population declines. Traditionally, such exercises focus on extrinsic factors, such as those related to the environment, habitat, predators, and competitors because many examples of their potential effect on populations exist (Caughley 1994, Newton 1998). However, as populations decline to small size or are brought into captivity where extrinsic influences are minimized, it has also been suggested that identifying the intrinsic traits that influence individual and population fitness should provide additional levers with which to model populations more precisely and manage them more successfully. It has also become apparent that, in a few species, the intrinsic traits of individuals interact with extrinsic factors to influence survival, reproductive success, and population growth. Recent examples include inbreeding and phenotype–environment interactions that influence survival and destabilize populations (Keller et al. 1994, 2001, Forchhammer et al. 2001, Coulson et al. 2001). These and other observations suggest that the extrinsic–intrinsic dichotomy is too simplistic. The identification of reliable indexes of individual quality may improve predictions of population response to alternate management tactics by identifying which individuals are likely to contribute more to population growth.

One approach to characterizing individual quality has been to identify phenotypic indexes of the relative fitness of individuals in populations. However, little has been done to estimate the repeatability of these indexes, the magnitude of their impact on individual fitness, or their cumulative impact on populations. Overall, the idea that easily observable indexes of individual quality exist and are sufficiently reliable as a basis for management decisions is an attractive one, but it is also fraught with practical and theoretical problems that will probably limit its application in conservation. In particular, demonstrating convincing interactions between individual variation in fitness and population trends in the wild will require very large data sets, careful analysis, and, in many cases, experimentation. Here, I described statistically factors accounting for approximately 76% of the observed variation in the annual reproductive success of female song sparrows over 25 years, but I was unable to detect a reliable index of individual quality likely to be markedly influential of population growth. It remains possible that a closer focus on survival, LRS, or species with a longer life span will provide more convincing examples of the application of individual quality to conservation. At this point, however, the application of these ideas to conserved populations remains uncertain.

Part V

Conclusion

17.

Where Do We Go from Here?

Marco Festa-Bianchet

The preceding chapters examined how animal behavior may (or may not) contribute to wildlife management and to the conservation of biodiversity. Human activities have affected the abundance and distribution of many species and drastically altered ecosystems, sometimes in radical ways through habitat destruction, other times in more subtle ways by altering species compositions or changing the sex/age structure of populations. Although much of the emphasis was on current conservation problems and on some spectacular failures to use available knowledge of animal behavior, most chapters also related some success stories or pointed to ways to reduce our future impact on biodiversity. In this concluding chapter I offer a few reflections about how the study of animal behavior may make a greater contribution to conservation in the future.

I suggested in chapter 1 that the main contribution of animal behavior to conservation likely lies in improving the management of populations. Most of the chapters support this viewpoint. Joel Berger and collaborators (chapter 9) also point out the community implications of some behaviors. Several chapters are also concerned with how artificial barriers affect connectivity between different habitat patches or seasonal ranges. In addition, mating behavior can dramatically affect the pattern of genetic diversity within a population, especially in the small populations that are typical of endangered species. It is clear that animal behavior should always be taken into consideration by wildlife managers interested in either protecting populations or exploiting them in a sustainable way. Although knowledge of population ecology and habitat requirements is very important to managing populations, in many cases behavior is the primary force responsible for changes in number, sex/age structure, or genetic diversity.

Two themes emerged repeatedly in this book: the importance of individual behavioral differences and the limited ability of animals to modify their behavior to deal with humanmade alterations to their environment. Individual differences are important because they affect how animal populations react to human developments or to conservation strategies. Conservation biologists must seek to preserve those individual differences. Individual differences in behavior are also important in predicting how animals may react to humanmade changes in their environment. For example, Steele and Hogg's discussion in chapter 15 about persistent individual differences in timing of estrus is clearly important in predicting the impact of climate change. It is not unreasonable to suspect that if winters continue to become shorter in the Northern Hemisphere, those bighorn ewes that tend to come into estrus earlier (and therefore give birth earlier) will be advantaged over those that typically conceive a few days later, given that in the study population the timing of birth has a strong effect on lamb survival (Festa-Bianchet 1988b).

We still know very little, however, about how genetic diversity and individual differences in behavior, both genetically derived and environmental, affect individual reproductive performance and ultimately population dynamics. Recent advances in population ecology show that differences in survival and reproductive parameters among sex/age classes are strong enough to affect population growth radically (Coulson et al. 2001), and it is likely that future research will also underline the importance of differences among cohorts and genotypes (Gaillard, Festa-Bianchet, and Yoccoz 2001). Clearly, a greater understanding of the role of individual differences in population dynamics would be of great utility in predicting how wildlife populations may react to human impacts.

Some chapters explored the role of learning in modifying animal behavior, and others emphasized the potential for behavior to evolve under natural or artificial selection. This knowledge could be put to good use in some cases; for example, in favoring the development of behaviors that reduce wildlife conflicts with humans. I pointed out in chapter 12 that numerous behavioral differences exist between the European and North American populations of several species, and that some of those differences may be due to the greater impact of humans on evolution of European than of North American animals (Martin and Clobert 1996).

In a few special cases, greater knowledge of animal behavior may point to management strategies to discourage or modify behaviors that increase the negative effects of humans on wildlife, such as aggressiveness in large carnivores, or crop raiding by herbivores. Increasingly, negative conditioning is used as an alternative to killing or transplants to discourage aggressive or other undesirable behaviors by bears in Canadian national parks. Knowledge of marine turtle swimming behavior was instrumental in the development of turtle excluding devices to decrease the number of turtles killed as bycatch during fishing operations (chapter 4). Similarly, studies of behavior of wildlife near major roads are useful in determining the best location and characteristics of underpasses and overpasses to reduce roadkills (Clevenger and Waltho 2000).

As remarked in many of the chapters, however, conservation will benefit much more from the modification of those human activities that lead to habitat destruction or species loss, than from seeking to modify animal behavior. For example, a short-term management solution to dog-killing behavior by cougars may be an aversive-conditioning technique to make attempted killing of domestic pets a very unpleasant experience. If successful, aversive conditioning may give the appearance that the problem has been solved. I suggest, however, that the real problem is not the killing of domestic dogs, but the suburbanization of cougar habitat in much of western North America. Preventive management is much more likely to be successful than remedial management: wildlife habitat is often not compatible with residential developments. Animals do not need wildlife managers to improve their lot, but people need wildlife managers to limit their negative impact on animal populations.

Many of the contributing authors, after discussing the science of animal behavior, pointed out how the application of scientific knowledge to conservation requires cooperation with other scientists, land managers, and elected officials. It is remarkable how many of the authors of chapters in this book either collaborate actively with conservation or management agencies, have been employed by such agencies, or are otherwise directly involved in wildlife conservation. For example, Morris Gosling led a successful program to eradicate exotic mammals from Britain; Joel Berger and John Hogg are employed by conservation nongovernmental organizations (NGOs) in the United States; Jon Swenson has worked as a wildlife manager in both the United States and Norway; and I currently chair the Committee on the Status of Endangered Wildlife in Canada. Many other chapter authors have volunteered time and expertise to assist in the solution of conservation problems. Research in animal behavior is most effectively translated into improvements in conservation when it involves a cooperative effort among academic researchers, government officials, and, often, environmental NGOs or land-use groups.

Cooperation with government agencies that control land use and wildlife management is particularly important for studies of animal behavior that seek to understand the relationships between individual variation, natural and artificial selection, and population dynamics under different management regimes. These studies require a long-term commitment because they must take environmental variability into account (Gaillard et al. 2000). Several chapters in this book underline the value of long-term studies of marked individuals to advance our knowledge of animal behavior and then to better prepare us to apply such knowledge to wildlife conservation. Typically, large-scale management experiments can only be conducted with the cooperation and support of the government agencies that control the land base. Long-term studies of specific populations require a secure long-term access to a study area, and it is often through the committed partnership of government departments that a secure access can be obtained.

A picture of a female mountain goat, her yearling daughter and male kid graces the cover of this book. The picture was taken at Caw Ridge, Alberta, where, with my students and collaborators, I've been studying mountain goats since 1988. Nanny #64 was tagged as a yearling in 1990 and as I write these words she's somewhere on Caw Ridge with her new male kid, probably being watched by a graduate student. During 14 years of research on mountain goats, we have learned a lot about their behavior. Many of our findings have practical implications for the management of this species. We have shown that goats are highly susceptible to helicopter harassment (Côté 1996), and our work has led to changes in guidelines for helicopter use over mountain goat habitat in many North American jurisdictions. Our findings about individual differences in female reproductive success (Côté and Festa-Bianchet 2001a,b,c) have been taken into account in the formulation of management plans for this species. Most of the financial support for the mountain goat research has come from the Natural Sciences and Engineering Research Council of Canada, but the study would have been impossible without the collaboration and support of the Alberta Fish and Wildlife Division. Clearly, this is an example of a very desirable partnership for wildlife conservation, and it is very similar to the arrangements existing, for example, between Jean-Michel Gaillard and the Office Nationale de la Chasse in France, Jack Hogg and the National Bison Range in the United States, Joel Berger and several American conservation and wildlife management agencies, or Jon Swenson and wildlife management agencies in Norway and Sweden.

Cooperative research efforts involving university-based researchers, government agencies, and NGOs also provide excellent opportunities for training graduate students. Long-term studies allow access to databases that can be incorporated in graduate student projects to address fundamental questions in animal behavior and conservation. Working alongside academics, government biologists, and a variety of other professionals also provides students with experience in both the fundamental and the applied aspects of research in animal behavior, and allows the development of skills that will be very useful in their future career. A conservation biologist that is able to consider the economic and political implications of alternative management strategies will be much more effective than one that knows the biology very well but has had no exposure to the realities of conservation. There is a need for conservation biologists because of human activities driven by economics and politics.

Research on animal behavior is often seen as a purely academic exercise. When right-wing politicians complain about government funding of fundamental research, they often cite titles of research in animal behavior as examples of wasted money. Yet, in this book we have demonstrated the importance of knowledge of animal behavior for wildlife conservation. Students of animal behavior have not been very effective in communicating the practical importance of their research. As pointed out in many chapters, the onus is on behavioral ecologists to communicate with other biologists and land-use managers to ensure that their research results are applied to conservation. Clearly, all conservation biologists and wildlife managers need to do a better job of communicating the economic value of conservation because so many members of the general public still see environmental protection as an alternative to economic development, not as a way to ensure greater economic opportunities (much less survival!) in the future (Lewis and Alpert 1997, Chichilnisky and Heal 1998, Armsworth and Roughgarden 2001). Cooperation with economists is essential to ensuring that a proper valuation of biodiversity is incorporated in economic and political analyses. In many cases, we are wasting biodiversity because we are unaware of its value.

Finally, I call attention to the enormously important role played by the media in our society. If we want to get our ideas across to the general public, we have to be able to do so in a way that can be easily understood by most people. Students of animal behavior are in a particularly enviable position in terms of communicating with the general public (compared to, say, students of inbreeding or extinction risk) because behavior is possibly the best characteristic of wildlife that can be used to communicate a message of conservation through the mass media—nature shows typically involve animals doing something. With this opportunity, however, comes a great responsibility: although it can be made suitable for a general audience, the message must have its basis in scientific evidence.

Literature Cited

Aanes R, Andersen R. 1996. The effects of sex, time of birth, and habitat on the vulnerability of roe deer fawns to red fox predation. Canadian Journal of Zoology 74:1857–1865.

Aars J, Ims RA. 1999. The effect of habitat corridors on rates of transfer and interbreeding between vole demes. Ecology 80:1648–1655.

Abe, EL 1996. Tusklessness amongst the Queen Elizabeth National Park elephants, Uganda. Pachyderm 22: 46–47.

Åberg J, Swenson JE, Andrén H. 2000. The dynamics of hazel grouse (*Bonasia bonasa*) occurrence in habitat fragments. Canadian Journal of Zoology 78:352–358.

Agrell J, Wolff JO, Ylönen H. 1998. Counter-strategies to infanticide in mammals: costs and consequences. Oikos 83:507–517.

Akçakaya HR, Burgman MA, Ginzburg LR. 1999. Applied population ecology: principles and computer exercises. Sunderland, MA: Sinauer Associates.

Alatalo RV, Lundberg A, Bjorklund M. 1982. Can the song of male birds attract other males? An experiment with the pied flycatcher *Ficedula hypoleuca*. Bird Behaviour 4:42–45.

Albon SD, Clutton-Brock TH, Guinness FE. 1987. Early development and population dynamics in red deer. II: Density-independent effects of cohort variation. Journal of Animal Ecology 56:69–81.

Alerstam T, Hedenstrom A. 1998. The development of bird migration theory. Journal of Avian Biology 29:343–369.

Alvarez F. 1994. Bone density and breaking stress in relation to consistent fracture position in fallow deer antlers. Acta Vertebrata 21:15–24.

Alverson WS, Waller DM. 1997. Deer populations and the widespread failure of hemlock regeneration in northern forests. In: The science of overabundance (McShea WJ, Underwood HB, Rappole JH, eds.). Washington, DC: Smithsonian Institution Press; 280–297.

Ammon E, Stacey PB. 1997. Avian nest success in a western montane riparian meadow: possible effects of past livestock grazing. Condor 99:7–13.

Andersen R, Linnell JDC. 1998. Ecological correlates of mortality in roe deer fawns in a predator-free environment. Canadian Journal of Zoology 76:1217–1225.

Anderson JL. 1986. Restoring a wilderness: The reintroduction of wildlife to an African national park. International Zoo Yearbook 24/25:192–199.

Andersson DI, Hughes D. 1996. Muller's ratchet decreases fitness of a DNA-based microbe. Proceedings of the National Academy of Sciences USA 93:906–907.

Andersson M. 1994. Sexual selection. Princeton, NJ: Princeton University Press.

Andreassen HP, Halle S, Ims RA. 1996. Optimal width of movement corridors for root voles: Not too narrow and not too wide. Journal of Applied Ecology 33:63–70.

Andrén H. 1995. Effects of landscape composition on predation rates at habitat edges. In: Mosaic landscapes and ecological processes (Hansson L, Fahrig L, Merriam G, eds.). New York, NY: Chapman and Hall; 225–255.

Anholt BR. 1997. How should we test for the role of behaviour in population dynamics? Evolutionary Ecology 11:633–640.

Anthony LL, Blumstein DT. 2000. Integrating behaviour into wildlife conservation: the multiple ways that behaviour can reduce N_e. Biological Conservation 95:303–315.

Apollonio M. 1998. Relationships between mating system, spatial behaviour, and genetic variation in ungulates, with special reference to European cervids. Acta Theriologica 5:155–162.

Apollonio M, Festa-Bianchet M, Mari F. 1989. Correlates of copulatory success in a fallow deer lek. Behavioral Ecology and Sociobiology 25:89–97.

Apollonio M, Hartl GB. 1993. Are biochemical–genetic variation and mating systems related in large mammals? Acta Theriologica 38(suppl 2):175–185.

Apollonio M, Mauri L, Bassano B. 1997. Reproductive strategies of male alpine ibex in Gran Paradiso National Park. Abstracts of the 2nd World Conference on Mountain Ungulates 1997, Saint-Vincent:12.

Arcese P. 1987. Age, intrusion pressure and defence against floaters by territorial male song sparrows. Animal Behaviour 35:773–784.

Arcese P. 1994. Horn symmetry and harem size in oribi. Animal Behaviour 48: 1485–1488.

Arcese P, Keller LF, Cary JR. 1997. Why hire a behaviorist into a conservation or management team? In: Behavioral approaches to conservation in the wild (Clemmons JR, Buchholz R, eds.). Cambridge: Cambridge University Press; 48–71.

Arcese P, Smith JNM. 1985. Phenotypic correlates and ecological consequences of dominance in the song sparrow. Journal of Animal Ecology 54:817–830.

Arcese P, Smith JNM. 1988. Effects of population density and supplemental food on reproduction in song sparrows. Journal of Animal Ecology 57:119–136.

Arcese P, Smith JNM. 1999. Impacts of nest predators and brood parasites on the production on North American passerines. In: Proceedings of the 22nd International Ornithological Congress, Durban, August 1998 (Adams N, Slotow R, eds.). Durban: University of Durban; 2953–2966.

Arcese P, Smith JNM, Hatch MI. 1996. Nest depredation by cowbirds and its effect on passerine demography. Proceedings of the National Academy of Sciences USA 93:4608–4611.

Arcese P, Smith JNM, Hochachka WM, Rogers CM, Ludwig D. 1992. Stability, regulation and the determination of abundance in an insular population of song sparrows. Ecology 73:805–822.

Arcese P, Sogge MK, Marr AB, Patton MS. 2002. Song sparrow. In: The birds of

North America (Pool A, Gill F, eds.). Philadelphia: Philadelphia Academy of Natural Sciences.

Armsworth PR, Roughgarden JE. 2001. An invitation to ecological economics. Trends in Ecology and Evolution 16:229–234.

Arnold GW, Maller RA. 1977. Effects of nutritional experience in early and adult life on the performance and dietary habits of sheep. Applied Animal Ethology 3:5–26.

Arnold KE. 2000. Group mobbing behaviour and nest defence in a cooperatively breeding Australian bird. Ethology 106:385–393.

Arnqvist G, Rowe L. 2002. Antagonistic coevolution between the sexes in a group of insects. Nature 415:787–789.

Arnqvist G, Edvardsson M, Friberg U, Nilsson T. 2000. Sexual conflict promotes speciation in insects. Proceedings of the National Academy of Sciences USA 97:10460–10464.

Atkinson CT, Woods KL, Dusek RJ, Sileo LS, Iko WM. 1995. Wildlife disease and conservation in Hawaii: pathogenicity of avian malaria (*Plasmodium relictum*) in experimentally infected Iiwi (*Vestiaria coccinea*). Parasitology 111(suppl):S59–S69.

Badyaev AV, Etges WJ, Faust JD, Martin TE. 1998. Fitness correlates of spur length and spur asymmetry in male wild turkeys. Journal of Animal Ecology 67:845–852.

Baer B, Schmid-Hempel P. 1999. Experimental variation in polyandry affects parasite loads and fitness in a bumble-bee. Nature 397:151–154.

Bagenal TB. 1978. Aspect of fish fecundity. In: Methods of assessment of ecology of freshwater fish production (Gerking SD, ed.). Oxford: Blackwell; 75–101.

Bailey DW, Gross JE, Laca EA, Rittenhouse LR, Coughenour MB, Swift DM, Sims PL. 1996. Mechanisms that result in large herbivore distribution patterns. Journal of Range Management 49:386–400.

Bailey TN. 1993. The African leopard: ecology and behavior of a solitary felid. New York: Columbia University Press.

Baker RR. 1978. The evolutionary ecology of animal migration. London: Hodder and Stoughton.

Bakker TCM, Pomiankowski A. 1995. The genetic-basis of female mate preferences. Journal of Evolutionary Biology 8:129–171.

Balazs GH. 1994. Homeward bound: satellite tracking of Hawaiian green turtles from nesting beaches to foraging pastures. In: Proceedings 13th International Symposium on Sea Turtle Biology and Conservation (Schroeder BA, Witherington BE, eds.). NOAA Technical Memo NMFS-SEFSC-341:205–208.

Balazs GH, Ellis DM. 2000. Satellite telemetry of migrant male and female green turtles breeding in the Hawaiian Islands. In: Proceedings 18th International Symposium on Sea Turtle Biology and Conservation–Suppl. (Abreu-Grobois FA, Briseño-Dueñas R, Márquez R, Sarti L, eds.). NOAA Technical Memo NMFS-SEFSC-436:3–5.

Ballard WB, Spraker TH, Taylor KP. 1991. Causes of neonatal moose calf mortality in southcentral Alaska. Journal of Wildlife Management 45:335–342.

Ballard WB, Whitman JS, Gardner CL. 1987. Ecology of an exploited wolf population in south-central Alaska. Wildlife Monographs 98:1–54.

Ballard WB, Whitlaw HA, Wakeling BF, Brown RL, deVos JC, Wallace MC. 2000. Survival of female elk in northern Arizona. Journal of Wildlife Management 64:500–504.

Balloux F, Goudet J, Perrin N. 1998. Breeding system and genetic variance in the monogamous shrew, *Crocidura russula*. Evolution 52:1230–1235.

Balmford A. 1996. Extinction filters and current resilience: the significance of past selection pressures for conservation biology. Trends in Ecology and Evolution 11:193–196.

Bannerot SP, Fox WW, Powers JE. 1987. Reproductive strategies and the management of snappers and groupers in the Gulf of Mexico and Caribbean. In: Tropical snappers and groupers: biology and fisheries management (Polovina JJ, Ralston D, eds.). Boulder, CO: Westview Press; 561–603.

Barbraud C, Lormee H, LeNeve A. 2000. Body size and determinants of laying date variation in the snow petrel *Pagodroma nivea*. Journal of Avian Biology 31:295–302.

Barraclough TG, Harvey PH, Nee S. 1995. Sexual selection and taxonomic diversity in passerine birds. Proceedings of the Royal Society B 259:211–215.

Bart J. 1995. Acceptance criteria for using individual-based models to make management decisions. Ecological Applications 5:411–420.

Bartmann RM, White GC, Carpenter LH. 1992. Compensatory mortality in a Colorado mule deer population. Wildlife Monographs 121:1–39.

Bartos L, Madlafousek J. 1994. Infanticide in a seasonal breeder: the case of red deer. Animal Behaviour 47:217–219.

Basset P, Balloux F, Perrin N. 2001. Testing demographic models of effective population size. Proceedings of the Royal Society of London, Series B 268:311–317.

Basson PA, Norval AG, Hofmeyr JM, Ebedes H, Schultz RA. 1982. Antelope and poisonous plants. I: Gifblaar containing monofluoroacetate. Madoqua 13:59–70.

Baur B, Baur A. 1997. Random mating with respect to relatedness in the simultaneously hermaphroditic land snail *Arianta arbustorum*. Invertebrate Biology 116:294–298.

Beavers SC, Cassano ER. 1996. Movements and dive behavior of a male sea turtle (*Lepidochelys olivacea*) in the eastern tropical Pacific. Journal of Herpetology 30:97–104.

Beck MW. 1996. On discerning the cause of late Pleistocene megafaunal extinctions. Paleobiology 22:91–103.

Becker PH, Wendeln H, Gonzalez-Solis J. 2001. Population dynamics, recruitment, individual quality and reproductive strategies in common terns *Sterna hirundo* marked with transponders. Ardea 89:241–252.

Beissinger SR. 1997. Integrating behavior into conservation biology: potentials and limitations. In: Behavioral approaches to conservation in the wild (Clemmons J, Buchholz R, ed.). Cambridge: Cambridge University Press; 23–47.

Beissinger SR, Westphal MI. 1998. On the use of demographic models of population viability in endangered species management. Journal of Wildlife Management 62:821–841.

Bekoff M. 1977. Coyotes. New York: Academic Press.

Bélisle M, Desrochers A. 2002. Gap-crossing decisions by forest birds: an empirical basis for parameterizing spatially-explicit, individual-based models. Landscape Ecology 17:219–231.

Bélisle M, Desrochers A, Fortin M-J. 2001. Influence of forest cover on the movements of forest birds: a homing experiment. Ecology 82:1893–1904. ·

Bellemain E. 2001. Etude du système d'appariement de l'ours de Scandinavie. Thesis, Université Claude Bernard, Villeurbanne, France.

Ben-David MR, Bowyer T, Duffy LK, Roby DD, Scheel DM. 1998. Social behavior and ecosystem processes: river otter latrines and nutrient dynamics of terrestrial vegetation. Ecology 79:2567–2571.

Benton TG, Grant A, Clutton-Brock TH. 1995. Does environmental stochasticity matter? Analysis of red deer life-histories on Rhum. Evolutionary Ecology 9:559–574.

Berger J. 1998. Future prey: some consequences of the loss and restoration of mammalian carnivores. In: Behavioral ecology and conservation biology (Caro TM, ed.). New York: Oxford University Press; 80–100.

Berger J. 1999. Anthropogenic extinction of top carnivores and interspecific animal behaviour: implications of the rapid decoupling of a web involving wolves, bears, moose, and ravens. Proceedings of the Royal Society of London, Series B 266:2261–2267.

Berger J, Cunningham C. 1994. Bison: mating and conservation in small populations. New York: Columbia University Press.

Berger J, Gompper ME. 1999. Sex ratios in extant ungulates: products of contemporary predation or past life histories? Journal of Mammalogy 80:1084–1113.

Berger J, Swenson JE, Persson I-L. 2001. Recolonizing carnivores and naive prey: conservation lessons from Pleistocene extinctions. Science 291:1036–1039.

Berger J, Testa JW, Roffe T, Montfort SL. 1999. Conservation endocrinology: a noninvasive tool to understand relationships between carnivore colonization and ecological carrying capacity. Conservation Biology 13:980–989.

Berger J, Stacey PB, Bellis L, Johnson L. 2001. A mammalian predator–prey disequilibrium: how the extinction of grizzly bears and wolves affects the diversity of neotropical avian migrants. Ecological Applications 11:947–960.

Berry HH. 1997. Historical review of the Etosha Region and its subsequent administration as a National Park. Madoqua 20:3–12.

Berthold P. 1993. Bird migration. Oxford: Oxford University Press.

Bérubé C, Festa-Bianchet M, Jorgenson JT. 1999. Individual differences, longevity, and reproductive senescence in bighorn ewes. Ecology 80:2555–2565.

Bevers M, Hof J, Uresk DW, Schenbeck GL. 1997. Spatial optimization of prairie dog colonies for black-footed ferret recovery. Operations Research 45:495–507.

Biederbeck HH, Boulay MC, Jackson DH. 2001. Effects of hunting regulations on bull elk survival and age structure. Wildlife Society Bulletin 29:1271–1277.

Billard R. 1985. Artificial insemination of salmonids. In: Salmonid reproduction: an international symposium (Iwamoto RN, Sower S, eds.). Seattle: Washington Sea Grant Program; 116–128.

Birkhead TR. 1991. The magpies. London: T and AD Poyser.

Birkhead T. 2000. Promiscuity. London: Faber and Faber.

Bjärvall A, Sandegren F. 1987. Early experiences with the first radio-marked brown bears in Sweden. International Conference on Bear Research and Management 7:9–12.

Blanco G, De la Puente J. 2002. Multiple elements of the black-billed magpie's tail correlate with variable honest information on quality in different age/sex classes. Animal Behaviour 63:217–225.

Blumstein JC, Daniel AS, Griffin S, Evans SC. 2000. Insular tammar wallabies (*Macropus eugenii*) respond to visual but not acoustic cues from predators. Behavioral Ecology 11:528–535.

Boertje RD, Valkenburg P, McNay M. 1996. Increases in moose, caribou, and wolves following wolf control. Journal of Wildlife Management 60:474–489.

Boitani L. 2000. Action plan for the conservation of wolves (*Canis lupus*) in Europe. http://www.nature.coe.int/cp20/tpvs23e.htm.

Boitani L. 2001. Carnivore introductions and invasions: their success and management options. In: Carnivore conservation (Gittleman JL, Funk SM, MacDonald D, Wayne RK). Cambridge: Cambridge University Press; 123–144.

Bolten AB, Bjorndal KA, Martins HR, Dellinger T, Biscoito MJ, Encalada SE, Bowen BW. 1998. Transatlantic developmental migrations of loggerhead sea turtles demonstrated by mtDNA sequence analysis. Ecological Applications 8:1–7.

Boorman L, Goss-Custard JD, McGrorty S. 1989. Climatic change, rising sea level and the British coast. ITE Research Publication 1. London: HMSO.

Borries C, Launhardt K, Epplen C, Epplen JT, Winkler P. 1999. DNA analyses support the hypothesis that infanticide is adaptive in langur monkeys. Proceedings of the Royal Society of London, Series B 226:901–904.

Both C. 1998. Density dependence of reproduction: from individual optimisation to population dynamics. Ph.D. diss., University of Utrecht.

Boulinier T, Danchin E. 1997. The use of conspecific reproductive success for breeding patch selection in terrestrial migratory species. Evolutionary Ecology 11:505–517.

Bouteiller C, Perrin N. 2000. Individual reproductive success and effective population size in the greater white-toothed shrew, *Crocidura russula.* Proceedings of the Royal Society of London, Series B 267:701–705.

Boutin S. 1990. Food supplementation experiments with terrestrial vertebrates: patterns, problems and the future. Canadian Journal of Zoology 68:203–220.

Boutin S. 1992. Predation and moose population dynamics; a critique. Journal of Wildlife Management 56:116–127.

Boving PS, Post E. 1997. Vigilance and foraging behaviour of female caribou in relation to predation risk. Rangifer 17:55–63.

Bowden DC, White GC, Bartmann RM. 2000. Optimal allocation of sampling effort for monitoring a harvested mule deer population. Journal of Wildlife Management 64:1013–1024.

Bowen BW, Karl SA. 1997. Population genetics, phylogeography, and molecular evolution. In: The biology of sea turtles (Lutz PE, Musick JA, eds.). Boca Raton, FL: CRC Press, 29–50.

Bowen BW, Abreu-Grobois FA, Balazs GH, Kamezaki N, Limpus CJ, Ferl RJ. 1995. Trans-Pacific migrations of the loggerhead turtle demonstrated with mithocondrial DNA markers. Proceedings of the National Academy of Sciences USA 92:3731–3734.

Bowyer RT, Van Vallenberghe V, Kie JG, Maier JAK. 1999a. Birth-site selection by Alaskan moose: maternal strategies for coping with a risky environment. Journal of Mammalogy 80:1070–1083.

Bowyer RT, Nicholson MC, Molvar E, Faro JB. 1999b. Moose on Kalgin Island: are density-dependent processes related to harvest? Alces 35:73–89.

Bowyer RT, Stewart KM, Kie JG, Gasaway WC. 2001. Fluctuating asymmetry in antlers of Alaskan moose: size matters. Journal of Mammology 82:814–824.

Boyce MS. 1992. Population viability analysis. Annual Review of Ecology and Systematics 23:481–506.

Boyce MS. 1999. Ecological-process management and ungulates: Yellowstone's conservation paradigm. Wildlife Society Bulletin 26:391–398.

Boyce MS, Sinclair ARE, White GC. 1999. Seasonal compensation of predation and harvesting. Oikos 87:419–426.

Boyd DK, Ream RR, Pletscher DH, Fairchild MW. 1994. Prey taken by colonizing wolves and hunters in the Glacier National Park areas. Journal of Wildlife Management 58:289–295.

Bradbury RB, Payne RJH, Wilson JD, Krebs JR. 2001. Predicting population responses to resource management. Trends in Evolution and Ecology 16:440–445.

Bradley JS, Wooller RD, Skira IJ. 2000. Intermittent breeding in the short-tailed shearwater *Puffinus tenuirostris.* Journal of Animal Ecology 69:639–650.

Brashares JS, Arcese P, Sam MK. 2001. Human demography and reserve size predict wildlife extinction in West Africa. Proceedings of the Royal Society of London, Series B 268:2473–2478.

Brault S, Caswell H. 1993. Pod-specific demography of killer whales (*Orcinus orca*). Ecology 74:1444–1454.

Breitenmoser U. 1998. Large predators in the Alps: the fall and rise of man's competitors. Biological Conservation 83:279–289.

Breitenmoser U, Haller UH. 1993. Patterns of predation by reintroduced European lynx in the Swiss Alps. Journal of Wildlife Management 57:135–144.

Breslow NE, Clayton DG. 1993. Approximate inference in generalized linear mixed models. Journal of the American Statistical Association 88:9–22.

British Columbia Ministry of Environment, Lands and Parks. 1995. Conservation of grizzly bears in British Columbia, background report. Victoria.

Brooks PM, Macdonald IAW. 1983. The Hluhluwe-Umfolozi Reserve: an ecological case history. In: Management of large mammals in African conservation areas (Owen-Smith RN, ed.). Pretoria: Haum; 51–77.

Brotons L, Desrochers A, Turcotte Y. 2001. Food hoarding behaviour of black-capped chickadees (*Poecile atricapilla*) in relation to forest edges. Oikos, 95:511–519.

Brown DE. 1985. The grizzly in the Southwest. Norman: University of Oklahoma Press.

Brown JH, Kodric-Brown A. 1977. Turnover rates in insular biogeography: effect of immigration on extinction. Ecology 58:445–449.

Brown JL. 1997. A theory of mate choice based on heterozygosity. Behavioral Ecology 8:60–65.

Brown JL, Eklund A. 1994. Kin recognition and the major histocompatibility complex: an integrative review. American Naturalist 143:435–461.

Bruton MN. 1995. Have fishes had their chips? The dilemma of threatened fishes. Environmental Biology of Fishes 43:1–27.

Bryant EH, Reed DH. 1999. Fitness decline under relaxed selection in captive populations. Conservation Biology 13:665–669.

Bryant EH, Vackus VL, Clark ME, Reed DH. 1999. Experimental tests of captive breeding for endangered species. Conservation Biology 13:1487–1496.

Bryk AS, Raudenbush SW. 1992. Hierarchical linear models. London: Sage.

Buchanan KL, Evans MR, Goldsmith AR, Bryant DM, Rowe LV. 2001. Testosterone influences basal metabolic rate in male house sparrows: a new cost of dominance signalling? Proceedings of the Royal Society of London, Series B 268:1337–1344.

Buckland ST, Goudie IBJ, Borchers DL. 2000. Wildlife population assessment: past developments and future directions. Biometrics 56:1–12.

Burke DM, Nol E. 1998. Influence of food abundance, nest-site habitat, and forest fragmentation on breeding ovenbirds. Auk 115:96–104.

Burley N. 1982. Reputed band attractiveness and sex manipulation in zebra finches. Science 215:423–424.

Burley N. 1986. Sexual selection for aesthetic traits in species with biparental care. American Naturalist 127:415–445.

Byers JA. 1997. American pronghorn: social adaptations and ghosts of predators past. Chicago: University of Chicago Press.

Byers JA, Hogg JT. 1995. Environmental effects on prenatal growth rate in pronghorn and bighorn: further evidence for energy constraint on sex-biased maternal expenditure. Behavioral Ecology 6:451–457.

Byers JA, Kitchen DW. 1988. Mating system shift in a pronghorn population. Behavioural Ecology and Sociobiology 22:355–360.

Cagnolaro L, Rosso D, Spagnesi M, Venturi B. 1974. Investigation on the wolf (*Canis lupus*) distribution in Italy and in Canton Ticino and Canton Grigioni (Switzerland). Ricerche Biologia della Selvaggina 59:1–75. [Italian]

Cam E, Monnat JY. 2000. Stratification based on reproductive state reveals contrasting patterns of age-related variation in demographic parameters in the kittiwake. Oikos 90:560–574.

Campbell BM, Sithole B, Frost P. 2000. CAMPFIRE experiences in Zimbabwe. Science 287:42–43.

Caro TM. 1994. Cheetahs of the Serengeti Plains. Chicago: University of Chicago Press.

Caro TM. 1998a. How do we refocus behavioral ecology to address conservation issues more directly? In: Behavioral ecology and conservation biology (Caro TM). New York: Oxford University Press; 557–565.

Caro TM (ed.). 1998b. Behavioral ecology and conservation biology. New York: Oxford University Press.

Caro TM. 1999. The behaviour–conservation interface. Trends in Ecology and Evolution 14:366–369.

Caro TM, Durant SM. 1995. The importance of behavioral ecology for conservation biology: examples from Serengeti carnivores. In: Serengeti II: Dynamics, Management and Conservation of an Ecosystem (Sinclair ARE, Arcese P, eds.). Chicago: University of Chicago Press; 451–472.

Caro TM, Laurenson MK. 1994. Ecological and genetic factors in conservation: a cautionary tale. Science 263:485–486.

Carr A. 1984. The sea turtle: so excellent a fishe. 2nd ed. Austin: University of Texas Press.

Carr A. 1987. New perspectives on the pelagic stage of sea turtle development. Conservation Biology 1:103–121.

Carr A, Carr MH, Meylan AB. 1978. The ecology and migrations of sea turtles. VII: The West Caribbean green turtle colonies. Bulletin of the American Museum of Natural History 162:1–46.

Carter PE, Rypstra AL. 1995. Top-down effects in soybean agroecosystems: spider density affects herbivore damage. Oikos 72:433–439.

Castelli G. 1935. The brown bear in Venetian Region. Trento: Hunting Association of Trento Province; pp. 171. [Italian]

Castle MD, Christensen BM. 1990. Hematozoa of wild turkeys from the midwestern United States: translocation of wild turkeys and its potential role in the introduction of *Plasmodium kempi*. Journal of Wildlife Diseases 26:180–185.

Caswell H. 1978. A general formula for the sensitivity of population growth rate to changes in life history parameters. Journal of Theoretical Biology 14:215–230.

Caswell H. 2000. Prospective and retrospective perturbation analyses: their roles in conservation biology. Ecology 81:619–627.

Caswell H. 2001. Matrix population models: construction, analysis and interpretation. Sunderland, MA: Sinauer Associates.

Catchpole CK. 1986. Song repertoires and reproductive success in the great reed warbler (*Acrocephalus arundinaceus*). Behavioral Ecology and Sociobiology 19:439–445.

Catry P, Ruxton GD, Ratcliffe N, Hamer KC, Furness RW. 1999. Short-lived repeatabilities in long-lived great skuas: implications for the study of individual quality. Oikos 84:473–479.

Caughley G. 1977. Analysis of vertebrate populations. Chichester: Wiley and Sons.

Caughley G. 1994. Directions in conservation biology. Journal of Animal Ecology 63:215–244.

Caughley G, Gunn A. 1996. Conservation biology in theory and practice. Cambridge, MA: Blackwell Science.

Caughley G, Sinclair ARE. 1994. Wildlife ecology and management. Boston: Blackwell Science Publications.

Charlesworth B. 1994. Evolution in age-structured populations. Cambridge: Cambridge University Press.

Cheeseman CL, Cresswell WJ, Harris S, Mallinson PJ. 1988. Comparison of dispersal and other movements in two badger (*Meles meles*) populations. Mammal Review 18:51–59.

Cheney DL, Seyfarth RM. 1990. How monkeys see the world. Chicago: University of Chicago Press.

Cheng IJ. 2000. Post-nesting migrations of green turtles (*Chelonia mydas*) at Wan-An Island, Penghu Archipelago, Taiwan. Marine Biology 137:747–754.

Chesser RK. 1983. Genetic variability within and among populations of the black-tailed prairie dog. Evolution 37:320–331.

Chesser RK. 1991a. Gene diversity and female philopatry. Genetics 127:437–447.

Chesser RK. 1991b. Influence of gene flow and breeding tactics on gene diversity within populations. Genetics 129:573–583.

Chesser RK. 1998a. Heteroplasmy and organelle gene dynamics. Genetics 150:1309–1327.

Chesser RK. 1998b. Relativity of behavioral interactions in socially structured populations. Journal of Mammalogy 79:713–724.

Chesser RK, Baker RJ. 1996. Effective sizes and dynamics of uniparentally and diparentally inherited genes. Genetics 144:1225–1235.

Chesser RK, Rhodes OE, Smith MH. 1996. Conservation. In: Population processes in ecological space and time (Rhodes OE, Chesser RK, Smith MH, eds.). Chicago: University of Chicago Press; 237–252.

Chesser RK, Willis KB, Mathews NE. 1994. Impacts of toxicants on population dynamics and gene diversity in avian species. In: Wildlife toxicology and population modeling (Kendall RJ, Lacher TE, Jr, eds.). Boca Raton, FL: CRC Press; 171–188.

Chesser RK, Rhodes OE, Sugg DW, Schnabel A. 1993. Effective sizes for subdivided populations. Genetics 135:1221–1232.

Chichilnisky G, Heal J. 1998. Economic returns from the biosphere. Nature 391:629–630.

Chitty D. 1967. The natural selection of self-regulatory behavior in animal populations. Proceedings Ecological Society of Australia 2:51–78.

Chitty D. 1999. Do lemmings commit suicide? Oxford: Oxford University Press.

Choquenot D, Bowman D. 1998. Marsupial megafauna, Aborigines and the overkill hypothesis: application of predator–prey models to the question of Pleistocene extinction in Australia. Global Ecology and Biogeography Letters 7:167–180.

Christians JK, Evanson M, Aiken JJ. 2001. Seasonal decline in clutch size in European starlings: a novel randomization test to distinguish between the timing and quality hypotheses. Journal of Animal Ecology 70:1080–1087.

Claereboudt C. 1999. Fertilization success in spatially distributed populations of benthic free-spawners: a simulation model. Ecological Modelling 121: 221–233.

Clark JD, Smith KG. 1994. A demographic comparison of two black bear populations in the interior highlands of Arkansas. Wildlife Society Bulletin 22: 593–603.

Clarke FM, Faulkes CG. 1999. Kin discrimination and female mate choice in the naked mole-rat *Heterocephalus glaber*. Proceedings of the Royal Society of London, Series B 266:1995–2002.

Clemmons JR, Buchholz R. 1997. Behavioral approaches to conservation in the wild. Cambridge: Cambridge University Press.

Clevenger AP, Waltho N. 2000. Factors influencing the effectiveness of wildlife underpasses in Banff National Park, Alberta, Canada. Conservation Biology 14:47–56.

Clutton-Brock TH. 1988. Reproductive success. Chicago: University of Chicago Press.

Clutton-Brock TH. 1989. Mammalian mating systems. Proceedings of the Royal Society of London, Series B 236:339–372.

Clutton-Brock TH, Albon SD. 1989. Red deer in the highlands. Oxford: BSP Professional Books.

Clutton-Brock TH, Albon SD, Guinness FE. 1988. Reproductive success in male and female red deer. In: Reproductive success (Clutton-Brock TH, ed.). Chicago: University of Chicago Press; 325–343.

Clutton-Brock TH, Guinness FE, Albon SD. 1982. Red deer: behavior and ecology of two sexes. Chicago: University of Chicago Press.

Clutton-Brock TH, Lonergan ME. 1994. Culling regimes and sex ratio biases in Highland red deer. Journal of Applied Ecology 31:521–527.

Clutton-Brock TH, Rose KE, Guinness FE. 1997. Density-related changes in sexual selection in red deer. Proceedings of the Royal Society of London, Series B 264:1509–1516.

Cobley ND, Croxall, JP, Prince PA. 1998. Individual quality and reproductive performance in the grey-headed albatross *Diomedea chrysostoma.* Ibis 140: 315–322.

Cockerham CC. 1967. Group inbreeding and coancestry. Genetics 56:89–104.

Cockerham CC. 1969. Variance of gene frequencies. Evolution 23:72–84.

Cockerham CC. 1973. Analysis of gene frequencies. Genetics 74:679–700.

Coleman FC, Koenig CC, Collins LA. 1996. Reproductive styles of shallow-water groupers (*Pisces:* Serranidae) in the eastern Gulf of Mexico and the consequences of fishing spawning aggregations. Environmental Biology of Fishes 47:129–141.

Coltman DW, Smith JA, Bancroft DR, Pilkington J, MacColl ADC, Clutton-Brock TH, Pemberton JM. 1999a. Density-dependent variation in lifetime breeding success and natural and sexual selection in Soay rams. American Naturalist 154:730–746.

Coltman DW, Pilkington JG, Smith JA, Pemberton JM. 1999b. Parasite-mediated selection against inbred Soay sheep in a free-living, island population. Evolution 53:1259–1267.

Coltman DW, Festa-Bianchet M, Jorgenson JT, Strobeck C. 2002. Age-dependent sexual selection in bighorn rams. Proceedings of the Royal Society of London, Series B 269:165–172.

Conner MM, White GC. 1999. Effects of individual heterogeneity in estimating the persistence of small populations. Natural Resource Modeling 12:109–120.

Connor KJ, Ballard WB, Dilworth T, Anions D. 2000. Changes in structure of a boreal forest community following intense herbivory by moose. Alces 36:111–132.

Cooper SM, Owen-Smith N. 1986. Effects of plant spinescence on large mammalian herbivores. Oecologia 68:446–455.

Cooper SM, Owen-Smith N, Bryant JP. 1988. Foliage acceptability to browsing ruminants in relation to seasonal changes in the leaf chemistry of woody plants in a South Africa savannah. Oecologia 75:336–342.

Cope ED. 1896. The Primary Factors of Organic Evolution. Chicago: Open Court.

Coss RG. 1991. Evolutionary persistence of ground squirrel anti-snake behavior: reflections on Burto's commentary. Ecological Psychology 5:171–194.

Côté IM, Vinyoles D, Reynolds JD, Doadrio I, Perdices A. 1999. Potential impacts of gravel extraction on Spanish populations of river blennies (*Salaria fluviatilis*). Biological Conservation 87:359–367.

Côté SD. 1996. Mountain goat responses to helicopter disturbance. Wildlife Society Bulletin 24:681–685.

Côté SD, Festa-Bianchet M. 2001a. Birthdate, mass and survival in mountain goat kids: effects of maternal characteristics and forage quality. Oecologia 127:230–238.

Côté SD, Festa-Bianchet M. 2001b. Reproductive success in female mountain goats: the influence of maternal age and social rank. Animal Behaviour 62:173–181.

Côté SD, Festa-Bianchet. M. 2001c. Offspring sex ratio in relation to maternal age and social rank in mountain goats. Behavioral Ecology and Sociobiology 49:260–265.

Côté SD, Festa-Bianchet M. 2001d. Life-history correlates of horn asymmetry in mountain goats. Journal of Mammology 82:389–400.

Côté SD, Festa-Bianchet M, Smith KG. 1998. Horn growth in mountain goats (*Oreamnos americanus*). Journal of Mammalogy 79:406–414.

Coulson T, Catchpole EA, Albon SD, Morgan BJT, Pemberton JM, Clutton-Brock TH, Crawley MJ, Grenfell BT. 2001. Age, sex, density, winter weather and population crashes in Soay sheep. Science 292:1528–1531.

Courchamp F, Clutton-Brock T, Grenfell B. 1999. Inverse density dependence and the Allee effect. Trends in Ecology and Evolution 14:405–410.

Couturier MJ. 1954. L'ours brun (*Ursus arctos* L.). Grenoble: Author.

Couturier MJ. 1962. Le bouquetin des Alpes. Grenoble: Arthaud Ed.

Cowlishaw, G. 1999. Predicting the pattern of decline of African primate diversity: an extinction debt from historical deforestation. Conservation Biology 13:1183–1193.

Crabtree RL, Sheldon JW. 1999. Coyotes and canid coexistence in Yellowstone. In: Carnivores in ecosystems; the Yellowstone experience (Clark TW, Curlee AP, Minta SC, Kareiva PM, eds.). New Haven: Yale University Press; 127–164.

Craighead FC. 1979. Track of the grizzly. San Francisco: Sierra Club Books.

Craighead JJ, Hornocker MG, Craighead FC. 1969. Reproductive biology of young female grizzly bears. Journal of Reproductive Fertility Supplement 6:447–475.

Craighead JJ, Sumner JS, Mitchell JA. 1995. The grizzly bears of Yellowstone: their ecology in the Yellowstone ecosystem 1959–1992. Washington, DC: Island Press.

Craighead L, Paetkau D, Reynolds HV, Vyse ER, Strobeck C. 1995. Microsatellite analysis of paternity and reproduction in arctic grizzly bears. Journal of Heredity 86:225–261.

Crim LW, Glebe BD. 1990. Reproduction. In: Methods for fish biology (Schreck CB, Moyle PB, eds.). Bethesda, MD: American Fisheries Society; 529–553.

Crockett CM Sekulic R. 1984. Infanticide in red howler monkeys (*Alouatta seniculus*). In: Infanticide: comparative and evolutionary perspectives (Hausfater G, Hrdy SB, eds.). New York: Aldine de Gruyter; 173–191.

Crodgington HS, Siva-Jothy MT. 2000. Genital damage, kicking and early death: the battle of the sexes takes a sinister turn in the bean weevil. Nature 407:855–856.

Crooks KR, Soulé ME. 1999. Mesopredator release and avifaunal extinctions in a fragmented system. Nature 400:563–566.

Crouse D. 1999. The WTO shrimp/turtle case. Marine Turtle Newsletter 83:1–3.

Csermely D. 1996. Anti-predator behavior in lemurs; evidence of an extinct eagle on Madagascar or something else? International Journal of Primatology 17:349–354.

Cunningham EJA, Russell AF. 2000. Egg investment is influenced by male attractiveness in the mallard. Nature 404:74–77.

Dagg AI. 1999. Infanticide by male lions hypothesis: a fallacy influencing research into human behavior. American Anthropologist 100:940–950.

Dahier T, Lequette B. 1997. Le loup *Canis lupus* dans le Massif du Mercantour (France): gestion des dommages occasionnés aux Ongulés domestiques. Bulletin de la Société Neuchateloise des Sciences Naturelles 120(2):19–26.

Dahle B, Swenson JE. In press. Home ranges in adult Scandinavian brown bears *Ursus arctos*: effect of population density, mass, sex, reproductive status and habitat type. Journal of Zoology.

Daldoss G. 1981. Sulle orme dell'orso. Trento: Edizioni Temi; pp. 250.

Davis-Born R, Wolff JO. 2000. Age- and sex-specific responses of the gray-tailed vole, *Microtus canicaudus*, to connected and unconnected habitat patches. Canadian Journal of Zoology 78:864–870.

DeGraaf RM. 1995. Nest predation rates in managed and reserved extensive northern hardwood forests. Forest Ecology and Management 79:227–234.

Dehn MM. 1990. Vigilance for predators: detection and dilution effects. Behavioral Ecology and Sociobiology 26:337–342.

de Kroon H, Plaisier A, Groenendael Jv, Caswell H. 1986. Elasticity: the relative contribution of demographic parameters to population growth rate. Ecology 67:1427–2431.

Dellinger T, Freitas C. 2000. Movements and diving behaviour of pelagic stage loggerhead sea turtles in the North Atlantic: preliminary results obtained through satellite telemetry. Proceedings 19th International Symposium on Sea Turtle Biology and Conservation (Kalb HJ, Wibbels T, eds.). NOAA Technical Memo NMFS-SEFSC-443:155–157.

Delope F, Møller AP. 1993. Female reproductive effort depends on the degree of ornamentation of their mates. Evolution 47:1152–1160.

Delorme D. 1989. L'effet observateur: une source de biais lors de l'application de l'indice kilométrique d'abondance (IKA) pour le dénombrement de chevreuils (*Capreolus capreolus*). Gibier Faune Sauvage 6:309–314.

Derocher AE, Stirling I. 1998. Maternal investment and factors affecting offspring size in polar bears (*Ursus maritimus*). Journal of Zoology 245:253–260.

Derocher AE, Taylor M. 1994. Density-dependent population regulation of polar bears. In: Density-dependent population regulation of brown, black, and polar bears (Taylor M, ed.). International Conference on Bear Research and Management 3:25–37.

Desrochers A, Fortin M-J. 2000. Understanding avian responses to forest boundaries: a case study with chickadee winter flocks. Oikos 91:376–384.

Desrochers A, Hannon SJ. 1997. Gap crossing decisions by dispersing forest songbirds during the post-fledging period. Conservation Biology 11: 1204–1210.

Desrochers A, Magrath RD. 1993. Environmental predictability and remating in European blackbirds. Behavioral Ecology 4:271–275.

Desrochers A, Hannon SJ, Bélisle M, St. Clair CC. 1999. Movement of songbirds in fragmented forests: Can we "scale up" from behaviour to explain occupancy patterns in the landscape? Proceedings of the 22nd International Ornithological Congress, Durban (Adams NJ, Slotow RH, eds.). Johannesburg: BirdLife South Africa; 2447–2464.

Diggle PJ, Liang K-Y, Zeger SL. 1994. The analysis of longitudinal data. Oxford: Oxford Science Publications, Clarendon Press.

Dingle H. 1996. Migration. New York: Oxford University Press.

Direction Regional de l'Environnement Midi-Pyrénées. Reintroduction of brown bear (*Ursus arctos*) in central Pyrénées: preliminary results (18/05 to 06/12/1996). Convegno Life per la salvaguardia dell'orso bruno, Tarvisio, dicembre 1996. Tarvisio.

Dobkin DS, Rich AC, Pyle WH. 1998. Habitat and avifaunal recovery from livestock grazing in a riparian meadow system of the northwestern Great Basin. Conservation Biology 12:209–221.

Dobson AJ. 1989. Introduction to statistical modelling. London: Chapman and Hall.

Dobson AP, Lyles AM. 1989. The population dynamics and conservation of primate populations. Conservation Biology 3:362–380.

Dobson AP, Lyles AM. 2000. Black-footed ferret recovery. Science 288:985–988.

Dobson AP, Poole JH. 1998. Conspecific aggregation and conservation biology. In: Behavioral ecology and conservation biology (Caro TM). New York: Oxford University Press; 193–208.

Dobson FS. 1982. Competition for mates and predominant juvenile male dispersal in mammals. Animal Behaviour 30:1183–1192.

Dobson FS. 1998. Social structure and gene dynamics in mammals. Journal of Mammalogy 79:667–670.

Dobson FS, Smith AT, Wang XG. 1998. Social and ecological influences on dispersal and philopatry in the plateau pika (*Ochotona curzoniae*). Behavioral Ecology 9:622–635.

Dobson FS, Smith AT, Wang XG. 2000. The mating system and gene dynamics of plateau pikas. Behavioural Processes 51:101–110.

Dobson FS, Chesser RK, Hoogland JL, Sugg DW, Foltz DW. 1997. Do black-tailed prairie dogs minimize inbreeding? Evolution 51:970–978.

Dobson FS, Chesser RK, Hoogland JL, Sugg DW, Foltz DW. 1998. Breeding groups and gene dynamics in a socially-structured population of prairie dogs. Journal of Mammalogy 79:671–680.

Dolby AS, Grubb TC. 1999. Effects of winter weather on horizontal vertical use of isolated forest fragments by bark-foraging birds. Condor 101:408–412.

Doligez B, Danchin E, Clobert J, Gustafsson L. 1999. The use of conspecific reproductive success for breeding habitat selection in a non-colonial, hole-nesting species, the collared flycatcher. Journal of Animal Ecology 68:1193–1206.

Donovan TM, Jones PW, Annand EM, Thompson FR, III. 1997. Variation in local-scale edge effects: mechanisms and landscape context. Ecology 78:2064–2075.

Draper NR, Smith H. 1998. Applied regression analysis. New York: Wiley.

Drent RH, Daan S. 1980. The prudent parent: energetic adjustments in avian breeding. Ardea 68:225–252.

Drolet B, Desrochers A, Fortin M-J. 1999. Are songbirds affected by landscape structure in exploited boreal forest? Condor 101:699–704.

Dufour KW, Weatherhead PJ. 1998. Bilateral symmetry and social dominance in captive male red-winged blackbirds. Behavioural Ecology and Sociobiology 42:71–76.

Dunham KM. 1980. The feeding behaviour of a tame impala. African Journal of Ecology 18:253–257.

Dunham KM. 1997. Population growth of mountain gazelles reintroduced into central Arabia. Biological Conservation 81:205–214.

Dunk JR, Cain SL, Reid ME, Smith RN. 1994. A high breeding density of common ravens in northwestern Wyoming. Northwest Naturalist 75:70–73.

Durant SM. 2000a. Dispersal patterns, social organization and population viability. In: Behaviour and conservation (Gosling LM, Sutherland WJ, eds.). Cambridge: Cambridge University Press; 172–197.

Durant SM. 2000b. Predator avoidance, breeding experience, and reproductive success in endangered cheetah, *Acinonyx jubatus*. Animal Behaviour 60:121–130.

du Toit JT. 1990. Feeding height stratification among African browsing ruminants. African Journal of Ecology 28:55–61.

du Toit JT. 1995. Sexual segregation in kudu: sex differences in competitive ability, predation risk or nutritional needs? South African Journal of Wildlife Research 25:127–132.

du Toit JT, Owen-Smith N. 1989. Body size, population metabolism and habitat specialization among African large herbivores. American Naturalist 133:736–740.

Eadie J, Sherman P, Semel B. 1998. Conspecific brood parasitism, population dynamics, and the conservation of cavity-nesting birds. In: Behavioral ecology and conservation biology (Caro TM). New York: Oxford University Press; 306–340.

Ebensperger LA. 1998. Strategies and counterstrategies to infanticide in mammals. Biological Review 73:321–346.

Ebensperger LA. 2001. No infanticide in the hystricognath rodent, *Octodon degus*: does ecology play a role? Acta Ethologica 3:89–93.

Eberhard WG. 1996. Female control: sexual selection by cryptic female choice. Princeton: Princeton University Press.

Eberhardt LL. 1991. Models of ungulate population dynamics. Rangifer 7:24–29.

Eckert KL. 1995. Anthropogenic threats to sea turtles. In: Biology and conservation of sea turtles (Bjorndal KA, ed.). Washington, DC: Smithsonian Institution Press; 611–612.

Eckert SA, Sarti L. 1997. Distant fisheries implicated in the loss of the world's largest leatherback nesting population. Marine Turtle Newsletter 78:2–7.

Edwards J. 1977. Learning to eat by following the mother in moose calves. American Midland Naturalist 96:229–232.

Ehrlich PR, Wilson EO. 1991. Biodiversity studies: science and policy. Science 253:758–762.

Eisenberg JF. 1981. The mammalian radiations. Chicago: University of Chicago Press.

Elgmork K. 1988. Reappraisal of the brown bear status in Norway. Biological Conservation 46:163–168.

Ellenberg H. 1978. Zur Populationsökologie des Rehes (*Capreolus capreolus*) in Mitteleuropa. Spixiana Zeitschrift für Zoologie (suppl 2):1–211.

Elowe KD, Dodge WE. 1989. Factors affecting black bear reproductive success and cub survival. Journal of Wildlife Management 53:962–968.

Elwood RW, Ostermeyer MC. 1984. Infanticide by male and female Mongolian gerbils: ontogeny, causation, and function. In: Infanticide: comparative and evolutionary perspectives (Hausfater G, Hrdy SB, eds.). New York: Aldine de Gruyter; 367–386.

Emlen JM. 1966. The role of time and energy in food preference. American Naturalist 100:611–617.

Endler JA. 1997. Light, behavior, and conservation of forest-dwelling organisms. In: Behavioral approaches to conservation in the wild (Clemmons JR, Buchholz R, eds.). Cambridge: Cambridge University Press; 329–355.

Engen F, Folstad I. 1999. Cod courtship song: a song at the expense of dance? Canadian Journal of Zoology 77:542–550.

Engen S, Bakke Ø, Islam A. 1998. Demographic and environmental stochasticity: concepts and definitions. Biometrics 16:840–846.

Erickson G. 1988. Permit auctions: the good, the bad and the ugly. Biennial Symposium of the Northern Wild Sheep and Goat Council 6:47–53.

Ericsson G, Wallin K. 2001. Age-specific moose (*Alces alces*) mortality in a predator-free environment: evidence for senescence in females. Écoscience 8:157–163.

Ericsson G, Wallin K, Ball JP, Broberg M. 2001. Age-related reproductive effort and senescence in free-ranging moose, *Alces alces*. Ecology 82:1613–1620.

Ernest HB, Rubin ES, Boyce WM. 2002. Fecal DNA analysis and risk assessment of mountain lion predation of bighorn sheep. Journal of Wildlife Management 66:75–85.

Estes JA, Rathbun GB, Vanblaricom GR. 1993. Paradigms for managing carnivores: the case of the sea otter. Symposia of the Zoological Society of London 65:307–320.

Estes JA, Tinker MT, Doak DF. 1998. Killer whale predation on sea otters linking oceanic and nearshore ecosystems. Science 282:473–475.

Fahrig L. 1997. Relative effects of habitat loss and fragmentation on population extinction. Journal of Wildlife Management 61:603–610.

Fahrmeir L, Tutz G. 1994. Multivariate statistical modelling based on generalized linear models. New York: Springer.

Faivre B, Preault M, Thery M, Secondi J, Patris B, Cézilly F. 2001. Breeding strategy and morphological characters in an urban population of blackbirds, *Turdus merula*. Animal Behaviour 61:969–974.

Falconer DS. 1981. Introduction to quantitative genetics. 2nd ed. London: Longman.

Faulkes CG, Abbott DH, Mellor AL. 1990. Investigation of genetic diversity in wild colonies of naked mole rats (*Heterocephalus glaber*) by DNA fingerprinting. Journal of Zoology 221:87–97.

Ferson S, Akçakaya HR. 1990. RAMAS/age, user manual, modeling fluctuations in age-structured populations. Setauket, NY: Exeter Software.

Festa-Bianchet M. 1986. Bighorn ram survival and harvest in southwestern Alberta. Symposium Northern Wild Sheep and Goat Council 5:102–109.

Festa-Bianchet M. 1988a. Age-specific reproduction of bighorn ewes in Alberta, Canada. Journal of Mammalogy 69:157–160.

Festa-Bianchet M. 1988b. Birthdate and lamb survival in bighorn lambs (*Ovis canadensis*). Journal of Zoology 214:653–661.

Festa-Bianchet M. 1989. Survival of male bighorn sheep in southwestern Alberta. Journal of Wildlife Management 53:259–263.

Festa-Bianchet M, Gaillard J-M, Jorgenson JT. 1998. Mass- and density-dependent reproductive success and reproductive costs in a capital breeder. American Naturalist 152:367–379.

Festa-Bianchet M, Jorgenson JT. 1998. Selfish mothers: reproductive expenditure and resource availability in bighorn ewes. Behavioral Ecology 9:144–150.

Festa-Bianchet M, Jorgenson JT, Réale D. 2000. Early development, adult mass, and reproductive success in bighorn sheep. Behavioral Ecology 11:633–639.

Festa-Bianchet M, Jorgenson JT, Lucherini M, Wishart WD. 1995. Life-history consequences of variation in age of primiparity in bighorn ewes. Ecology 76:871–881.

Festa-Bianchet M, Jorgenson JT, Bérubé CH, Portier C, Wishart WD. 1997. Body mass and survival of bighorn sheep. Canadian Journal of Zoology 75:1372–1379.

Fischer J, Lindenmayer DB. 2000. An assessment of the published results of animal relocations. Biological Conservation 96:1–11.

Fisher RA. 1930. The genetical theory of natural selection. Oxford: Clarendon Press.

Fitzsimmon NN, Buskirk SW, Smith MH. 1995. Population history, genetic variability, and horn growth in bighorn sheep. Conservation Biology 9:314–323.

Fiumera AC, Parker PG, Fuerst PA. 2000. Effective population size and maintenance of genetic diversity in captive-bred populations of a Lake Victoria cichlid. Conservation Biology 14:886–892.

Fleming IA, Gross MR. 1993. Breeding success of hatchery and wild coho salmon (*Oncorhynchus kisutch*) in competition. Ecological Applications 3:230–245.

Fleming IA, Jonsson B, Gross MR. 1994. Phenotypic divergence of sea-ranched, farmed, and wild salmon. Canadian Journal of Fisheries and Aquatic Sciences 51:2808–2824.

Flowers MA, Graves BM. 1997. Juvenile toads avoid chemical cues from snake predators. Animal Behaviour 53:641–646.

Folstad I, Karter AJ. 1992. Adaptive or nonadaptive immunosuppression by sex hormones? American Naturalist 139:603–622.

Forchhammer MC, Clutton-Brock TH, Lindstrom J, Albon SD. 2001. Climate and population density induce long-term cohort variation in a northern ungulate. Journal of Animal Ecology 70:721–729.

Fowler CW. 1987. A review of density dependence in populations of large mammals. In: Current mammalogy (Genoways HH, ed.). New York: Plenum; 401–441.

Fowler K, Whitlock MC. 2002. Environmental stress, inbreeding, and the nature of phenotypic and genetic variance in *Drosophila melanogaster*. Proceedings of the Royal Society of London, Series B 269:677–683.

Fox GA, Kendall BE. 2002. Demographic stochasticity and the variance reduction effect. Ecology 83:1928–1934.

Frame LH, Fanshawe JH. 1990. African wild dog *Lycaon pictus*: a survey of status and distribution 1985–88. Unpublished report.

Frank LG, Holekamp KE, Smale L. 1995. Dominance, demography and reproductive success of female spotted hyenas. In: Serengeti II: research, management, and conservation of an ecosystem (Sinclair ARE, Arcese P, eds.). Chicago: University of Chicago Press.

Frankham R. 1996. Effective population size/adult population size ratios in wildlife: a review. Genetical Research 66:95–107.

Frankham R. 1999. Quantitative genetics in conservation biology. Genetical Research 74:237–244.

Frankham R, Manning H, Margan SH, Briscoe DA. 2000. Does equalization of family sizes reduce genetic adaptation to captivity? Animal Conservation 3:357–363.

Franzmann AW, Schwartz CC. 1985. Moose twinning rates: a possible condition assessment. Journal of Wildlife Management 49:394–396.

Franzmann AW, Schwartz CC. 1986. Black bear predation on moose calves in highly productive versus marginal moose habitats on the Kenai Peninsula, Alaska. Alces 22:139–153.

Frati F, Lovari S, Hartl GB. 2000. Does protection from hunting favour genetic uniformity in the red fox? Zeitschrift für Säugetierkunde 65:76–83.

French J. 1994. Wildlife telemetry by satellite. Endeavour 18:32–37.

French Ministry of Agriculture and Fisheries. 2000. Action Plan for the preservation of pastoralism and the wolf in the Alpine Chain. Document presented at the Bern Convention Meeting of Oslo 22–24 June 2000. Pp. 10.

Frisch RE. 1984. Body fat, puberty and fertility. Biological Reviews of the Cambridge Philosophical Society 59:161–188.

Fritts SH. 1983. Record dispersal by a wolf from Minnesota. Journal of Mammalogy 64:166–167.

Fritts SH, Mech LD. 1981. Dynamics, movements and feeding ecology of a newly protected wolf population in northwestern Minnesota. Wildlife Monographs 80:1–79.

Fritts SH, Paul WJ, Mech LD. 1984. Movements of translocated wolves in Minnesota. Journal of Wildlife Management 48:709–721.

Fritts SH, Paul WJ, Mech, LD. 1985. Can relocated wolves survive? Wildlife Society Bulletin 13:459–463.

Frost SK. 1981. Food selection in young naive impala. South African Journal of Zoology 16:123–124.

Fryxell JM, Hussell DJT, Lambert AB, Smith PC. 1991. Time lags and population fluctuations in white-tailed deer. Journal of Wildlife Management 55:377–385.

Fuller TK. 1989. Population dynamics of wolves in north-central Minnesota. Wildlife Monographs 105:1–41.

Gaillard JM. 1988. Contribution á la dynamique des populations de grands mammifères: l'exemple du chevreuil (*Capreolus capreolus*). Ph.D. thesis, Université de Lyon.

Gaillard JM, Delorme D, Jullien JM. 1993. Effects of cohort, sex, and birth date on body development of roe deer *Capreolus capreolus* fawns. Oecologia 94:57–61.

Gaillard JM, Festa-Bianchet M, Yoccoz NG. 1998. Population dynamics of large herbivores: variable recruitment with constant adult survival. Trends in Ecology and Evolution 13:58–63.

Gaillard JM, Festa-Bianchet M, Yoccoz NG. 2001. Not all sheep are equal. Science 292:1499–1500.

Gaillard JM, Pontier D, Allainé D, Lebreton JD, Trouvilliez J, Clobert J. 1989. An analysis of demographic tactics in birds and mammals. Oikos 56:59–76.

Gaillard JM, Sempéré AJ, Boutin J-M, Laere GV, Boisaubert B. 1992. Effects of age and body weight on the proportion of females breeding in a population of roe deer (*Capreolus capreolus*). Canadian Journal of Zoology 70:1541–1545.

Gaillard JM, Delorme D, Boutin JM, Laere GV, Boisaubert B, Pradel R. 1993. Roe deer survival patterns: a comparative analysis of contrasting populations. Journal of Animal Ecology 62:778–791.

Gaillard JM, Delorme D, Boutin JM, Laere GV, Boisaubert B. 1996. Body mass of roe deer fawns during winter in two contrasting populations. Journal of Wildlife Management 60:29–36.

Gaillard JM, Delorme D, Laere GV, Duncan P, Lebreton JD. 1997. Early survival in roe deer: causes and consequences of cohort variation in two contrasted populations. Oecologia 112:502–513.

Gaillard JM, Andersen R, Delorme D, Linnell JDC. 1998a. Family effects on growth and survival of juvenile roe deer. Ecology 79:2878–2889.

Gaillard JM, Liberg O, Andersen R, Hewison AJM, Cederlund G. 1998b. Population dynamics of roe deer. In: The European roe deer: the biology of success (Andersen R, Duncan P, Linnell JDC, eds.). Oslo: Scandinavian University Press; 309–335.

Gaillard JM, Festa-Bianchet M, Yoccoz NG, Loison A, Toïgo C. 2000. Temporal variation in fitness components and population dynamics of large herbivores. Annual Review of Ecology and Systematics 31:367–393.

Gammel M. 2002. Dominance rank in male fallow deer (*Dama dama* L.): ranking methods, ranking conditions, adult phenotypic correlates and early-life effects. Ph.D. thesis, National University of Ireland, Dublin.

Garshelis DL. 1994. Density-dependent population regulation of brown bears. In: Density-dependent population regulation of brown, black, and polar bears (Taylor M, ed.). International Conference on Bear Research and Management, No. 3:3–14.

Gasaway WG, Boertje RD, Grangaard DV, Kellyhouse DG, Stephenson RO, Larsen DG. 1992. The role of predation in limiting moose at low densities in Alaska and Yukon and implications for conservation. Wildlife Monographs 120:1–59.

Gates JE, Gysel LW. 1978. Avian nest dispersion and fledgling success in field-forest ecotones. Ecology 59:871–883.

Gauthier D, Michallet J, Villaret JC, Rivet A. 1994. Taille et composition des groupes sociaux dans six populations de Bouquetins des Alpes. Travaux Scientifiques du Parc national de la Vanoise 18:101–124.

Gauthier G, Pradel R, Menu S, Lebreton J-D. 2001. Assessing seasonal survival rate of greater snow geese and the effect of hunting in presence of dependence in sighting probabilities. Ecology 82:3105–3119.

Gavin TA. 1989. What's wrong with the questions we ask in wildlife research? Wildlife Society Bulletin 17:345–350.

Geist V. 1971. Mountain sheep. Chicago: University of Chicago Press.

Geist V. 1994. Wildlife conservation as wealth. Nature 368:491–492.

Genov PV, Gigantesco P, Massei G. 1998. Interactions between black-billed magpie and fallow deer. Condor 100:177–179.

Georgii B. 1980. Home range patterns of female red deer (*Cervus elaphus*) in the Alps. Oecologia 47:278–285.

Georgii B, Schroeder W. 1983. Home range and activity patterns of male red deer in the Alps. Oecologia 58:238–248.

Gese EM, Crabtree RL. 1996. Social and nutritional factors influencing the dispersal of resident coyotes. Animal Behaviour 52:1025–1043.

Getz W, Fortmann L, Cumming D, Du Toit J, Martin R. 2000. CAMPFIRE experiences in Zimbabwe–Response. Science 287:43.

Giacometti M, Bassano B, Peracino V, Ratti P. 1997. Die konstitution des Alpensteinbockes *(Capra i. ibex* L.) in Abhängigkeit von Geschlecht, Alter, Herkunft und Jahreszeit in Graubünden (Schweiz) und im Parco Nazionale Gran Paradiso (Italien). Zeitschrift Jagdwisserchaft 43:24–34.

Gibbons EF, Durrant BS, Demarest J. 1995. Conservation of endangered species in captivity. Albany: State University of New York Press.

Gibbs JP, Faaborg J. 1990. Estimating the viability of ovenbird and Kentucky warbler populations in forest fragments. Conservation Biology 4:193–196.

Gibson RM, Bradbury JW. 1985. Sexual selection in lekking sage grouse: phenotypic correlates of male mating success. Behavioural Ecology and Sociobiology 18:117–123.

Gil D, Graves J, Hazon N, Wells A. 1999. Male attractiveness and differential testosterone investment in zebra finch eggs. Science 286:126–128.

Gill RMA. 1990. Monitoring the status of European and North American cervids: the global environment monitoring system (GEMS). London: UNEP.

Gillingham S, Lee PC. 1999. The impact of wildlife-related benefits on the conservation attitudes of local people around the Selous Game Reserve, Tanzania. Environmental Conservation 26:218–228.

Gilpin ME, Hanski I. 1991. Metapopulation dynamics: empirical and theoretical investigations. London: Academic.

Gilpin ME, Soulé ME. 1986. Minimum viable populations: processes of species extinction. In: Conservation biology: the science of scarcity and diversity (Soulé ME, ed.). Sunderland, MA: Sinauer Associates; 19–34.

Ginsberg JR, Milner-Gulland EJ. 1994. Sex-biased harvesting and population dynamics in ungulates: implications for conservation and sustainable use. Conservation Biology 8:157–166.

Girard I. 2000. Dynamique des populations et expansion géographique du bouquetin des Alpes *(Capra ibex)* dans le Parc national de la Vanoise. Ph.D. thesis, Université de Savoie.

Girard I, Toïgo C, Gaillard JM, Gauthier D, Martinot JP. 1999. Patron de survie chez le bouquetin des Alpes *(Capra ibex ibex)* dans le Parc National de la Vanoise. Revue d'Écologie 54:235–251.

Gitschlag GR. 1996. Migration and diving behavior of Kemp's ridley (Garman) sea turtles along the U.S. southeastern Atlantic coast. Journal of Experimental Marine Biology and Ecology 205:115–135.

Godfrey MH. 1996. Ethology and sea turtle conservation. Trends in Ecology and Evolution 11:433–434.

Godley BJ, Broderick AC, Hays GC. 2001. Nesting of green turtles (*Chelonia mydas*) at Ascension Island, South Atlantic. Biological Conservation 97:151–158.

Godley BJ, Broderick AC, Frauenstein R, Glen F, Hays GC. 2002. Reproductive seasonality and sexual dimorphism in green turtles. Marine Ecology Progress Series 226:125–133.

Gonzalez G, Sorci G, Smith LC, de Lope F. 2001. Testosterone and sexual signalling in male house sparrows (*Passer domesticus*). Behavioral Ecology and Sociobiology 50:557–562.

Goossens B, Graziani L, Waits LP, Farand E, Magnolon S, Coulon J, Jacques M-C, Taberlet P, Allainé D. 1998. Extra-pair paternity in the monogamous alpine marmot revealed by nuclear DNA microsatellite analysis. Behavioral Ecology and Sociobiology 43:281–288.

Gordon IJ. 1991. Ungulate re-introductions: the case of the scimitar-horned oryx. Symposium of the Zoological Society of London 62:217–240.

Gordon IJ, Illius AW. 1988. Incisor arcade structure and diet selection in ruminants. Functional Ecology 2:15–22.

Gordon IJ, Illius AW. 1994. The functional significance of the browser-grazer dichotomy in African ruminants. Oecologia 98:167–175.

Gorman M. 1999. Oryx go back to the brink. Nature 398:190.

Gosling LM. 1986. The evolution of mating strategies in male antelopes. In: Ecological aspects of social evolution (Rubenstein DI, Wrangham RW). Princeton: Princeton University Press.

Gosling LM, Baker SJ. 1989. Demographic consequences of differences in the ranging behavior of male and female coypus. In: Mammals as pests (Putman RJ, ed.). London: Chapman and Hall; 155–167.

Gosling LM, Petrie M. 1990. Lekking in topi: a consequence of satellite behaviour by small males at hot spots. Animal Behaviour 40:272–287.

Gosling LM, Roberts SC. 2001. Scent-marking by male mammals: cheat-proof signals to competitors and mates. Advances in the Study of Animal Behaviour 30:169–217.

Gosling LM, Roberts SC, Thornton EA, Andrew MJ. 2000. Life history costs of olfactory status signalling in mice. Behavioural Ecology and Sociobiology 48:328–332.

Gosling LM, Sutherland WJ (eds.). 2000. Behaviour and conservation. Cambridge: Cambridge University Press.

Goss-Custard JD, Sutherland WJ. 1997. Individual behaviour, populations and conservation. In: Behavioural ecology (Krebs JR, ed.). Oxford: Blackwell Science; 373–395.

Goss-Custard JD, Caldow RWG, Clarke RT, Durell SEA, Urfi AJ, West, AD. 1995a. Consequences of winter habitat loss and change to populations of wintering migratory birds: predicting the local and global effects from studies of individuals. Ibis 137(suppl.):S56–S66.

Goss-Custard JD, Caldow RWG, Clarke RT, Durell SEA, Sutherland WJ. 1995b. Deriving population parameters from individual variations in foraging behaviour. I: Empirical game theory distribution model of oystercatchers

Haematopus ostralegus feeding on mussels *Mytilus edulis*. Journal of Animal Ecology 64:265–276.

Goss-Custard JD, Stillman RA, West AD, Caldow RWG, McGrorty S. 2002. Carrying capacity in overwintering migratory birds. Biological Conservation 105:27–41.

Gossow H, Stadlmann I. 1985. Red deer management and the use of overwintering enclosures. Transactions 18th IUGB Congress, Brussels; 133.

Grafen A. 1988. On the uses of data on lifetime reproductive success. In: Reproductive success (Clutton-Brock TH, ed.). Chicago: University of Chicago Press; 454–470.

Grafen A. 1990. Biological signals as handicaps. Journal of Theoretical Biology 144:517–546.

Grahn M, Langesfors A, von Schantz T. 1998. The importance of mate choice in improving viability in captive populations. In: Behavioral ecology and conservation biology (Caro TM, ed.). New York: Oxford University Press.

Greene C, Umbanhowar J, Mangel M, Caro T. 1998. Animal breeding systems, hunter selectivity, and consumptive use in wildlife conservation. In: Behavioral ecology and conservation biology (Caro T, ed.). New York: Oxford University Press; 271–305.

Greenwood PJ. 1980. Mating systems, philopatry and dispersal in birds and mammals. Animal Behaviour 28:1140–1162.

Greenwood PJ. 1993. The ecology and conservation management of geese. Trends in Ecology and Evolution 8:307–308.

Greenwood PJ, Harvey PH. 1982. The natal and breeding dispersal of birds. Annual Review of Ecology and Systematics 13:1–21.

Griffin AS, Blumstein AT, Evans CS. 2000. Training captive-bred or translocated animals to avoid predators. Conservation Biology 14:1317–1326.

Griffiths B, Scott JM, Carpenter JW, Reed C. 1989. Translocation as a species conservation tool: status and strategy. Science 245:472–480.

Grinnell J, McComb K. 1996. Maternal grouping as a defence against infanticide by males: evidence from field playback experiments on African lions. Behavioral Ecology 7:55–59.

Gulland J. 1977. The analysis of data and development of models. In: Fish population dynamics (Gulland J, ed.). New York: John Wiley; 67–95.

Gustafsson L. 1986. Lifetime reproductive success and heritability: empirical support for Fisher's fundamental theorem. American Naturalist 128:761–764.

Haddad NM. 1999a. Corridor and distance effects on interpatch movements: a landscape experiment with butterflies. Ecological Applications 9:612–622.

Haddad NM. 1999b. Corridor use predicted from behaviours at habitat boundaries. American Naturalist 153:215–227.

Haffer J. 1997. Alternative models of vertebrate speciation in Amazonia: an overview. Biodiversity and Conservation 6:451–476.

Hamilton PH. 1986. Status of the leopard in Kenya with reference to sub-Saharan Africa. In: Cats of the world: biology, conservation and management (Miller SD, Everett DD, eds.). Washington DC: National Wildlife Federation.

Hamilton WD, Axelrod R, Tanese R. 1990. Sexual reproduction as an adaptation to resist parasites (a review). Proceedings of the National Academy of Sciences USA 87:3566–3573.

Hamilton WD, Zuk M. 1982. Heritable true fitness and bright birds: a role for parasites. Science 218:384–387.

Hannon SJ, Cotterill SE. 1998. Nest predation in aspen woodlots in an agricultural area in Alberta: the enemy from within. Auk 115:16–25.

Hanski IA. 1997. Metapopulation dynamics: from concepts and observations to predictive models. In: Metapopulation biology: ecology, genetics and evolution (Hanski IA, Gilpin ME, eds.). London: Academic Press; 69–91.

Hanski IA, Gilpin M (eds.). 1997. Metapopulation dynamics: ecology, genetics and evolution. London: Academic Press.

Harborne JB. 1988. Introduction to ecological biochemistry. 3rd ed. London: Academic Press.

Harcourt AH, Stewart KJ, Fossey D. 1976. Male emigration and female transfer in wild mountain gorilla. Nature 263:226–227.

Harestad AS, Bunnell FL. 1979. Home range and body weight: a reevaluation. Ecology 60:389–402.

Hargis CD, Bissonette JA, Turner DC. 1999. The influence of forest fragmentation and landscape pattern on American martens. Journal of Applied Ecology 36:157–172.

Harrington FH. 1978. Ravens attracted to wolf howling. Condor 80:236–237.

Harrington R, Owen-Smith N, Viljoen PC, Biggs HC, Mason DR, Funston P. 1999. Establishing the causes of the roan antelope decline in the Kruger National Park, South Africa. Biological Conservation 90:69–78.

Harris RB, Allendorf FW. 1989. Genetically effective population size of large mammals, an assessment of estimators. Conservation Biology 3:181–191.

Harris RB, Wall WA, Allendorf FW. 2002. Genetic consequences of hunting: what do we know and what should we do? Wildlife Society Bulletin 30:634–643.

Harris S, White PCL. 1992. Is reduced affiliative rather than increase agonistic behaviour associated with dispersal in red foxes? Animal Behaviour 44:1085–1089.

Harrison DJ. 1992. Dispersal characteristics of juvenile coyotes in Maine. Journal of Wildlife Management 56:128–138.

Hartl DL. 1988. A primer of population genetics. 2nd ed. Sunderland, MA: Sinauer Associates.

Hartl GB, Lang G, Klein F, Willing R. 1991. Relationships between allozymes, heterozygosity and morphological characters in red deer (*Cervus elaphus*), and the influence of selective hunting on allele frequency distribution. Heredity 66:343–350.

Hartl GB, Klein F, Willing R, Apollonio M, Lang G. 1995. Allozymes and the genetics of antler development in red deer (*Cervus elaphus*). Journal of Zoology 237:83–100.

Hartley MJ, Hunter ML. 1998. A meta-analysis of forest cover, edge effects, and artificial nest predation rates. Conservation Biology 12:465–469.

Harvey PH, Zammuto RM. 1985. Patterns of mortality and age at first reproduction in natural populations of mammals. Nature 315:319–320.

Hasselquist D, Bensch S, vonSchantz T. 1996. Correlation between male song repertoire, extra-pair paternity and offspring survival in the great reed warbler. Nature 381:229–232.

Hasselquist D, Wasson MF, Winkler DW. 2001. Humoral immunocompetence correlates with date of egg-laying and reflects work load in female tree swallows. Behavioral Ecology 12:93–97.

Hatch MI. 1996. Factors influencing nest failure in the song sparrow. Master of Science thesis, University of Wisconsin–Madison.

Hayes RD, Gunson JR. 1995. Status and management of wolves in Canada. In: Ecology and conservation of wolves in a changing world (Carbyn LN, Fritts SH, Seip DR, eds.). Edmonton, AB: Canadian Circumpolar Institute.

Hays GC, Broderick AC, Glen F, Godley BJ, Nichols WJ. 2001. The movements and submergence behaviour of male green turtles at Ascension Island. Marine Biology 139:395–399.

Hays GC, Marsh R. 1997. Estimating the age of juvenile loggerhead sea turtles in the North Atlantic. Canadian Journal of Zoology 75:40–46.

Hedenstrom A, Alerstam T. 1996. Skylark optimal flight speeds for flying nowhere and somewhere. Behavioral Ecology 7:121–126.

Hedrick PW, Kalinowski ST. 2000. Inbreeding depression in conservation biology. Annual Reviews of Ecology and Systematics 31:139–162.

Hedrick PW, Miller PS. 1994. Rare alleles, MHC and captive breeding. In: Conservation genetics (Loeschke V, Tomiuk J, Jain SK, eds.). Basel: Birkhaeuser; 187–204.

Heibert S, Stoddard P, Arcese P. 1989. Repertoire size, territory acquisition and reproductive success in the song sparrow. Animal Behaviour 37:266–273.

Heimer WE, Watson SM, Smith TC. 1984. Excess ram mortality in a heavily hunted Dall sheep population. Symposium Northern Wild Sheep and Goat Council 4:425–432.

Heinrich B. 1989. Ravens in winter. New York: Random House.

Hellgren EC, Vaughan MR. 1986. Home range and movements of winter-active black bears in the Great Dismal Swamp. International Conference on Bear Research and Management 7:227–234.

Henry JD. 1986. The red fox. Washington, DC: Smithsonian Institution Press.

Henshaw RE., Lockwood R, Shideler R, Stephenson RO. 1979. Experimental release of captive wolves. In: The behavior and ecology of wolves (Klinghammer E, ed.). New York: Garland STPM Press.

Herrero S, Hamer D. 1977. Courtship and copulation of a pair of grizzly bears, with comments on reproductive plasticity and strategy. Journal of Mammalogy 58:441–444.

Hessing P, Aumiller L. 1994. Observations of conspecific predation by brown bears, *Ursus arctos*, in Alaska. Canadian Field-Naturalist 108:332–336.

Hewison AJM. 1997. Evidence for a genetic component of female fecundity in British roe deer from studies of cranial morphometrics. Functional Ecology 11:508–517.

Hill JK, Thomas CD, Lewis OT. 1996. Effects of habitat patch size and isolation on dispersal by *Hesperia comma* butterflies: implications for metapopulation structure. Journal of Animal Ecology 65:725–735.

Hill WG. 1979. A note on effective population size with overlapping generations. Genetics 92:317–322.

Hindar K, Balstad T. 1994. Salmonid culture and interspecific hybridisation. Conservation Biology 8:881–882.

Hobson KA, Villard M-A. 1998. Forest fragmentation affects the behavioral response of American redstarts to the threat of cowbird parasitism. Condor 100:389–394.

Hochachka WM. 1993. Repeatable reproduction in song sparrows. Auk 110:603–613.

Hochachka WM, Smith JNM, Arcese P. 1989. Song sparrow. In: Lifetime reproduction in birds (Newton I, ed.). New York: Academic Press; 135–152.

Hoenig JM. 1983. Empirical use of longevity data to estimate mortality rates. Fishery Bulletin 82:898–903.

Hofmann RR. 1989. Evolutionary steps of ecophysiological adaptation and diversification of ruminants: a comparative review of their digestive system. Oecologia 78:443–457.

Hogg JT. 2000. Mating systems and conservation at large spatial scales. In: Vertebrate mating systems (Apollonio M, Festa-Bianchet M, Mainardi D, eds.). Singapore: World Scientific; 214–252.

Hogg JT, Forbes SH. 1997. Mating in bighorn sheep: frequent male reproduction via a high-risk "unconventional" tactic. Behavioral Ecology and Sociobiology 41:33–48.

Hogg JT, Hass CC, Jenni DA. 1992. Sex-biased maternal expenditure in Rocky Mountain bighorn sheep. Behavioral Ecology and Sociobiology 31:243–251.

Holekamp KE, Ogutu JO, Frank LG, Dublin HT, Smale L. 1993. Fission of a spotted hyena clan: consequences of prolonged female absenteeism and causes of female emigration. Ethology 93:285–299.

Holthausen RS, Raphael MG, Samson FB, Ebert D, Hiebert R, Menasco K. 1999. Population viability in ecosystem management. In: Ecological stewardship: a common reference for ecosystem management, vol. 2 (Szaro RC, Johnson NC, Sexton WT, Malk AJ, eds.). Oxford: Elsevier Science; 135–156.

Hoogland JL. 1992. Levels of inbreeding among prairie dogs. American Naturalist 139:591–602.

Hoogland JL. 1995. The black-tailed prairie dog—social life of a burrowing mammal. Chicago: University of Chicago Press.

Horak P, Ots I, Vellau H, Spottiswoode C, Møller AP. 2001. Carotenoid-based plumage coloration reflects hemoparasite infection and local survival in breeding great tits. Oecologia 126:166–173.

Horner MA, Powell RA. 1990. Internal structure of home ranges of black bears and analyses of home range overlap. Journal of Mammalogy 71:402–410.

Houston DB. 1968. The Shiras moose in Jackson Hole, Wyoming. National Park Service Technical Bulletin 1:1–110.

Houston DB. 1992. Willow–moose relationships in Grand Teton National Park: an update. Moose, WY: National Park Service; unpublished ms.

Hovey FW, McLellan BN. 1996. Estimating population growth of grizzly bears from the Flathead River drainage using computer simulations of reproductive and survival rates. Canadian Journal of Zoology 74:1409–1416.

Howard RS, Lively CM. 1994. Parasitism, mutation accumulation and the maintenance of sex. Nature 367:554–557.

Hrdy SB. 1979. Infanticide among animals: a review, classification, and examination of the implications for the reproductive strategies of females. Ethology and Sociobiology 1:13–40.

Hrdy SB, Hausfater G. 1984. Comparative and evolutionary perspectives on infanticide: introduction and overview. In: Infanticide: comparative and evolutionary perspectives (Hausfater G, Hrdy SB, eds.). New York: Aldine de Gruyter; xiii–xxxv.

Huber D, Roth HU. 1986. Home ranges and movements of brown bears in Plitvice Lakes National Park, Yugoslavia. International Conference on Bear Research and Management 6:93–97.

Huber D, Roth HU. 1993. Movements of European brown bears in Croatia. Acta Theriologica 38:151–159.

Hughes GR. 1996. Nesting of the leatherback turtle (*Dermochelys coriacea*) in Tongaland, KwaZulu-Natal, South Africa. 1963–1995. Chelonian Conservation and Biology 2:153–158.

Hughes GR, Luschi P, Mencacci R, Papi F. 1998. The 7000-km oceanic journey of a leatherback turtle tracked by satellite. Journal of Experimental Marine Biology and Ecology 229:209–217.

Hunter LTB, Skinner JD. 1998. Vigilance behaviour in African ungulates: the role of predation pressure. Behaviour 135;195–211.

Hunter ML. 1996. Fundamentals of conservation biology. London: Blackwell Science.

Huntsman GR, Schaaf WE. 1994. Simulation of the impact of fishing on reproduction of a protogynous grouper, the graysby. North American Journal of Fisheries Management 12:41–52.

Huston MD, DeAngelis D, Post W. 1988. New computer models unify ecological theory. BioScience 38:682–691.

Hutchings JA. 1996. Spatial and temporal variation in the density of northern cod and a review of hypotheses for the stock's collapse. Canadian Journal of Fisheries and Aquatic Sciences 53:943–962.

Hutchinson GE. 1959. Homage to Santa Rosalia, or why are there so many kinds of animals? American Naturalist 93:145–159.

Hykle D. 2000. The Convention on Migratory Species and marine turtle conservation. Marine Turtle Newsletter 87:1–3.

Illius AW, O'Connor TG. 2000. Resource heterogeneity and ungulate population dynamics. Oikos 89:283–294.

Istituto Nazionale per la Fauna Selvatica (INFS). 1998. Studio di fattibilitá per la reintroduzione dell'orso bruno (*Ursus arctos*) sulle Alpi Centrali. Relazioni Interne del Parco Adamello Brenta. Treviso. [Italian]

Jachmann H, Berry PS, Imahe H. 1995. Tusklessness in African elephants: a future trend. African Journal of Ecology 33:230–235.

Janson CH, van Schaik CP. 2000. The behavioral ecology of infanticide by males. In: Infanticide by males and its implications (van Schaik CP, Janson CH, eds.). Cambridge: Cambridge University Press; 469–494.

Jarman PJ. 1974. The social organisation of antelope in relation to their ecology. Behaviour 48:215–267.

Jennings S, Reynolds JD, Mills SC. 1998. Life history correlates of responses to fisheries exploitation. Proceedings of the Royal Society of London, Series B 265:333–339.

Jennions MD, Petrie M. 1997. Variation in mate choice and mating preferences: a review of causes and consequences. Biological Reviews 72:283–327.

Jennions MD, Petrie M. 2000. Why do females mate multiply? A review of the genetic benefits. Biological Reviews 75:21–64.

Johnstone RA. 1995. Sexual selection, honest advertisement and the handicap principle: reviewing the evidence. Biological Reviews 70:1–65.

Jokimäki J, Huhta E, Itämies J, Rahko P. 1998. Distribution of arthropods in relation to forest patch size, edge, and stand characteristics. Canadian Journal of Forest Research 28:1068–1072.

Jordan ND, Ride JP, Rudd JJ, Davies EM, FranklinTong VE, Franklin FCH. 2000. Inhibition of self-incompatible pollen in *Papaver rhoeas* involves a complex series of cellular events. Annual Botany 85:197–202.

Jordan WC, Bruford MW. 1998. New perspectives on mate choice and the MHC. Heredity 81:239–245.

Jorgenson JP, Redford KH. 1993. Humans and big cats as predators in the Neotropics. Symposia of the Zoological Society of London 65:367–390.

Jorgenson JT, Festa-Bianchet M, Wishart WD. 1993. Harvesting bighorn ewes: consequences for population size and trophy ram production. Journal of Wildlife Management 57:429–435.

Jorgenson JT, Festa-Bianchet M, Wishart WD. 1998. Effects of population density on horn development in bighorn rams. Journal of Wildlife Management 62:1011–1020.

Jorgenson JT, Festa-Bianchet M, Lucherini M, Wishart WD. 1993. Effects of body size, population density and maternal characteristics on age of first reproduction in bighorn ewes. Canadian Journal of Zoology 71:2509–2517.

Jorgenson JT, Festa-Bianchet M, Gaillard J-M, Wishart WD. 1997. Effects of age, sex, disease and density on survival of bighorn sheep. Ecology 78:1019–1032.

Judson OP. 1994. The rise of the individual-based model in ecology. Trends in Ecology and Evolution 9:9–14.

Kaczensky P (ed). 2000. Co-existence of brown bear and men in the cultural landscape of Slovenia. Unpublished final report of the Project Meoved.

Karns PD 1998. Population distribution, density, and trends. In: The ecology and management of the North American moose (Franzmann AW, Schwartz CC, eds.). Washington, DC: Smithsonian Institution Press; 124–140.

Katajisto JK. 2001. Bed site selection of female brown bears (*Ursus arctos*) as a counter-strategy to avoid sexually selected infanticide by males. Master of Science thesis, University of Helsinki.

Kauffman JB, Krueger WC. 1984. Livestock impacts on riparian ecosystems and streamside management implications. a review. Journal of Range Management 37:430–438.

Kaufman L. 1992. Catastrophic change in a secies-rich freshwater ecosystem: the lessons of Lake Victoria. BioScience 42:846–858.

Kay CE. 1994a. Aboriginal overkill: the role of native Americans in structuring western ecosystems. Human Nature 5:359–98.

Kay CE. 1994b. The impact of native ungulates and beaver on riparian communities in the Intermountain West. Natural Resources and Environmental Issues 1:23–24.

Kay CE. 1998. Are ecosystems structured from the top-down or bottom-up? Wildlife Society Bulletin 26:484–498.

Keane B, Creel SR, Waser PM. 1996. No evidence of inbreeding avoidance or inbreeding depression in a social carnivore. Behavioral Ecology 7:480–489.

Keller LF. 1998. Inbreeding and its fitness effects in an insular population of song sparrows (*Melospiza melodia*). Evolution 52:240–250.

Keller LF, Arcese P. 1998. No evidence for inbreeding avoidance in a natural population of song sparrows (*Melospiza melodia*). American Naturalist 152: 380–392.

Keller LF, Waller DM. 2002. Inbreeding effects in wild populations. Trends in Ecology and Evolution 17:230–241.

Keller LF, Arcese P, Smith JNM, Hochachka WM, Stearns SC. 1994. Selection against inbred song sparrows during a natural population bottleneck. Nature 372:356–357.

Keller LF, Jeffery KJ, Arcese P, Beaumont MA, Hochachka WM, Smith JNM, Bruford M. 2001. Immigration and the ephemerality of natural population bottlenecks: evidence from molecular markers. Proceedings of the Royal Society of London, Series B 268:1387–1394.

Keller LF, Grant PR, Grant BR, Petren K. 2002. Environmental conditions affect the magnitude of inbreeding depression in survival of Darwin's finches. Evolution 56:1229–1239.

Kempenaers B, Verheyen GR, Vandenbroeck M, Burke T, Vanbroeckhoven C, Dhondt AA. 1992. Extra-pair paternity results from female preference for high-quality males in the blue tit. Nature 357:494–496.

Kendall BE, Gordon FA. 2002. Variation among individuals and reduced demographic stochasticity. Conservation Biology 16:109–116.

Khuri AI, Mathew T, Sinha BK. 1998. Statistical tests for mixed linear models. New York: Wiley.

Kimball RT, Ligon JD, MerolaZwartjes M. 1997. Fluctuating asymmetry in red junglefowl. Journal of Evolutionary Biology 10:441–457.

Kirkpatrick M. 1993. The evolution of size and growth in harvested natural populations. In: The exploitation of evolving resources (Stokes TK, McGlade JM, Law R, eds.). Berlin: Springer-Verlag; 145–154.

Kleiman DG. 1989. Reintroduction of captive mammals for conservation. BioScience 39:152–161.

Knight RR, Eberhardt LL. 1985. Population dynamics of Yellowstone grizzly bears. Ecology 66:323–334.

Knopf FL, Cannon RW. 1982. Structural resilience of a willow riparian community to changes in grazing practices. In: Wildlife–Livestock Relationship Symposium, 10 (JM Peek, Dalke PD, eds.). Moscow: University of Idaho; 198–207.

Kock R, Wambua J, Mwanzia J, Fitzjohn T, Manyibe T, Kambe S, Lergoi D. 1999. African hunting dog translocation from Mount Kenya (Timau) to Tsavo West National Park Kenya 1996–1998. Nairobi: World Wildlife Fund; unpublished report.

Kohlmann SG. 1999. Adaptive fetal sex allocation in elk: evidence and implications. Journal of Wildlife Management 63:1109–1117.

Kokko H, Lindström J , Ranta E. 2001. Life histories and sustainable harvesting. In: Conservation of exploited species (Reynolds JD, Mace GM, Redford KH, Robinson JG, eds.). Cambridge: Cambridge University Press; 301–322.

Kokko H, Mackenzie A, Reynolds JD, Lindström J, Sutherland WJ. 1999. Measures of inequality are not equal. American Naturalist 154:358–382.

Komdeur J, Huffstadt A, Prast W, Castle G, Mileto R, Wattel J. 1995. Transfer experiments of Seychelles warblers to new islands: changes in dispersal and helping behaviour. Animal Behaviour 49:695–708.

Komers PE, Curman CP. 2000. The effect of demographic characteristics on the success of ungulate re-introductions. Biological Conservation 93:187–193.

Kondrashov AS. 1993. Classification of hypotheses on the advantage of amphimixis. Journal of Heredity 84:372–387.

Kotiaho JS, Alatalo RV, Mappes J, Nielsen MG, Parri S, Rivero A. 1998. Energetic costs of size and sexual signalling in a wolf spider. Proceedings of the Royal Society of London, Series B 265:2203–2209.

Kozakiewicz M. 1993. Habitat isolation and ecological barriers: the effect on small mammal populations and communities. Acta Theriologica 38:1–30.

Krebs CJ. 2000. Hypothesis testing in ecology. In: Research techniques in animal ecology: controversies and consequences (Boitani L, Fuller TK, eds.). New York: Columbia University Press; 1–14.

Kristoffersson M. 2003. The choice of bed: a factor to avoid sex, rapes and murder? A study of habitat selection of female brown bears (*Ursus arctos*) as a possible counter strategy to avoid infanticide. Undergraduate thesis, Swedish University of Agricultural Sciences, Umeå.

Kruuk LEB, Clutton-Brock TH, Slate J, Pemberton JM, Brotherstone S, Guinness FE. 2000. Heritability of fitness in a wild mammal population. Proceedings of the National Academy of Sciences USA 97:698–703.

Kuitunen M, Mäkinen M. 1993. An experiment on nest site choice of the common treecreeper in fragmented boreal forest. Ornis Fennica 70:163–167.

Kunkel KE, Pletscher DH. 1999. Species-specific dynamics of cervids in a multi-predator system. Journal of Wildlife Management 63:1082–1093.

Kunkel KE, Ruth TK, Pletscher DH, Hornocker MG. 1999 Winter prey selection by wolves and cougars in and near Glacier National Park, Montana. Journal of Wildlife Management 63:901–910.

Lacy RC. 1995. Clarification of genetic terms and their use in the management of captive populations. Zoo Biology 14:565–577.

Lacy RC. 2000. Should we select genetic alleles in our conservation breeding programs? Zoo Biology 19:279–282.

Landa A, Tufto J, Franzén R, Bø T, Lindén M, Swenson JE. 1998. Active wolverine *Gulo gulo* dens as a minimum population estimator in Scandinavia. Wildlife Biology 4:159–168.

Landa A, Linnell JDC, Lindén M, Swenson JE, Røskaft E, Moksnes A. 2000. Conservation of Scandinavian wolverines in ecological and political landscapes. In: Mustelids in a modern world, management and conservation aspects of small carnivore:human interactions (Griffiths HI, ed.). Leiden: Backhuys Publishers; 1–20.

Lande R. 1995. Mutation and conservation. Conservation Biology 9:782–791.

Lande R. 1998a. Demographic stochasticity and Allee effect on a scale with isotropic noise. Oikos 83:353–358.

Lande R. 1998b. Anthropogenic, ecological and genetic factors in extinction. In: Conservation in a changing world (Mace GM, Balmford A, Ginsberg JR, eds.). Cambridge: Cambridge University Press; 29–51.

Landry J-M. 1997. Distribution potentielle du loup *Canis lupus* dans trois Canton Alpins Suisses: première analyses. Bulletin de la Société Neuchateloise des Sciences Naturelles 120:105–116.

Langvatn R, Loison A. 1999. Consequences of harvesting on age structure, sex ratio and population dynamics of red deer *Cervus elaphus* in central Norway. Wildlife Biology 5:213–223.

Langvatn R, Albon SD, Burkey T, Clutton-Brock TH. 1996. Climate, plant phenology and variation in age of first reproduction in a temperate herbivore. Journal of Animal Ecology 65:653–670.

Laurian C, Ouellet JP, Courtois R, Breton L, Onge SS. 2000. Effects of intensive harvesting on moose reproduction. Journal of Applied Ecology 37:515–531.

Law R. 2001. Phenotypic and genetic changes due to selective exploitation. In: Conservation of exploited species (Reynolds JD, Mace GM, Redford KH, Robinson JG, eds.). Cambridge: Cambridge University Press; 323–342.

Leal M. 1999. Honest signalling during prey–predator interactions in the lizard *Anolis cristatellus*. Animal Behaviour 58:521–526.

Lebreton JD, Millier C. 1982. Modèles dynamiques déterministes en biologie. Paris: Masson.

LeCount AL. 1983. Evidence of wild black bears breeding while raising cubs. Journal of Wildlife Management 47:264–267.

LeCount AL. 1987. Causes of black bear cub mortality. International Conference on Bear Research and Management 7:75–82.

Lee BY, Nelder JA. 1996. Hierarchical generalized linear models. Journal of the Royal Statistical Society, Series B 58:619–678.

Leech DI, Hartley IR, Stewart IRK, Griffith SC, Burke T. 2001. No effect of parental quality or extrapair paternity on brood sex ratio in the blue tit (*Parus caeruleus*). Behavioural Ecology 12:674–680.

Legendre S, Clobert J, Møller AP, Sorci G. 1999. Demographic stochasticity and social mating system in the process of extinction of small populations: the case of passerines introduced to New Zealand. American Naturalist 153:449–463.

Lens L, van Dongen S, Wilder CM, Brooks TM, Matthysen E. 1999. Fluctuating asymmetry increases with habitat disturbance in seven bird species of a fragmented afrotropical forest. Proceedings of the Royal Society of London, Series B 266:1241–1246.

Leoni G. 1995. Red deer in the Canton Ticino. Ufficio Caccia e Pesca. Lugano: Dipartimento del Territorio Bellinzona. [Italian]

Leopold A. 1933. Game management. New York: Charles Scribner's Sons.

Lercel BA, Kaminski RM, Cox RR, Jr. 1999. Mate loss in winter affects reproduction of mallards. Journal of Wildlife Management 63:621–629.

Leslie PH. 1945. On the use of matrices in population mathematics. Biometrika 33:182–212.

Leslie PH. 1966. The intrinsic rate of increase and the overlap of successive generations in a population of guillemot (*Uria aalge*). Journal of Animal Ecology 35:291–301.

Lessells CM, Boag PT. 1987. Unrepeatable repeatabilities: a common mistake. Auk 104:116–121.

Leuthold W. 1971. A note on the formation of food habits in young antelopes. East African Wildlife Journal 9:154–156.

Levin PS, Tolimieri N, Nicklin M, Sale PF. 2000. Integrating individual behavior and population ecology: the potential for habitat-dependent population regulation in a reef fish. Behavioral Ecology 11:565–571.

Levins R. 1969. Some demographic and genetic consequences of environmental heterogeneity for biological control. Bulletin of the Entomological Society of America 15:237–240.

Levitan DR, Sewell MA, Chia FS. 1992. How distribution and abundance influence fertilization success in the sea urchin *Strongylocentrotus franciscanus*. Ecology 73:248–254.

Lewis DM, Alpert P. 1997. Trophy hunting and wildlife conservation in Zambia. Conservation Biology 11:59–68.

Liberg O, Johansson A, Wahlström LK, Axen AH. 1996. Mating tactics and success in male roe deer. In: Territorial dynamics and marking behaviour in male roe deer (Johansson A.). Ph.D. thesis, University of Stockholm.

Liberg O, Johansson A, Andersen R, Linnell JDC. 1998. Mating systems, mating tactics and the function of male territory in roe deer. In: The European roe deer: the biology of success (Andersen R, Duncan P, Linnell JDC, eds.). Stockholm: Scandinavian University Press; 221–256.

Lidicker WZ, Koenig WD. 1996. Responses of terrestrial vertebrates to habitat edges and corridors. In: Metapopulations and wildlife conservation (McCullough DR, ed.). Washington DC: Island Press; 85–109.

Liermann M, Hilborn R. 1997. Depensation in fish stocks: a hierarchic Bayesian meta-analysis. Canadian Journal of Fisheries and Aquatic Science 54: 1976–1984.

Liew HC, Chan EH, Papi F, Luschi P. 1995. Long-distance migration of green turtles from Redang Island, Malaysia: the need for regional cooperation in sea turtle conservation. In: Proceedings International Congress Chelonian Conservation, Gonfaron, France (Deveraux B, Pritchard PCH, eds.). Gonfaron; 73–74.

Liley D. 2000. Predicting the consequences of human disturbance, predation and sea level rise for ringed plover populations. Ph.D. thesis, University of East Anglia.

Lima SL. 1987. Vigilance while feeding and its relation to the risk of predation. Journal of Theoretical Biology 124:303–316.

Lima SL. 1998. Nonlethal effects in the ecology of predator–prey interactions. Bioscience 48:25–34.

Lima SL, Zollner PA. 1996. Towards a behavioral ecology of ecological landscapes. Trends in Ecology and Evolution 11:131–135.

Limpus CJ. 1995. Global overview of the status of marine turtles: a 1995 viewpoint. In: Biology and conservation of sea turtles (Bjorndal KA, ed.). Washington, DC: Smithsonian Institution Press; 605–610.

Lin X, Breslow NE. 1996. Bias correction in generalized linear mixed models with multiple components of dispersion. Journal of the American Statistical Association 91:1007–1016.

Lindzey FG, Barber KR, Peters RD, Meslow EC. 1986. Responses of a black bear population to a changing environment. International Conference on Bear Research and Management 6:57–63.

Linnell JDC, Aanes R, Andersen R. 1995. Who killed Bambi? The role of predation in the neonatal mortality of temperate ungulates. Wildlife Biology 1:209–223.

Lizaso JLS, Goni R, Renones O, Charton G, Galzin R, Bayle JT, Jerez PS, Ruzafa AP, Maitland PS. 1995. The conservation of freshwater fish: past and present experience. Biological Conservation 72:259–270.

Lisazo JLS, Goni R, Renones O, Charton G, Galzin R, Bayle JT, Jeres PS, Ruzafa AP, Ramos AA. 2000. Density dependence in marine protected populations: a review. Environmental Conservation 27:144–158.

Logan KA, Sweanor LL. 2001. Desert puma: evolutionary ecology and conservation of an enduring carnivore. Washington DC: Island Press.

Lohmann KJ, Witherington BE, Lohmann CMF, Salmon M. 1997. Orientation, navigation and natal beach homing in sea turtles. In: The biology of sea turtles (Lutz PL, Musick JA, eds.). Boca Raton, FL: CRC Press; 107–136.

Loison A, Gaillard J-M, Jullien J-M. 1996. Demographic patterns after an epizootic of keratoconjunctivitis in a chamois population. Journal of Wildlife Management 60:517–527.

Loison A, Festa-Bianchet M, Gaillard JM, Jorgenson JT, Jullien J-M. 1999a. Age-specific survival in five populations of ungulates: evidence of senescence. Ecology 80:2539–2554.

Loison A, Gaillard JM, Pélabon C, Yoccoz NG. 1999b. What factors shape sexual size dimorphism in ungulates? Evolutionary Ecology Research 1:611–633.

Łomnicki A. 1978. Individual differences between animals and the natural regulation of their numbers. Journal of Animal Ecology 47:461–475.

Łomnicki A. 1988. Population ecology of individuals. Princeton University Press.

Long JC. 1986. The allelic correlation structure of Gainj- and Kalam-speaking people. I: The estimation and interpretation of Wright's F-statistics. Genetics 112:629–647.

Long JC, Romero FC, Urbanek M, Goldman D. 1998. Mating patterns and gene dynamics of an American Indian population isolate. Journal of Mammalogy 79:681–691.

Longford NT. 1993. Random coefficient models. Oxford: Clarendon Press.

Lovari S, Cosentino R. 1986. Seasonal habitat selection and group size of the Abruzzo chamois (*Rupicapra pyrenaica ornata*). Bollettino di Zoologia 53:73–78.

Luccarini S, Mauri L. 2000. Red deer in Susa Valley, spatial behavior and habitat use. Torino Province: Game and Wildlife Service. [Italian]

Lukefahr SD, Jacobson HA. 1998. Variance component analysis and heritability of antler traits in white-tailed deer. Journal of Wildlife Management 62:262–268.

Luschi P, Papi F, Liew H, Chan E, Bonadonna F. 1996. Long-distance migration and homing after displacement in the green turtle (*Chelonia mydas*): a satellite tracking study. Journal of Comparative Physiology 178A:447–452.

Luschi P, Hays GC, Del Seppia C, Marsh R, Papi F. 1998. The navigational feats of green sea turtles migrating from Ascension Island investigated by satellite telemetry. Proceedings of the Royal Society of London, Series B 265:2279–2284.

Lutcavage ME, Plotkin P, Witherington BE, Lutz PE. 1997. Human impacts and sea turtle survival. In: The biology of sea turtles (Lutz PE, Musick JA, eds.). Boca Raton, FL: CRC Press; 387–410.

Lynch M, Walsh B. 1998. Genetics and the analysis of quantitative traits. Sunderland, MA: Sinauer Associates.

MacArthur RH, Pianka ER. 1966. On optimal use of a patchy environment. American Naturalist 100:603–609.

Mace GM, Balmford A, Ginsberg J. 1999. Conservation in a changing world. Cambridge: Cambridge University Press.

Mace RD, Waller JS. 1997. Spatial and temporal interaction of male and female grizzly bears in northwestern Montana. Journal of Wildlife Management 61:39–52.

Machlis L, Dodd PWD, Fentress JC. 1985. The pooling fallacy: problems arising when individuals contribute more than one observation to the data set. Zeitschrift für Tierpsychologie 68:201–214.

Machtans CS, Villard M-A, Hannon SJ. 1996. Use of riparian buffer strips as movement corridors by forest birds. Conservation Biology 10:1366–1379.

Madsen T, Shine R, Olsson M, Wittzell H. 1999. Conservation biology: restoration of an inbred adder population. Nature 402:34–35.

Maehr DS. 1997. The comparative ecology of bobcat, black bear and Florida panther in South Florida. Bulletin of the Florida Museum of Natural History 40:1–176.

Magome DT. 1991. Habitat selection and the feeding ecology of the sable antelope in Pilanesberg National Park, Bophuthatswana. Master of Science dissertation, University of the Witwatersrand, Johannesburg.

Magoun AJ. 1985. Population characteristics, ecology and management of wolverines in northwestern Alaska. Ph.D. thesis, University of Alaska, Fairbanks.

Maillard D, Gaultier P, Boisaubert B. 1999. Revue de l'utilisation des différentes méthodes de suivi des populations de chevreuils (*Capreolus capreolus*) en France. Bulletin Mensuel de l'ONC 244:30–37.

Maitland PS. 1995. The conservation of freshwater fish: past and present experience. Biological Conservation 72:259–270.

Markusson E, Folstad I. 1997. Reindeer antlers: visual indicators of individual quality? Oecologica 110:501–507.

Marr AB, Keller LF, Arcese P. 2002. Heterosis and outbreeding depression in descendants of natural immigrants to an inbred population of song sparrows. Evolution 56:131–142.

Marshall JT, Dedrick KG. 1994. Endemic song sparrows and yellowthroats of San Francisco Bay. Studies in Avian Biology 15:316–327.

Martin TE, Clobert J. 1996. Nest predation and avian life-history evolution in

Europe versus North America: a possible role of humans? American Naturalist 147:1028–1046.

Matthysen E. 1999. Nuthatches (*Sitta europaea*) in forest fragments: demography of a patchy population. Oecologia 119:501–509.

Matthysen E, Currie D. 1996. Habitat fragmentation reduces disperser success in juvenile nuthatches *Sitta europaea*: evidence from patterns of territory establishment. Ecography 19:67–72.

Mattson DJ. 1997. Use of ungulates by Yellowstone grizzly bears. Biological Conservation 81:161–177.

Mattson DJ, Knight RR, Blanchard BM. 1987. The effects of development and primary roads on grizzly bear habitat use in Yellowstone National Park, Wyoming. International Conference on Bear Research and Management 7:259–273.

Mattson DJ, Knight RR, Blanchard BM. 1992. Cannibalism and predation on black bears by grizzly bears in the Yellowstone ecosystem, 1975–1990. Journal of Mammalogy 73:422–425.

Maynard Smith J. 1978. The evolution of sex. Cambridge: Cambridge University Press.

McAdam AG, Millar JS. 1999. Breeding lay young-of-the-year female deer mice: why weight? Ecoscience 6:400–405.

McCollin D. 1998. Forest edges and habitat selection in birds: a functional approach. Ecography 21:247–260.

McComb K, Pusey AE, Packer C, Grinnell J. 1993. Female lions can detect potentially infanticidal males from their roars. Proceedings of the Royal Society of London, Series B 252:59–64.

McCorquodale SM. 1999. Movements, survival, and mortality of black-tailed deer in the Klickitat Basin of Washington. Journal of Wildlife Management 63:861–871.

McCullagh P, Nelder JA. 1992. Generalized linear models. London: Chapman and Hall.

McCulloch CE. 1997. Maximum likelihood algorithms for generalised linear mixed models. Journal of the American Statistical Association 92:162–170.

McCullough DR. 1981. Population dynamics of the Yellowstone grizzly bear. In: Dynamics of large mammal populations (Fowler CW, Smith TD, eds.). New York: John Wiley and Sons; 173–196.

McCullough DR. 1996. Metapopulations and wildlife conservation. Washington DC: Island Press.

McElligott AG, Hayden TJ. 2000. Lifetime mating success, sexual selection and life history of fallow bucks (*Dama dama*). Behavioral Ecology and Sociobiology 48:203–210.

McLain DK. 1993. Cope's rule, sexual selection and the loss of ecological plasticity. Oikos 68:490–500.

McLain DK, Boulton MP, Redfearn TP. 1995. Sexual selection and the risk of extinction of birds on oceanic islands. Oikos 74:27–34.

McLean IG. 1997. Conservation and the ontogeny of behavior. In: Behavioral approaches to conservation in the wild (Clemmons JR, Buchholz R, eds). Chicago: University of Chicago Press; 132–156.

McLean IG, Lundie-Jenkins G, Jarman PJ. 1996. Teaching an endangered mammal to recognize predators. Biological Conservation 75:51–62.

McLellan BN. 1994. Density-dependent population regulation of brown bears. In: Density-dependent population regulation of brown, black, and polar bears (Taylor M, ed.). International Conference on Bear Research and Management 3:15–24.

McLellan BN, Hovey FW. 2001. Natal dispersal of grizzly bears. Canadian Journal of Zoology 79:838–844.

McLellan BN, Shackleton DM. 1988. Grizzly bears and resource extraction industries: effects of roads on behaviour, habitat use, and demography. Journal of Applied Ecology 25:451–460.

McLellan BN, Hovey FW, Mace RD, Woods JG, Carney DW, Gibeau ML, Wakkinen WL, Kasworm WF. 1999. Rates and causes of grizzly bear mortality in the interior mountains of British Columbia, Alberta, Montana, Washington, and Idaho. Journal of Wildlife Management 63:911–920.

McNamara JM, Houston AI. 1996. State-dependent life histories. Nature 380:215–221.

McNutt JW. 1996. Sex-biased dispersal in African wild dogs, *Lycaon pictus*. Animal Behaviour 52:1067–1077.

McShea WJ, Rappole JH. 1997. Herbivores and the ecology of forest understory birds. In: The science of overabundance (McShea WJ, Underwood HB, Rappole JH, eds.). Washington, DC: Smithsonian Institution Press; 298–309.

McShea WJ, Underwood HB, Rappole JH (eds.). 1997. The science of overabundance: deer ecology and population management. Washington, DC: Smithsonian Institution Press.

Mech LD. 1970. The wolf. Minneapolis: University of Minnesota Press.

Meriggi A, Lovari S. 1996. A review of wolf predation in southern Europe: does the wolf prefer wild prey to livestock? Journal of Applied Ecology 33:1561–1567.

Merila J, Andersson M. 1999. Reproductive effort and success are related to haematozoan infections in blue tits. Ecoscience 6:421–428.

Merila J, Gustafsson L. 1996. Temporal stability and microgeographic homogeneity of heritability estimates in a natural bird population. Journal of Heredity 87:199–204.

Merila J, Hemborg C. 2000. Fitness and feather wear in the collared flycatcher *Ficedula albicollis*. Journal of Avian Biology 31:504–510.

Messier F. 1991. The significance of limiting and regulating factors on the demography of moose and white-tailed deer. Journal of Animal Ecology 60:371–393.

Messier F. 1994. Ungulate population models with predation: a case study with the North American moose. Ecology 75:478–488.

Messier F, Taylor MK, Ramsay MA. 1992. Seasonal activity patterns of female polar bears (*Ursus maritimus*) in the Canadian Arctic as revealed by satellite telemetry. Journal of Zoology 226:219–229.

Michod RE, Levin BR. 1987. The evolution of sex: an examination of current ideas. Sunderland, MA: Sinauer Associates.

Milinski M, Bakker TCM. 1992. Costs influence sequential mate choice in sticklebacks, *Gasterosteus aculeatus*. Proceedings of the Royal Society of London, Series B 250:229–233.

Miller B, Reading RP, Forrest S. 1996. Prairie night: black-footed ferrets and the recovery of endangered species. Washington, DC: Smithsonian Institution Press.

Miller JD. 1997. Reproduction in sea turtles. In: The biology of sea turtles (Lutz PE, Musick JA, eds.). Boca Raton, FL: CRC Press; 51–82.

Miller RB. 1957. Have the genetic patterns of fishes been altered by introductions or by selective fishing? Journal of the Fisheries Research Board of Canada 14:797–806.

Miller SD. 1990a. Population management of bears in North America. International Conference on Bear Research and Management 8:357–373.

Miller SD. 1990b. Impact of increased bear hunting on survivorship of young bears. Wildlife Society Bulletin 18:462–467.

Miller SD, White GC, Sellers RA, Reynolds HV, Schoen JW, Titus K, Barnes VG, Smith RB, Nelson RR, Ballard WB, Schwartz CC. 1997. Brown and black bear density estimation in Alaska using radiotelemetry and replicated mark-resight techniques. Wildlife Monographs 133:1–155.

Mills MGL, Ellis S, Woodroffe R, Maddock A, Stander P, Rasmussen G, Pole A, Fletcher P, Bruford M, Wildt D, Macdonald DW, Seal US. 1998. Population and habitat viability assessment for the African wild dog (*Lycaon pictus*) in southern Africa. Apple Valley, MN: IUCN/SSC Conservation Breeding Specialist Group.

Milner JM, Pemberton JM, Brotherstone S, Albon SD. 2000. Estimating variance components and heritabilities in the wild: a case study using the "animal model" approach. Journal of Evolutionary Biology 13:804–813.

Milner-Gulland EJ, Coulson TN, Clutton-Brock TH. 2000. On harvesting a structured ungulate population. Oikos 88:592–602.

Ministry of Agriculture and Fisheries and Ministry for Regional Planning and the Environment. 2000. Action Plan for the preservation of pastoralism and the wolf in the Alpine Chain. Document T-PVS 21, Council of Europe:1–10.

Mitchell B, Lincoln GA. 1973. Conception dates in relation to age and condition in two populations of red deer in Scotland. Journal of Zoology 171:141–152.

Mladenoff DJ, Sickley TA, Wydeven AP. 1999. Predicting gray wolf landscape recolonization: logistic regression models vs. new field data. Ecological Applications 9:37–44.

Moehlmann PD. 1987. Ecology of cooperation in canids. In: Ecological aspects of social evolution (Rubenstein DI, Wrangham RW, eds.). Princeton: Princeton University Press; 64–86.

Moehlman PD, Amato G, and Runyoro V. 1996. Genetic and demographic threats to the black rhinoceros population in the Ngorongoro Crater. Conservation Biology 10:1107–1114.

Møller AP. 1994. Sexual selection and the barn swallow. Oxford: Oxford University Press; 1–365.

Møller AP. 2000. Sexual selection and conservation. In: Behavior and Conservation (Gosling LM, Sutherland WJ, eds.). Cambridge: Cambridge University Press; 161–171.

Møller AP, Alatalo RV. 1999. Good-genes effects in sexual selection. Proceedings of the Royal Society of London, Series B 266:85–91.

Møller AP, Christe P, Lux E. 1999. Parasitism, host immune function, and sexual selection. Quarterly Review of Biology 74:3–20.

Møller AP, Cuervo JJ. 1998. Speciation and feather ornamentation in birds. Evolution 52:859–869.

Møller AP, Legendre S. 2001. Allee effect, sexual selection and demographic stochasticity. Oikos 92:27–34.

Monfort SL, Schwartz CC, Wasser SK. 1993. Monitoring reproductive in captive moose using urinary and fecal steroid metabolites. Journal of Wildlife Management 57:400–407.

Montgomery ME, Ballou JD, Nurthen RK, England PR, Briscoe DA, Frankham R. 1997. Minimizing kinship in captive breeding programs. Zoo Biology 16:377–389.

Moorcroft PR, Albon SD, Pemberton JM, Stevenson IR, Clutton-Brock TH. 1996. Density-dependent selection in a fluctuating ungulate population. Proceedings of the Royal Society of London, Series B 263:31–38.

Morellet N, Champely S, Gaillard JM, Ballon P, Boscardin Y. 2001. The browsing index: a new tool to monitor deer populations from browsing pressure. Wildlife Society Bulletin 29:1243–1252.

Morowitz HJ. 1991. Balancing species preservation and economic considerations. Science 253:752–754.

Morreale SJ, Standora EA, Spotila JR, Paladino FV. 1996. Migration corridor for sea turtles. Nature 384:319–320.

Morris W, Doak D, Groom M, Kareiva P, Fieberg J, Gerber L, Murphy P, Thompson D. 1999. A practical handbook for population viability analysis. Washington, DC: The Nature Conservancy.

Mortimer JA, Balazs GH. 2000. Post-nesting migrations of hawksbill turtles in the Granitic Seychelles and implications for conservation. In: Proceedings 19th International Annual Symposium on Sea Turtle Biology and Conservation (Kalb HJ, Wibbels T, eds.). NOAA Technical Memo NMFS-SEFSC-443:22–26.

Muller HJ. 1932. Some genetic aspects of sex. American Naturalist 66:118–138.

Murcia C. 1995. Edge effects in fragmented forests: implications for conservation. Trends in Ecology and Evolution 10:58–62.

Murie A. 1940. Ecology of the coyote in Yellowstone. Fauna Series 4:1–206.

Murie A. 1981. The grizzlies of Mount McKinley. USDA National Park Service Monograph Series Number 14.

Murphy KM, Ross PI, Hornocker MG. 1999. The ecology of anthropogenic influences on cougars. In: Carnivores in ecosystems: the Yellowstone experience (Clark TW, Curlee AP, Minta SC, Kareiva PM, eds.). New Haven: Yale University Press.

Musick JA, Limpus CJ. 1997. Habitat utilization and migration in juvenile sea turtles. In: The biology of sea turtles (Lutz PL, Musick JA, eds.). Boca Raton, FL: CRC Press; 137–164.

Mustoni A. 2000. La reintroduzione dell'orso nelle Alpi Centrali. Trento: Rel. Int. Parco Naturale Adamello Brenta; pp. 134.

Myers JH, Krebs CJ. 1971. Genetic, behavioural and reproductive attributes of dispersing field voles, *Microtus pennsylvanicus* and *Microtus ochrogaster*. Ecological Monographs 41:53–78.

Myre TY. 2000. Strategies for female brown bear (*Ursus arctos*) to avoid infanticide: activity patterns. Thesis, Agricultural University of Norway, Ås.

Mysterud I. 1973. Behaviour of the brown bear (*Ursus arctos*) at moose kills. Norwegian Journal of Zoology 21:267–272.

Nagelkerken WP. 1979. Biology of the graysby, *Epinephelus cruentatus*, of the coral reef of Curacao. Studies of the Fauna of Curaçao and Other Caribbean Islands 61:1–118.

National Research Council. 1990. Decline of the sea turtles: causes and prevention. Washington, DC: National Academy Press.

Nee S, May RM. 1992. Dynamics of metapopulations: habitat destruction and competitive coexistence. Journal of Animal Ecology 61:37–40.

Nelson RK. 1983. Make prayers to the raven. Chicago: University of Chicago Press.

Newmark WD. 1996. Insularization of Tanzanian parks and the local extinction of large mammals. Conservation Biology 10:1549–1556.

Newton I. 1998. Population limitation in birds. London: Academic Press.

Nice MM. 1937. Studies in the life history of the song sparrow, vol. 1.

Nichols WJ, Resendiz A, Seminoff JA, Resendiz B. 2000. Transpacific migration of a loggerhead turtle monitored by satellite telemetry. Bulletin of Marine Sciences 67:937–947.

Nievergelt B. 1966. Der Alpensteinbock (*Capra ibex* L.) in seinem Lebensraum: ein ökologischer Vergleich. Mammalia Depicta. Hamburg and Berlin: Verlag P. Parey; 1–86.

Nievergelt B. 1967. Steinwild und Steinwildhege. Schweizer Naturschutz 33:154–157.

Nol E, Smith JNM. 1985. Effects of age and breeding experience on seasonal reproductive success in the song sparrow. Journal of Animal Ecology 56:301–314.

Nomura T. 1999. Effective population size in supportive breeding. Conservation Biology 13:670–672.

Noon BR, McKelvey KS. 1996. Management of the spotted owl: a case history in conservation biology. Annual Review of Ecology and Systematics 27:135–162.

Noon BR, Lamberson RH, Boyce MS, Irwin LL. 1999. Population viability analysis: a primer on its principal technical concepts. In: Ecological stewardship: a common reference for ecosystem management, vol. 2 (Szaro RC, Johnson NC, Sexton WT, Malk AJ, eds.). Oxford: Elsevier Science; 87–134.

Norrdahl K, Korpimaki E. 2002. Changes in individual quality during a 3-year population cycle of voles. Oecologia 130:239–249.

Norris DR, Stutchbury BJM. 2001. Extraterritorial movements of a forest songbird in a fragmented landscape. Conservation Biology 15:729–736.

Novaro AJ, Funes MC, Walker RS. 2000. Ecological extinction of native prey of a carnivore assemblage in Argentine Patagonia. Biological Conservation 92:25–33.

Novellie PA, Knight M. 1994. Repatriation and translocation of ungulates into South African national parks: an assessment of past attempts. Koedoe 37:115–119.

Novellie PA, Millar PS, Lloyd PH. 1996. The use of VORTEX simulation models in a long-term programme of reintroduction of an endangered large mammal, the Cape mountain zebra. Acta Oecologia 17:657–671.

Nowak RM. 1991. Walker's mammals of the world, vol. 2. 5th ed. Baltimore and London: Johns Hopkins University Press.

Nowell K, Jackson P. 1996. Wild cats: status survey and conservation action plan. Gland, Switzerland: IUCN.

Noyes JH, Johnson BK, Bryant LD, Findholt SL, Thomas JW. 1996. Effects of bull age on conception dates and pregnancy rates of cow elk. Journal of Wildlife Management 60:508–517.

Nunney L. 1993. The influence of mating system and overlapping generations on effective population size. Evolution 47:1329–1341.

Nunney L. 1996. The influence of variation in female fecundity on effective population size. Biological Journal of the Linnean Society 59:411–425.

Nunney L. 1999. The effective size of a hierarchically structured population. Evolution 53:1–10.

Nunney L, Campbell A. 1993. Assessing viable population size: demography meets population genetics. Trends in Ecology and Evolution 8:234–239.

Nunney L, Elam DR. 1994. Estimating the effective population size of conserved populations. Conservation Biology 8:175–184.

Ober C, Weitkamp LR, Cox N, Dytch H, Kostyu D, Elias S. 1997. HLA and mate choice in humans. American Journal of Human Genetics 61:497–504.

Oedekoven MA, Joern A. 2000. Plant quality and spider predation affects grasshoppers (Acrididae): food-quality-dependent compensatory mortality. Ecology 81:66–77.

Oftedal OT. 1984. Milk composition, milk yield and energy output at peak lactation. Symposium of the Zoological Society of London 51:33–85.

Ohmart RD. 1994. The effects of human-induced changes in the avifauna of western riparian habitats. Studies in Avian Biology 15:273–285.

Okarma H. 1993. Status and management of the wolf in Poland. Biological Conservation 66:153–158.

Olsen, TL. 1993. Infanticide in brown bears, *Ursus arctos*, at Brooks River, Alaska. Canadian Field-Naturalist 107:92–94.

Opdam P, Rijsdijk G, Hustings F. 1985. Bird communities in small woods in an agricultural landscape: effects of area and isolation. Biological Conservation 34:333–352.

Orians GH, Cochran PA, Duffield JW, Fuller TK, Gutierrez RJ, Hanemann MW, James FJ, Karieva P, Kellert SR, Klein D, McLellan BN, Olson PD, Yaska G. 1997. Wolves, bears, and their prey in Alaska: biological and social challenges in wildlife management. Washington, DC: National Academy Press.

Osti F. 1988. L'orso bruno in Trentino: presenza e distribuzione nel quinquennio 1982–1986. Natura alpina. Anno 39. N. 3–4.51–55. [Italian]

Osti F. 1994. The brown bear (*Ursus arctos* L.) in Trentino, Italy: geographical distribution and population size during 1987–91. Atti dei Convegno—L'Orso bruno nelle Regioni di Alpe, Ljubljana 29–30.06.1992. Ljubljana; 103–109.

Osti F. 1999. L'orso bruno nel Trentino: distribuzione, biologia, ecologia e protezione della specie. Trento: Edizioni Arca. [Italian]

Ott L, Longnecker MT. 2000. Introduction to statistical methods and data analysis. Pacific Grove, CA: Brooks/Cole.

Owens IPF, Bennett PM. 1994. Mortality costs of parental care and sexual dimorphism in birds. Proceedings of the Royal Society of London, Series B 257:1–8.

Owens IPF, Bennett PM. 1995. Ancient ecological diversification explains life-history variation among living birds. Proceedings of the Royal Society of London, Series B 261:227–232.

Owen-Smith N. 1979. Assessing the forage efficiency of a large herbivore, the kudu. South African Journal of Wildlife Research 9:102–110.

Owen-Smith N. 1982. Factors influencing the consumption of plant products by large herbivores. In: The ecology of tropical savannas (Huntley BJ, Walker BH, eds.). Berlin and Hamburg: Springer Verlag; 359–404.

Owen-Smith N. 1985. Niche separation among African ungulates. In: Species and speciation (Vrba ES, ed.). Transvaal Museum Monograph, no. 4. Pretoria: Transvaal Museum; 167–171.

Owen-Smith RN. 1988. Megaherbivores: the influence of very large body size on ecology. Cambridge: Cambridge University Press.

Owen-Smith N. 1989. Morphological factors and their consequences for resource partitioning among African savanna ungulates: a simulation modeling approach. In: Patterns in the structure of mammalian communities (Morris

DW, Abramsky Z, Fox BJ, Willig ML, eds.). Lubbock: Texas Tech Museum; 155–165.

Owen-Smith N. 1993. Evaluating optimal diet models for an African browsing ruminants, the kudu: how constraining are the assumed constraints? Evolutionary Ecology 7:499–524.

Owen-Smith N. 1994. Foraging responses of kudus to seasonal changes in food resources: elasticity in constraints. Ecology 75:1050–1062.

Owen-Smith N. 2002. Adaptive herbivore ecology: from resources to populations in variable environments. Cambridge: Cambridge University Press.

Owen-Smith N, Cooper SM. 1987. Palatability of woody plants to browsing ungulates in a South African savanna. Ecology 68:319–331.

Owen-Smith N, Cooper SM. 1989. Nutritional ecology of a browsing ruminant, the kudu, through the seasonal cycle. Journal of Zoology, London 219:29–43.

Owen-Smith N, Novellie P. 1982. What should a clever ungulate eat? American Naturalist 119:151–178.

Packer C, Pusey AE. 1983a. Adaptations of female lions to infanticide by incoming males. American Naturalist 121:716–728.

Packer C, Pusey AE. 1983b. Male takeovers and female reproductive parameters: a simulation of oestrous synchrony in lions (*Panthera leo*). Animal Behavior 31:334–340.

Packer C, Pusey AE. 1984. Infanticide in carnivores. In: Infanticide: comparative and evolutionary perspectives (Hausfater G, Hrdy SB, eds.). New York: Aldine de Gruyter; 31–42.

Packer C, Scheel D, Pusey AE. 1990. Why lions form groups: food is not enough. American Naturalist 136:1–19.

Paine RT. 1966. Food web complexity and species diversity. American Naturalist 100:65–75.

Palmer AR. 2000. Quasireplication and the contract of error: lessons from sex ratios, heritabilities and fluctuating asymmetry. Annual Review of Ecology and Systematics 31:441–480.

Papi F, Luschi P. 1996. Pinpointing "Isla Meta": the case of sea turtles and albatrosses. Journal of Experimental Biology 199:65–71.

Papi F, Luschi P, Crosio E, Hughes GR. 1997. Satellite tracking experiments on the navigation ability and migratory behaviour of the loggerhead turtle *Caretta caretta*. Marine Biology 129:215–220.

Papi F, Luschi P, Akesson S, Capogrossi S, Hays GC. 2000. Open-sea migration of magnetically disturbed sea turtles. Journal of Experimental Biology 203:3435–3443.

Parker GA, Partridge L. 1998. Sexual conflict and speciation. Philosophical Transactions of the Royal Society, Series B 353:261–274.

Parker PG, Waite TA. 1997. Mating systems, effective population size, and conservation of natural populations. In: Behavioral approaches to conservation in the wild (Clemmons JR, Buchholz R, eds.). Cambridge: Cambridge University Press; 243–261.

Passerin d'Entreves P. 2000. Les Chasses Royales in Valle d'Aosta (1850–1919). Torino: Umberto Allemandi Ed.

Pastor JB. Dewey J, Naiman RJ, McInnes PF, Cohen Y. 1993. Moose browsing and soil fertility in the boreal forests of Isle Royale National Park. Ecology 74:467–480.

Paton PWC. 1994. The effect of edge on avian nest success: how strong is the evidence? Conservation Biology 8:17–26.

Patton JL, Feder FH. 1981. Microspatial genetic heterogeneity in pocket gophers: non-random breeding and drift. Evolution 35:912–920.

Pauly D. 1980. On the interrelationships between natural mortality, growth parameters and mean environmental temperature in 175 fish stocks. Journal du Conseil International pour l'Exploration de la Mer 39:195–212.

Pearson AM. 1975. The northern interior grizzly bear *Ursus arctos*. Canadian Wildlife Service Report Series, no. 34. Ottawa, ON: Canadian Wildlife Service.

Peek JM. 1998. Habitat relationships. In: Ecology and management of the North American moose (Franzmann, AW, Schwartz CC, eds.). Washington, DC: Smithsonian Institution Press; 351–376.

Pelabon C, van Breukelen L. 1998. Asymmetry in antler size in roe deer (*Capreolus capreolus*): an index of individual and population conditions. Oecologia 116:1–8.

Pelabon C, Joly P. 2000. What, if anything, does visual asymmetry in fallow deer antlers reveal? Animal Behaviour 59:193–199.

Pellew RA. 1984. Food consumption and energy budgets of the giraffe. Journal of Applied Ecology 21:141–159.

Penn D, Potts W. 1998. MHC-disassortative mating preferences reversed by cross-fostering. Proceedings of the Royal Society of London, Series B 265:1299–1306.

Penn DJ, Potts WK. 1999. The evolution of mating preferences and major histocompatibility complex genes. American Naturalist 153:145–164.

Persson J, Willebrand T, Landa A, Andersen R, Segerström P. In press. The role of intraspecific predation in the survival of juvenile wolverines *Gulo gulo*. Wildlife Biology.

Peters A, Michiels NK. 1996. Evidence for lack of inbreeding avoidance by selective simultaneous hermaphrodite. Invertebrate Biology 115:99–103.

Petersen CW, Warner RR, Cohen S, Hess HC, Sewell AT. 1992. Variable pelagic fertilization success: implications for mate choice and spatial patterns of mating. Ecology 73:391–401.

Peterson RO. 1977. Wolf ecology and prey relationships on Isle Royale. National Park Service Science Monograph Series 11. Washington, DC: U.S. Government Printing Office.

Peterson RO. 1995. The wolves of Isle Royale: a broken balance. Minocqua, WI: Willow Creek Press.

Peterson RO. 1999. Wolf–moose interaction on Isle Royale: the end of natural regulation? Ecological Applications 9:10–16.

Petrie M, Doums C, Møller AP. 1998. The degree of extra-pair paternity increases with genetic variability. Proceedings of the National Academy of Sciences USA 95:9390–9395.

Petrie M, Halliday T, Sanders C. 1991. Peahens prefer peacocks with elaborate trains. Animal Behavior 41:323–331.

Petrie M, Kempenaers B. 1998. Extra-pair paternity in birds: explaining variation between species and populations. Trends in Ecology and Evolution 13:52–58.

Petrie M, Lipsitch M. 1994. Avian polygyny is most likely in populations with high variability in heritable male fitness. Proceedings of the Royal Society of London, Series B 256:275–280.

Petrie M, Williams A. 1993. Peahens lay more eggs for peacocks with larger trains. Proceedings of the Royal Society of London, Series B 251:127–131.

Pettifor RA, Norris KJ, Rowcliffe M. 2000. Incorporating behaviour in predictive models for conservation. In: Behaviour and conservation (Gosling LM, Sutherland WJ, eds.). Cambridge: Cambridge University Press; 198–220.

Pettorelli N, Gaillard JM, Van Laere G, Duncan P, Kjellander P, Liberg O, Delorme D, Maillard D. 2002. Variations in adult body mass in roe deer: the effects of population density at birth and of habitat quality. Proceedings of the Royal Society of London, Series B 269:747–753.

Pfister CA. 1998. Patterns of variance in stage-structured populations: evolutionary predictions and ecological implications. Proceedings of the National Academy of Sciences USA 95:213–218.

Philips JA, Alberts AC. 1992. Naive ophipahus lizards recognize and avoid venomous snakes using chemical cues. Journal of Chemical Ecology 18: 1775–1783.

Phillips M. 1995. Conserving the red wolf. Canid News 3:13–17.

Phillips MK, Smith DW. 1996. The wolves of Yellowstone. Stillwater, MN: Voyageur Press.

Pielowski Z. 1984. Some aspects of population structure and longevity of field roe deer. Acta Theriologica 29:17–33.

Piersma T, Baker AJ. 2000. Life history characteristics and the conservation of migratory shorebirds. In: Behavior and conservation (Gosling LM, Sutherland WJ, eds.). Cambridge: Cambridge University Press; 105–124.

Pinheiro JC, Bates DM. 2000. Mixed-effects models in S and S-PLUS. New York: Springer.

Pokki J. 1981. Distribution, demography and dispersal of the field vole, *Microtus agrestis* (L.) in the Tvarminne archipelago, Finland. Acta Zoologica Fennica 164:1–48.

Policansky D. 1993. Fishing as a cause of evolution in fishes. In: The exploitation of evolving resources (Stokes TK, McGlade JM, Law R, ed.). Berlin: Springer-Verlag; 2–18.

Polis GA, Strong DR. 1996. Food web complexity and community dynamics. American Naturalist 147:813–846.

Polovina JJ, Kobayashi DR, Parker DM, Seki MP, Balazs GH. 2000. Turtles on the edge: movement of loggerhead turtles (*Caretta caretta*) along oceanic fronts, spanning longline fishing grounds in the central North Pacific. 1997–1998. Fisheries Oceanography 9:71–82.

Pomiankowski A, Møller AP. 1995. A resolution of the lek paradox. Proceedings of the Royal Society of London, Series B 260:21–29.

Poole JH. 1989. Mate guarding, reproductive success and female choice in African elephants. Animal Behaviour 37:842–849.

Pope TR. 1992. The influence of dispersal patterns and mating system on genetic differentiation within and between populations of the red howler monkey (*Alouatta seniculus*). Evolution 46:1112–1128.

Pope TR. 1998. Effects of demographic change on group kin structure and gene dynamics of red howling monkey populations. Journal of Mammalogy 79:692–712.

Post E, Langvatn R, Forchhamer MC, Stenseth N. 1999. Environmental variation shapes sexual dimorphism in red deer. Proceedings of the National Academy of Sciences USA 96:4467–4471.

Potts WK, Manning CJ, Wakeland EK. 1991. Mating patterns in seminatural populations of mice influenced by MHC genotype. Nature 352:619–621.

Potvin F, Jolicoeur H, Breton L, Lemieux R. 1992. Evaluation of an experimental wolf reduction and its impact on deer in Papineau-Labelle Reserve, Quebec. Canadian Journal of Zoology 70:1595–1603.

Poulle M, Dahier T, de Beaufort R, Durand C. 2000. Conservation du loup en France. Programme Life-Nature, Rapport final 1997–1999. Programme Life-Loup, French Ministry of Environment, Office National de la Chasse.

Powell RA. 1986. Black bear home range overlap in North Carolina and the concept of home range applied to black bears. International Conference on Bear Research and Management 7:235–242.

Power ME. 1992. Top-down and bottom-up forces in food webs: do plants have primacy? Ecology 73:733–746.

Primack RB. 1998. Essentials of conservation biology. 2nd ed. Sunderland, MA: Sinauer Associates.

Pritchard PCH. 1997. Evolution, phylogeny and current status. In: The biology of sea turtles (Lutz PE, Musick JA, eds.). Boca Raton, FL: CRC Press, 1–28.

Promberger C, Schroeder S. 1992. Wolves in Europe: current status and prospects. Oberammergau, Germany: Wildbiologische Gesellschaft München.

Promislow DEL, Harvey PH. 1990. Living fast and dying young: a comparative analysis of life-history variation among mammals. Journal of Zoology 220:417–437.

Promislow DEL, Montgomerie RD, Martin TE. 1992. Mortality costs of sexual dimorphism in birds. Proceedings of the Royal Society of London, Series B 250:143–150.

Provenza FD, Balph DF. 1987. Diet learning by domestic ruminants: theory, evidence and practical implications. Applied Animal Behavior Science 18:211–232.

Provenza FD, Pfister JA, Cheney CD. 1992. Mechanisms of learning in diet selection with reference to phytotoxicosis in herbivores. Journal of Range Management 45:36–45.

Pusey AE. 1987. Sex-biased dispersal and inbreeding avoidance in birds and mammals. Trends in Ecology and Evolution 2:295–299.

Pusey AE, Packer C. 1987. The evolution of sex-biased dispersal in lions. Behaviour 101:275–310.

Pusey AE, Packer C. 1994a. Infanticide in lions: consequences and counterstrategies. In: Infanticide and parental care (Parmigiani S, vom Saal FS, eds.). London: Harwood Academic; 277–299.

Pusey AE, Packer C. 1994b. Non-offspring nursing in social carnivores: minimizing the costs. Behavioral Ecology 5:362–374.

Pusey AE, Wolf M. 1996. Inbreeding avoidance in animals. Trends in Ecology and Evolution 11:201–206.

Pyare S, Berger J. 2003. Beyond demography and delisting: ecological recovery for Yellowstone's grizzly bears and wolves. Biological Conservation: In press.

Quinn JF, Dunham AE. 1983. On hypothesis testing in ecology and evolution. American Naturalist 122:602–617.

Ramsey FL, Schafer DW. 1997. The statistical sleuth. Belmont, CA: Duxbury Press.

Rauer G, Gutleb B. 1997. Der Braunbar im Österreich. Monographien, Band 88. Wien (Vienna).

Roth HU. 1983. Diel activity of a remnant population of European brown bears. International Conference on Bear Research and Management 5:223–229.

Read AF, Harvey PH. 1989. Life history differences among the eutherian radiations. Journal of Zoology 219:329–353.

Real LA. 1991. Search theory and mate choice. II: Mutual interaction, assortative mating, and equilibrium variation in male and female fitness. American Naturalist 138:901–917.

Réale D, Festa-Bianchet M, Jorgenson JT. 1999. Heritability of body mass varies with age and season in wild bighorn sheep. Heredity 83:526–532.

Redford KH. 1992. The empty forest. Bioscience 42:412–422.

Redford KH, Feinsinger P. 2001. The half-empty forest: sustainable use and the ecology of interactions. In: Conservation of exploited species (Reynolds J, Mace G, Redford KH, Robinson JH, eds.). Cambridge: Cambridge University Press.

Redpath SM. 1995. Habitat fragmentation and the individual: tawny owls *Strix aluco* in woodland patches. Journal of Animal Ecology 64:652–661.

Reed JM, Dobson AP. 1993. Behavioural constraints and conservation biology: conspecific attraction and recruitment. Trends in Ecology and Evolution 8:253–256.

Reichel JD, Wiles GJ, Glass PO. 1992. Island extinctions: the case of the endangered nightingale reedwarbler. Wilson Bulletin 104:44–54.

Reid JM, Bignal EM, Bignal S, McCracken DI, Monaghan P. 2003. Environmental variability, life-history covariation and cohort effects in the red-billed chough. Journal of Animal Ecology 72:36–46.

Reimchen TE, Nosil P. 2001. Lateral plate asymmetry, diet and parasitism in threespine stickleback. Journal of Evolutionary Biology 14:632–645.

Reynolds JD, Mace GM, Redford KH, Robinson JG (eds.). 2001. Conservation of exploited species. Cambridge: Cambridge University Press.

Reznick DA, Bryga H, Endler JA. 1990. Experimentally induced life-history evolution in a natural population. Nature 346:357–359.

Reznick DN. 1993. Norms of reaction in fishes. In: The exploitation of evolving resources (Stokes TK, McGlade JM, Law R, eds.). Berlin: Springer-Verlag; 72–90.

Reznikoff Etievant MF, Bonneau JC, Alcalay D, Cavelier B, Toure C, Lobet R, Netter A. 1991. HLA antigen-sharing in couples with repeated spontaneous abortions and the birthweight of babies in successful pregnancies. American Journal of Reproduction and Immunology 25:25–27.

Richardson P. 2000. Obstacles to objectivity: first impressions of a CITES CoP. Marine Turtle Newsletter 89:1–4.

Ricklefs RE. 1969. An analysis of nesting mortality in birds. Smithsonian Contributions to Zoology 9:1–48.

Rijnsdorp AD. 1993. Fisheries as a large-scale experiment on life-history evolution: disentangling phenotypic and genetic effects in changes in maturation and reproduction of North Sea Plaice, *Pleuronectes Platessa*. Oecologia 96:391–401.

Ripple WJ, Larsen EJ, Renkin RA, Smith DW. 2001. Trophic cascades among wolves, elk and aspen on Yellowstone National Park's northern Range. Biological Conservation 102:227–234.

Ritter JC, Bednekoff PA. 1995. Dry season water, female movements and male territoriality in springbok: preliminary evidence of waterhole-directed sexual selection. African Journal of Ecology 33:395–404.

Robbins CT, Spalinger DE, van Hoven W. 1995. Adaptations of ruminants to browse and grass diets: are anatomical-based browser-grazer interpretations valid? Oecologia 103:208–213.

Robinson SK, Thompson FR, Donovan TM, Whitehead DR, Faaborg J. 1995. Regional forest fragmentation and the nesting success of migratory birds. Science 267:1987–1990.

Rochet MJ. 1998. Short-term effects of fishing on life history traits of fishes. ICES Journal of Marine Science 55:371–391.

Rochet MJ, Cornillon PA, Sabatier R, Pontier D. 2000. Comparative analysis of phylogenetic and fishing effects in life history patterns of teleost fishes. Oikos 91:255–270.

Ross PI, Jalkotzy MG. 1992. Characteristics of a hunted population of cougars in southwestern Alberta. Journal of Wildlife Management 56:417–426.

Ross PI, Jalkotzy MG. 1996. Cougar predation on moose in southwestern Alberta. Alces 32:1–8.

Ross PI, Jalkotzy MG, Festa-Bianchet M. 1997. Cougar predation on bighorn sheep in southwestern Alberta during winter. Canadian Journal of Zoology 75:771–775.

Roth HU. 1983. Diel activity of a remnant population of European brown bears. International Conference on Bear Research and Management 5:223–229.

Roth HU, Osti E. 1979. Prime esperienze di radiolocalizzazione di due Orsi bruni del Trentino. Natura alpina 17:27–37.

Rothstein SI. 1990. A model system for coevolution: avian brood parasitism. Annual Review of Ecology and Systematics 21:481–508.

Roulin A, Richner H, Ducrest AL. 1998. Genetic, environmental, and condition-dependent effects on female and male ornamentation in the barn owl *Tyto alba*. Evolution 52:1451–1460.

Rowell CA. 1993. The effects of fishing on the timing of maturity in North Sea cod (*Gadus morhua* L.). In: The exploitation of evolving resources (Stokes TK, McGlade JM, Law R, eds.). Berlin: Springer-Verlag; 44–61.

Rülicke T, Chapuisat M, Homberger FR, Macas E, Wedekind C. 1998. MHC-genotype of progeny influenced by parental infection. Proceedings of the Royal Society of London, Series B 265:711–716.

Ruusila V, Poysa H, Runko P. 2001. Female wing plumage reflects reproductive success in common goldeneye *Bucephala clangula*. Journal of Avian Biology 32:1–5.

Ryman N, Jorde PE, Laikre L. 1995. Supportive breeding and variance effective population size. Conservation Biology 9:1619–1628.

Ryman N, Jorde PE, Laikre L. 1999. Supportive breeding and inbreeding effective number: reply to Nomura. Conservation Biology 13:673–676.

Ryman N, Laikre L. 1991. Effects of supportive breeding on the genetically effective population-size. Conservation Biology 5:325–329.

Saab VA, Bock CE, Rich TD, Dobkin DS. 1995. Livestock grazing effects in western North America. In: Population ecology and conservation of neotropical migrant birds (Martin TE, Finch DM, eds.). New York: Oxford University Press; 311–353.

Saccheri I, Kuussaari M, Kankare M, Vikman P, Fortelius W, Hanski I. 1998. Inbreeding and extinction in a butterfly metapopulation. Nature 392:491–494.

Sadovy Y. 1994. Grouper stocks of the western central Atlantic: the need for management and management needs. Proceedings of the Gulf and Caribbean Fisheries Institute 43:43–64.

Sadovy Y, Eklund AM. 1999. Synopsis of biological data on the Nassau grouper, *Epinephelus striatus* (Bloch, 1792), and the jewfish, *E. itajara* (Lichtenstein, 1822). NOAA Technical Report NMFS 146:1–65.

Sæther BE. 1990. Age-specific variation in reproductive performance in birds. Current Ornithology 7:251–283.

Sæther BE, Bakke O. 2000. Avian life history variation and contribution of demographic traits to the population growth rate. Ecology 81:642–653.

Sæther BE, Gordon IJ. 1994. The adaptive significance of reproductive strategies in ungulates. Proceedings of the Royal Society of London, Series B 256:263–268.

Sæther BE, Engen S, Swenson JE, Bakke Ø, Sandegren F. 1998. Viability of Scandinavian brown bear *Ursus arctos* populations: the effects of uncertain parameter estimates. Oikos 83:403–416.

Saino N, Bolzern AM, Møller AP. 1997. Immunocompetence, ornamentation and viability in male barn swallows (*Hirundo rustica*). Proceedings of the Royal Society of London, Series B 94:549–552.

Saino N, Møller AP. 1996. Sexual ornamentation and immunocompetence in the barn swallow. Behavioral Ecology 7:227–232.

Saino N, Calza S, Ninni P, Møller AP. 1999. Barn swallows trade survival against offspring condition and immunocompetence. Journal of Animal Ecology 68:999–1009.

Saltz D, Rowen M, Rubenstein DI. 2000. The effects of space use patterns of reintroduced Asiatic wild ass on effective population size. Conservation Biology 14:1852–1861.

Saltz D, Rubenstein DI. 1995. Population dynamics of a reintroduced Asiatic wild ass herd. Ecological Applications 5:327–335.

Santos M. 2001. Fluctuating asymmetry is nongenetically related to mating success in male *Drosophila buzzatii*. Evolution 55:2248–2256.

Sargeant AB, Allen SH. 1989. Observed interactions between coyotes and red foxes. Journal of Mammalogy 70:631–633.

Sargeant AB, Allen SH, Eberhardt RT. 1984. Red fox predation on breeding ducks in midcontinent North America. Wildlife Monographs 89:1–41.

Sargeant AB, Allen SH, Hastings SO. 1987. Spatial relations between sympatric coyotes and red foxes in North Dakota. Journal of Wildlife Management 51:285–293.

Sargeant GA, Ruff RL. 2001. Demographic response of black bears at Cold Lake, Alberta, to the removal of adult males. Ursus 12:59–68.

Sauer JR, Slade NA. 1987. Size-based demography of vertebrates. Annual Review of Ecology and Systematics 18:71–90.

Scandura M, Apollonio M, Mattioli L. 2001. Recent recovery of the Italian wolf population: a genetic investigation using microsatellites. Mammalian Biology 66:321–331.

Schlichting CD, Pigliucci M. 1998. Phenotypic evolution: a reaction norm perspective. Sunderland, MA: Sinauer Associates.

Schluter D, Smith JNM. 1986. Natural selection on beak and body size in the song sparrow. Evolution 40:221–231.

Schmidt KT. 1993. Winter ecology of nonmigratory alpine red deer. Oecologia 95:226–233.

Schmidt KT, Stien A, Albon SD, Guinness FE. 2001. Antler length of yearling red deer is determined by population density, weather and early life-history. Oecologia, 127:191–197.

Schmitz OJ, Sinclair ARE. 1997. Rethinking the role of deer in forest ecosystems.

In: The science of overabundance (McShea WJ, Underwood HB, Rappole JH, eds.). Washington, DC: Smithsonian Institution Press; 201–223.

Schneider RR, Wasel S. 2000. The effect of human settlement on the density of moose in northern Alberta. Journal of Wildlife Management 64:513–520.

Schultz CB. 1998. Dispersal behaviour and its implications for reserve design in a rare Oregon butterfly. Conservation Biology 12:284–292.

Schwabl H. 1993. Yolk is a source of maternal testosterone for developing birds. Proceedings of the National Academy of Sciences USA 90:11446–11450.

Schwabl H, Mock DW, Gieg JA. 1997. A hormonal mechanism for parental favouritism. Nature 386:231.

Schwartz OA, Armitage, KB. 1980. Genetic variation in social mammals: the marmot model. Science 207:665–667.

Schwartz CC, Franzmann AW. 1991. Interrelationships of black bears to moose and forest succession in the northern coniferous forest. Wildlife Monographs 113:1–58.

Schwartz CC, Monfort SL, Dennis PH, Hundertmark KH. 1995. Fecal progestagen concentration as an indicator of the estrous cycle and pregnancy in moose. Journal of Wildlife Management 59:580–583.

Schwarz CJ, Seber GAF. 2001. Estimating animal abundance: review 3. Statistical Science 1:134.

Searle SR, Casella G, McCulloch CE. 1992. Variance components. New York: Wiley.

Seber GAF. 1986. A review of estimating animal abundance. Biometrics 42:267–292.

Seehausen O. 2000. Explosive speciation rates and unusual species richness in haplochromine cichlid fishes: effects of sexual selection. Advances in Ecological Research 31:237–274.

Seehausen O, van Alphen JJM, Witte F. 1997. Cichlid fish diversity threatened by eutrophication that curbs sexual selection. Science 277:1808–1811.

Seidensticker J, Lahiri JE, Das KC, Wright A. 1976. Problem tiger in the Sundarbans. Oryx 11:267–273.

Servheen C, Herrero S, Peyton B. 1999. Bears: status survey and conservation action plan. Gland, Switzerland: IUCN/SSC Bear and Polar Bear Specialist Groups.

Shackleton DM. 1997. Wild sheep and goats and their relatives: status survey and conservation action plan for Caprinae. Gland, Switzerland: IUCN.

Shapiro DY. 1980. Serial female sex changes after simultaneous removal of males from social groups of a coral reef fish. Science 209:1136–1137.

Short RV, Baladan E. 1994. The differences between the sexes. Cambridge: Cambridge University Press.

Short J, Kinnear JE, Robley A. 2002. Surplus killing by introduced predators in Australia: evidence for ineffective antipredator adaptations in native prey species? Biological Conservation 103:283–301.

Short J, Bradshaw SD, Giles J, Prince RIT, Wilson GR. 1992. Reintroduction of macropods in Australia: a review. Biological Conservation 62:189–204.

Shun Z. 1997. Another look at the salamander mating data: a modified Laplace approximation approach. Journal of the American Statistical Association 92:341–349.

Shykoff JA, Møller AP. 1999. Fitness and asymmetry under different environmental conditions in the barn swallow. Oikos 86:152–158.

Sih A, Wooster DE. 1994. Prey behavior, prey dispersal, and predator impacts on stream prey. Ecology 75:1199–1207.

Sillero-Zubiri C, Gottelli D, Macdonald DW. 1996. Male philopatry, extra-pair copulations and inbreeding avoidance in the Ethiopian wolf (*Canis simensis*). Behavioral Ecology and Sociobiology 38:331–340.

Simberloff D, Farr JA, Cox J, Mehlman DW. 1992. Movement corridors: conservation bargains or poor investments? Conservation Biology 6:493–504.

Sinclair ARE. 1989. Population regulation in animals. In: Ecological concepts: the contribution of ecology to an understanding of the natural world. (Cherrett JM, ed.). Oxford: Blackwell Science; 197–241.

Sinclair ARE, Arcese P. 1995a. Population consequences of predation-sensitive foraging: the Serengeti wildebeest. Ecology 76:882–891.

Sinclair ARE, Arcese P. 1995b. Serengeti in the context of global conservation. In: Serengeti II: research, management and conservation of an ecosystem (Sinclair ARE, Arcese P, eds.). Chicago: University of Chicago Press; 31–46.

Slate J, Kruuk LEB, Marshall TC, Pemberton JM, Clutton-Brock TH. 2000. Inbreeding depression influences lifetime breeding success in a wild population of red deer (*Cervus elaphus*). Proceedings of the Royal Society of London, Series B 267:1657–1662.

Slatkin, M. 1987. Gene flow and the geographic structure of natural populations. Science 236:787–792.

Smith AT, Wang XG. 1991. Social relationships of adult black-lipped pikas (*Ochotona curzoniae*). Journal of Mammalogy 72:231–247.

Smith B. 1991. Hunt wisely: a guide to male-selective grizzly bear hunting. Extension Report. Whitehorse: Fish and Wildlife Branch, Yukon Renewable Resources.

Smith BL, Andersen SH. 1996. Patterns of neonatal mortality of elk in northwest Wyoming. Canadian Journal of Zoology 74:1229–1237.

Smith C, Reynolds JD, Sutherland WJ. 2000. Population consequences of reproductive decisions. Proceedings of the Royal Society of London, Series B 267:1327–1334.

Smith C, Reynolds JD, Sutherland WJ, Jurajda P. 2000. Adaptive host choice and avoidance of superparasitism in the spawning decisions of bitterling (*Rhodeus sericeus*). Behavioral Ecology and Sociobiology 48:29–35.

Smith CL. 1972. A spawning aggregation of Nassau grouper, *Epinephelus striatus* (Block). Transactions of the American Fisheries Society 101:225–261.

Smith JA, Wilson K, Pilkington JG, Pemberton JM. 1999. Heritable variation in resistance to gastrointestinal nematodes in an unmanaged mammal population. Proceedings of the Royal Society of London, Series B. 266:1283–1290.

Smith JNM. 1981a. Cowbird parasitism, host fitness and age of the host female in an island song sparrow population. Condor 83:152–161.

Smith JNM. 1981b. Does high fecundity reduce future survival in the song sparrow? Evolution 35:1142–1148.

Smith JNM. 1988. Determinants of lifetime reproductive success in the song sparrow. In: Reproductive success: studies of individual variation in contrasting breeding systems (Clutton-Brock TH, ed.). Chicago: University of Chicago Press; 154–172.

Smith JNM, Arcese P. 1994. Brown-headed cowbirds and an island population of song sparrows: a 16-year study. Condor 96:916–934.

Smith JLD, McDougal C. 1991. The contribution of variance in lifetime reproduction to effective population size in tigers. Conservation Biology 5:484–490.

Smuts GL. 1978. Effects of population reduction on the travels and reproduction of lions in Kruger National Park. Carnivore 1:61–72.

Sneddon LU, Swaddle JP. 1999. Asymmetry and fighting performance in the shore crab *Carcinus maenas*. Animal Behaviour 58:431–435.

Sockman KW, Schwabl H. 1999. Female kestrels hormonally regulate the survival of their offspring. American Zoology 39:369.

Soderquist TR. 1994. The importance of hypothesis testing in reintroduction biology: examples from the reintroduction of the carnivorous marsupial *Phascogale tapoatafa*. In: Reintroduction biology of Australian and New Zealand fauna. Serena, MT: Surrey Beaty and Sons; 156–164.

Sogge MK, van Riper C, III. 1988. Breeding biology and population dynamics of the San Miquel Island song sparrow (*Melospiza melodia miconyx*). Cooperative National Park Resources Studies Unit, Technical Report, no. 26.

Solberg EJ, Sæther BE, Strand O, Loison A. 1999. Dynamics of a harvested moose population in a variable environment. Journal of Animal Ecology 68:186–204.

Solberg EJ, Loison A, Sæther BE, Strand O. 2000. Age-specific harvest mortality in a Norwegian moose *Alces alces* population. Wildlife Biology 6:41–52.

Sommer V. 2000. The holy wars about infanticide. Which side are you on? and why? In: Infanticide by males and its implications (van Schaik CP, Janson CH, eds.). Cambridge: Cambridge University Press; 9–26.

Sorci G, Møller AP, Clobert J. 1998. Plumage dichromatism of birds predicts introduction success to New Zealand. Journal of Animal Ecology 67:263–269.

Soulé ME. 1987. Viable populations for conservation. Cambridge: Cambridge University Press.

Soulé ME, Wilcox BA, Holtby C. 1979. Benign neglect: a model of faunal collapse in the game reserves of East Africa. Biological Conservation 15:259–272.

Spaggiari J. 2000. Dynamique de la population de bouquetins des Alpes du Valbonnais-Oisans: Aspects démographiques et spatiaux. Ph.D. thesis, Université Lyon 1.

Spalinger DE, Cooper SM, Martin DJ, Shipley LA. 1997. Is social learning an important influence on foraging behavior in white-tailed deer? Journal of Wildlife Management 61:611–621.

Spalton JA. 1992. The Arabian oryx reintroduction project in Oman: 10 years on. In: Ongules/ungulates 91 (Spitz F, Janeau G, Gonzalez G, Aulagnier S, eds.). Paris-Toulouse: SFEPM-IRGM; 343–347.

Spiller DA, Schoener TW. 1994. Effects of top and intermediate predators in a terrestrial food web. Ecology 75:182–196.

Spotila JR, Dunham AE, Leslie AJ, Steyermark AC, Plotkin PT, Paladino FV. 1996. Worldwide population decline of *Dermochelys coriacea*: are leatherback turtles really going extinct? Chelonian Conservation and Biology 2:209–222.

Spotila J, Reina R, Steyermark A, Plotkin P, Paladino F. 2000. Pacific leatherback turtle faces extinction. Nature 405:529–530.

Spreadbury BR, Musil K, Musil J, Kaisner C, Kovak J. 1996. Cougar population characteristics in southeastern British Columbia. Journal of Wildlife Management 60:962–969.

Squibb RC. 1985. Mating success of yearling and older bull elk. Journal of Wildlife Management 49:744–750.

Stacey PB. 1995. Biodiversity of rangeland bird populations. In: Biodiversity of rangelands (West N, ed.). Logan: Utah State University Press; 33–41.

Stamps JA. 1988. Conspecific attraction and aggregation in territorial species. American Naturalist 131:329–347.

Stamps JA, Buechner M, Krishnan VV. 1987. The effects of edge permeability and habitat geometry on emigration from patches of habitat. American Naturalist 129:533–552.

Stander PE. 1990. A suggested management strategy for stock-raiding lions in Namibia. South African Journal of Wildlife Research 20:37–43.

Stanley Price MR. 1990. Animal reintroductions: the Arabian oryx in Oman. Cambridge: Cambridge University Press.

Starfield AM, Bleloch AL. 1991. Building models for conservation and wildlife management. 2nd ed. Edina, MN: Burgess International Group.

Starfield AM, Furniss PR, Smuts GL. 1981. A model of lion population dynamics as a function of social behavior. In: Dynamics of large mammal populations (Fowler CW, Smith TD, eds.). New York: John Wiley and Sons; 121–134.

Stearns SC. 1983. The influence of size and phylogeny on patterns of covariation among life history traits in the mammals. Oikos 41:173–187.

Stearns SC. 1987. The evolution of sex and its consequences. Basel: Birkhäuser.

Stearns SC. 1992. The evolution of life histories. Oxford: Oxford University Press.

Steele BM. 1996. A modified EM algorithm for estimation in generalized linear mixed models. Biometrics 52:1295–1310.

Stephens PA, Sutherland WJ. 1999. Consequences of the Allee effect for behavior, ecology and conservation. Trends in Ecology and Evolution 14:401–405.

Stephens PA, Sutherland WJ, Freckleton RP. 1999. What is the Allee effect? Oikos 87:185–190.

Stillman RA, Goss-Custard JD, West AD, Durell SEA le V, Caldow RWG,

McGrorty S, Clarke RT. 2000. Predicting to novel environments: tests and sensitivity of a behaviour-based population model. Journal of Applied Ecology 37:564–588.

Stillman RA, Goss-Custard JD, West AD, McGrorty S, Caldow RWG, Durell SEA, Norris KJ, Johnstone IG, Ens BJ, van der Meer J, Triplet P. 2001. Predicting oystercatcher mortality and population size under different regimes of shellfishery management. Journal of Applied Ecology 38:857–868.

Stockwell CA. 1991. Behavioural reactions of desert bighorn sheep to avian scavengers. Journal of Zoology 225:563–566.

Stoneburner DL. 1982. Satellite telemetry of loggerhead sea turtle movement in the Georgia Bight. Copeia 1982:400–408.

Storz, JF. 1999. Genetic consequences of mammalian social structure. Journal of Mammalogy 80:553–569.

Strandgaard H. 1967. Reliability of the Petersen method tested on a roe deer population. Journal of Wildlife Management 31:643–651.

Strandgaard H. 1972. The roe deer (*Capreolus capreolus*) population at Kalö and the factors regulating its size. Danish Review of Game Biology 7:1–205.

Stringham SF. 1980. Possible impacts of hunting on the grizzly/brown bear, a threatened species. International Conference on Bear Research and Management 4:337–349.

Stringham SF. 1983. Roles of adult males in grizzly bear population biology. International Conference on Bear Research and Management 5:140–151.

Stubsjøen T, Sæther BE, Solberg EJ, Heim M, Rolandsen CM. 2000. Moose (*Alces alces*) survival in three populations in northern Norway. Canadian Journal of Zoology 78:1822–1830.

Sugg DW, Chesser RK. 1994. Effective population sizes with multiple paternity. Genetics 137:1147–1155.

Sugg DW, Chesser RK, Dobson FS, Hoogland JL. 1996. Population genetics meets behavioral ecology. Trends in Ecology and Evolution 11:338–342.

Sutherland WJ. 1996. From individual behaviour to population ecology. Oxford: Oxford University Press.

Sutherland WJ. 1998. The importance of behavioural studies in conservation biology. Animal Behaviour 56:801–809.

Sutherland WJ. 2001. Sustainable exploitation: a review of principles and methods. Wildlife Biology 7:131–140.

Sutherland WJ, Dolman PM. 1994. Combining behaviour and population dynamics with applications for predicting consequences of habitat loss. Proceedings of the Royal Society of London, Series B 225:133–138.

Sutherland WJ, Gosling LM. 2000. Advances in the study of behaviour and their role in conservation. In: Behaviour and conservation (Gosling LM, Sutherland WJ, eds.). Cambridge: Cambridge University Press; 3–9.

SYSTAT. 1992. SYSTAT for Windows: statistics, version 5. Evanston, IL: SYSTAT, Inc.

Sweanor LL, Logan KA, Hornocker MG. 2000. Cougar dispersal patterns, metapopulation dynamics, and conservation. Conservation Biology 14: 798–808.

Swenson JE. 1982. Effects of hunting on habitat use by mule deer on mixed-grass prairie in Montana. Wildlife Society Bulletin 10:115–120.

Swenson JE. 1985. Compensatory reproduction in an introduced mountain goat population in the Absaroka Mountains, Montana. Journal of Wildlife Management 49:837–843.

Swenson JE, Dahle B, Sandegren F. 2001a. Bjørnens predasjon på elg. Trondheim: Norwegian Institute for Nature Research, Fagrapport 048.

Swenson JE, Dahle B, Sandegren F. 2001b. Intraspecific predation in Scandinavian brown bears older than cubs-of-the-year. Ursus 12:81–92.

Swenson JE, Sandegren F. 1999. Mistänkt illegal björnjakt i Sverige. In: Bilagor till Sammanhållen rovdjurspolitik: Slutbetänkande av Rovdjursutredningen. Stockholm: Statens Offentliga Utredningar; 146:201–206.

Swenson JE, Sandegren F, Söderberg A. 1998. Geographical expansion of an increasing brown bear population: evidence for presaturation dispersal. Journal of Animal Ecology 67:819–826.

Swenson JE, Wabakken P, Sandegren F, Bjärvall A, Franzén R, Söderberg A. 1995. The near extinction and recovery of brown bears in Scandinavia in relation to the bear management policies of Norway and Sweden. Wildlife Biology 1:11–25.

Swenson JE, Sandegren F, Söderberg A, Bjärvall A, Franzén R, Wabakken P. 1997. Infanticide caused by hunting of male bears. Nature 386:450–451.

Swenson JE, Sandegren F, Söderberg A, Heim M, Sørensen OJ, Bjärvall A, Franzén R, Wikan S, Wabakken P. 1999a. Interactions between brown bears and humans in Scandinavia. Biosphere Conservation 2:1–9.

Swenson JE, Wallin K, Ericsson G, Cederlund C, Sandegren F. 1999b. Effects of ear-tagging with radio-transmitters on survival of moose calves. Journal of Wildlife Management 63:354–358.

Swenson JE, Gerstl N, Dahle B, Zedrosser A. 2000. Action Plan for the Conservation of the Brown bear in Europe (*Ursus arctos*). Nature and Environment Series, 114.

Swenson JE, Sandegren F, Brunberg S, Segerström P. 2001. Factors associated with loss of brown bear cubs in Sweden. Ursus 12:69–80.

Swihart RK, Weeks HP, Easter-Pilcher AL, DeNicola AJ. 1998. Nutritional condition and fertility of white-tailed deer (*Odocoileus virginianus*) from areas with contrasting histories of hunting. Canadian Journal of Zoology 76:1932–1941.

Taberlet P, Swenson JE, Sandegren F, Bjärvall A. 1995. Localization of a contact zone between two highly divergent mitochondrial DNA Lineages of the brown bear (*Ursus arctos*) in Scandinavia. Conservation Biology 9:1255–1261.

Tarof SA, Ratcliffe LM. 2000. Pair formation and copulation behavior in least flycatcher clusters. Condor 102:832–837.

Tear TH, Mosley JC, Ables ED. 1997. Landscape-scale foraging decisions by reintroduced Arabian oryx. Journal of Wildlife Management 61:1142–1154.

Terborgh J. 1987. The big things that run the world: a sequel to E.O. Wilson. Conservation Biology 2:402–403.

Terborgh J, Estes JA, Paquet P, Ralls K, Boyd-Heger D, Miller BJ, Noss RF. 1999. The role of top carnivores in regulating terrestrial ecosystems. In: Continental conservation: scientific foundations of regional reserve networks (Soulé ME, Terborgh J, eds.). Washington, DC: Island Press; 39–64.

Tessier N, Bernatchez L, Wright JM. 1997. Population structure and impact of supportive breeding inferred from mitochondrial and microsatellite DNA analyses in land-locked Atlantic salmon *Salmo salar* L. Molecular Ecology 6:735–750.

Testa JW, Adams GP. 1998. Body condition and adjustments to reproductive effort in female moose (*Alces alces*). Journal of Mammalogy 79:1345–1354.

Testa JW, Becker EF, Lee GR. 2000. Temporal patterns in the survival of twin and single moose (*Alces alces*) calves in southcentral Alaska. Journal of Mammalogy 81:162–168.

Thingstad PG. 1999. Predicting autumn population sizes of tetraonid game birds from reproduction data of pied flycatcher *Ficedula hypoleuca*. Biological Conservation 87:143–148.

Thomas CD. 2000. Dispersal and extinction in fragmented landscapes. Proceedings of the Royal Society of London, Series B 267:139–145.

Thomas CD, Baguette M, Lewis OT. 2000. Butterfly movements and conservation in patchy landscapes. In: Behaviour and conservation (Gosling LM, Sutherland WJ, eds.). Cambridge: Cambridge University Press; 85–104.

Thomas CD, Hanski I. 1997. Butterfly metapopulations. In: Metapopulation dynamics: ecology, genetics and evolution (Hanski IA, Gilpin ME, eds.). London: Academic Press; 359–386.

Thornhill, NW (ed.). 1993. The natural history of inbreeding and outbreeding: theoretical and empirical perspectives. Chicago: University of Chicago Press.

Thouless CR, Sakwa J. 1995. Shocking elephants: fences and crop raiders in Laikipia District, Kenya. Biological Conservation 72:99–107.

Tilman D, May RM, Lehman CL, Nowak MA. 1994. Habitat destruction and the extinction debt. Nature 371:65–66.

Tinbergen N. 1951. The study of instinct. New York: Oxford University Press.

Tixier H, Maizerit C, Duncan P, Bertrand R, Poirel C, Roger M. 1998. Development of feeding selectivity in roe deer. Behavioural Processes 43:33–42.

Toïgo C. 1998. Stratégies biodémographiques et sélection naturelle chez le bouquetin des Alpes. Ph.D. thesis, Université Lyon 1.

Toïgo C, Galliard J-M, Michallet J. 1996. La taille des groupes: un bioindicateur de l'effectif des populations de bouquetin des Alpes (*Capra ibex ibex*)? Mammalia 60:463–472.

Toïgo C, Gaillard J-M, Michallet J. 1997. Adult survival of the sexually dimorphic Alpine ibex (*Capra ibex ibex*). Canadian Journal of Zoology 75:75–79.

Toïgo C, Gaillard JM, Michallet J. 1999. Cohort affects growth of males but not females in alpine ibex (*Capra ibex ibex*). Journal of Mammalogy 80:1021–1027.

Tremblay J-P, Crête M, Huot J. 1998. Summer foraging behaviour of eastern coyotes in rural versus forest landscape: a possible mechanism of source-sink dynamics. Écoscience 5:172–182.

Trivers, R. 1985. Social evolution. Menlo Park, CA: Benjamin/Cummings.

Trivers RL, Willard DE. 1973. Natural selection of parental ability to vary the sex ratio of offspring. Science 179:90–91.

Troyer WA, Hensel RJ. 1962. Cannibalism in brown bear. Animal Behaviour 10:231.

Tufto J, Sæther BE, Engen S, Swenson JE, Sandegren F. 1999. Harvesting strategies for conserving minimum viable populations based on World Conservation Union criteria: brown bears in Norway. Proceedings of the Royal Society of London, Series B 266:961–967.

Tuljapurkar SD. 1990. Population dynamics in variable environments. New York: Springer-Verlag.

Tuljapurkar S, Caswell H. 1996. Structured-population models in marine, terrestrial, and freshwater systems. New York: Chapman and Hall.

Turchin P. 1998. Quantitative analysis of movement: measuring and modeling population redistribution in animals and plants. Sunderland, MA: Sinauer Associates.

Turner MG, Gardner RH, O'Neill RV. 2001. Landscape ecology in theory and practice: pattern and process. New York: Springer-Verlag.

Twigg LE, King DR. 1991. The impact of fluoracetate-bearing vegetation on nature: Australian fauna: a review. Oikos 61:412–430.

Ulfstrand S. 1996. Behavioural ecology as a tool in conservation biology: an introduction. Oikos 77:183.

Van Ballenberghe V. 1983. Extraterritorial movements and dispersal of wolves in southcentral Alaska. Journal of Mammalogy 64:168–171.

van Horn MA, Gentry RM, Faaborg J. 1995. Patterns of Ovenbird (*Seiurus aurocapillus*) pairing success in Missouri forest tracks. Auk 112:98–106.

Van Laere G, Maillard D, Delorme D, Gaillard JM. 2001. Fiabilité de la méthode des approches sur secteurs pour le dénombrement des chevreuils en forêt tempérée de plaine. Mammalia 65:240–244.

van Noordwijk MA, van Schaik CP. 2000. Reproductive patterns in eutherian mammals: adaptations against infanticide? In: Infanticide by males and its implications (van Schaik CP, Janson CH, eds.). Cambridge: Cambridge University Press; 322–360.

van Schaik CP. 2000a. Vulnerability to infanticide by males: patterns among mammals. In: Infanticide by males and its implications (van Schaik CP, Janson CH, eds.). Cambridge: Cambridge University Press; 61–71.

van Schaik CP. 2000b. Infanticide by male primates: the sexual selection hypothesis revisited. In: Infanticide by males and its implications (van Schaik CP, Janson CH, eds.). Cambridge: Cambridge University Press; 27–60.

Van Vuren, D. 1998. Mammalian dispersal and reserve design. In: Behavioural ecology and conservation biology (Caro T, ed.). Oxford: Oxford University Press; 369–393.

Van Vuren D, Kuenzi AJ, Loredo I, Leider AL, Morrison ML. 1997. Translocation as a nonlethal alternative for managing California ground squirrels. Journal of Wildlife Management 61:351–359.

Veiga JP, Moreno J, Cordero PJ, Minguez E. 2001. Territory size and polygyny in the spotless starling: resource-holding potential or social inertia? Canadian Journal of Zoology 79:1951–1956.

Venter J, Hopkins ME. 1988. Use of a simulation model in the management of a lion population. South African Journal of Wildlife Research 18:126–130.

Verbeke G, Molenberghs G. 1997. Linear mixed models in practice. New York: Springer.

Verhulst S, Dieleman SJ, Parmentier HK. 1999. A tradeoff between innumocompetence and sexual ornamentation in domestic fowl. Proceedings of the National Academy of Sciences 96:4478–4481.

Villard M-A. 1998. On forest-interior species, edge avoidance, area-sensitivity, and dogmas in avian conservation. Auk 115:801–805.

Villard M-A, Martin PR, Drummond CG. 1993. Habitat fragmentation and pairing success in the ovenbird (*Seiurus aurocapillus*). Auk 110:759–768.

Villaret JC, Bon R. 1998. Social and spatial segregation in alpine ibex (*Capra ibex*) in Bargy, French Alps. Ethology 101:291–300.

Voelk FH. 1998. Schaelschaeden und Rotwildmanagement in Relation zu Jagdgesetz und Waldaufbau in Oesterreich. Alpine Umweltprobleme. Ergebnisse des Forschungsprojekts Achenkirch Teil XXXIV. Beiträge zur Umweltgestaltung, Band A 141. Berlin: Erich Schmidt Verlag; Pp. 514.

Voland E, Stephan, P. 2000. "The hate that love generated": sexually selected neglect of one's own offspring in humans. In: Infanticide by males and its implications (van Schaik CP, Janson CH, eds.). Cambridge: Cambridge University Press; 447–465.

von Hardenberg A, Bassano B, Peracino A, Lovari S. 2000. Male alpine chamois occupy territories at hotspots before the mating season. Ethology 106:617–630.

von Raesfeld, 1898. Die Hege in der freien Wildbahn. Berlin: Paul Parey.

von Schantz T, Bensch S, Grahn M, Hasselquist D, Wittzell H. 1999. Good genes, oxidative stress and condition-dependent sexual signals. Proceedings of the Royal Society of London, Series B 266:1–12.

Vosburgh TC, Irby LR. 1998. Effects of recreational shooting on prairie dog colonies. Journal of Wildlife Management 62:363–372.

Walker TA, Parmenter CJ. 1990. Absence of a pelagic phase in the life cycle of the flatback turtle, *Natator depressa* Garman. Journal of Biogeography 17:275–278.

Wang JL, Caballero A. 1999. Developments in predicting the effective size of subdivided populations. Heredity 82:212–226.

Warner RR. 1975. The adaptive significance of sequential hermaphroditism in animals. American Scientist 109:61–82.

Warner RR. 1984. Mating behavior and hermaphroditism in coral reef fishes. American Scientist 72:128–136.

Warren CD, Peek JM, Servheen GL, Zager P. 1996. Habitat use and movements of two ecotypes of translocated caribou in Idaho and British Columbia. Conservation Biology 10:547–553.

Waser PM. 1985. Does competition drive dispersal? Ecology 66:1170–1175.

Waser PM. 1996. Patterns and consequences of dispersal in gregarious carnivores. In: Carnivore behaviour, ecology and evolution, vol. 2 (Gittleman JL, ed.). Ithaca: Cornell University Press; 267–295.

Weaver JL. 1994. Ecology of wolf predation amidst high ungulate diversity in Jasper National Park, Alberta. Ph.D. diss., University of Montana, Missoula.

Wedekind C. 1994a. Handicaps not obligatory in sexual selection for resistance genes. Journal of Theoretical Biology 170:57–62.

Wedekind C. 1994b. Mate choice and maternal selection for specific parasite resistances before, during and after fertilization. Philosophical Transactions of the Royal Society of London B 346:303–311.

Wedekind C. 2002a. Manipulating sex ratios for conservation: short-term risks and long-term benefits. Animal Conservation 5:13–20.

Wedekind C. 2002b. Sexual selection and life-history decisions: implications for supportive breeding and the management of captive populations. Conservation Biology 16:1204–1211.

Wedekind C, Füri S. 1997. Body odour preferences in men and women: do they aim for specific MHC combinations or simply heterozygosity? Proceedings of the Royal Society of London, Series B 264:1471–1479.

Wedekind C, Müller R, Spicher H. 2001. Potential genetic benefits of mate selection in whitefish. Journal of Evolutionary Biology 14:980–986.

Wedekind C, Strahm D, Schärer L. 1998. Evidence for strategic egg production in a hermaphroditic cestode. Parasitology 117:373–382.

Wedekind C, Chapuisat M, Macas E, Rülicke T. 1996. Non-random fertilization in mice correlates with the MHC and something else. Heredity 77:400–409.

Wendeln H, Becker PH, Gonzalez-Solis J. 2000. Parental care of replacement clutches in common terns (*Sterna hirundo*). Behavioral Ecology and Sociobiology 47:382–392.

West AD, Goss-Custard JD, Stillman RA, Caldow RWG, Durell SEA, McGrorty S. 2002. Predicting the impacts of disturbance on wintering waders using a behaviour based individuals model. Biological Conservation 106:319–328.

Westemeier RL, Brawn JD, Simpson SA, Esker TL, Jansen RW, Walk JW, Kershner EL, Bouzat JL, Paige KN. 1998. Tracking the long-term decline and recovery of an isolated population. Science 266:217–273.

Western D. 1994. Ecosystem conservation and rural development: the case of Amboseli. In: Natural connections: perspectives in community-based conservation (Western D, Wright RM, eds.). Workshop proceedings. Airlie, VA; 15–52.

Western D, Ssemakula J. 1981. The future of the savannah ecosystems: ecological islands or faunal enclaves. African Journal of Ecology 19:7–19.

Westneat DF, Birkhead TR. 1998. Alternative hypotheses linking the immune system and mate choice for good genes. Proceedings of the Royal Society of London, Series B 265:1065–1073.

Westneat DF, Sherman PW. 1997. Density and extra-pair fertilizations in birds. Behavioral Ecology and Sociobiology 41:205–215.

Whitcomb RF, Robbins CS, Lynch JF, Whitcomb BL, Klimkiewicz MK, Bystrak D. 1981. Effects of forest fragmentation on avifauna of the eastern deciduous forest. In: Forest island dynamics in man-dominated landscapes (Burgess RL, Sharpe DM, eds.). New York: Springer-Verlag; 125–205.

Wielgus RB, Bunnell FL. 1994. Sexual segregation and female grizzly bear avoidance of males. Journal of Wildlife Management 58:405–413.

Wielgus RB, Bunnell FL. 1995. Tests of hypotheses for sexual segregation in grizzly bears. Journal of Wildlife Management 59:552–560.

Wielgus RB, Bunnell FL. 2000. Possible negative effects of adult male mortality on female grizzly bear reproduction. Biological Conservation 93:145–154.

Wielgus RB, Sarrazin F, Ferriere R, Clobert J. 2001. Estimating effects of adult male mortality on grizzly bear population growth and persistence using matrix models. Biological Conservation 98:293–303.

Wiens JA. 1995. Landscape mosaics and ecological theory. In: Mosaic landscapes and ecological processes (Hansson L, Fahrig L, Merriam G, eds.). London: Chapman and Hall; 1–26.

Wiens JA, Schooley RL, Weeks RD, Jr. 1997. Patchy landscapes and animal movements: do beetles percolate? Oikos 78:257–264.

Wiens JA, Crist TO, With KA, Milne BT. 1995. Fractal patterns of insect movement in microlandscape mosaics. Ecology 76:663–666.

Wiktander U, Olsson O, Nilsson SG. 2001. Age and reproduction in lesser spotted woodpeckers (*Dendrocopos minor*). Auk 118:624–635.

Wilcove DS. 1985. Nest predation in forest tracts and the decline of migratory songbirds. Ecology 66:1211–1214.

Williams GC. 1966. Natural selection, the cost of reproduction, and a refinement of Lack's principle. American Naturalist 100:687–690.

Williams GC. 1975. Sex and evolution. Princeton: Princeton University Press.

Williams GC. 1992. Natural selection: domains, levels, and challenges. New York: Oxford University Press.

Wilson EO. 1975. Sociobiology; the new synthesis. Cambridge, MA: Belknap Press.

Wilson EO. 1987. The little things that run the world. Conservation Biology 2:344–346.

Wilson EO. 1992. The diversity of life. Cambridge: Harvard University Press.

Wilson PJ, Grewal S, Lawford ID, Heal JNM, Granacki AG, Pennock D, Theberge JB, Theberge MT, Voigt DR, Waddell W, Chambers RE, Paquet PC, Goulet G, Cluff D, White BN. 2000. DNA profiles of the eastern Canadian wolf and the red wolf provide evidence for a common evolutionary history independent of the gray wolf. Canadian Journal of Zoology 78:2156–2166.

Witherington BE, Martin RE. 1996. Understanding, assessing, and resolving light-pollution problems on sea turtle nesting beaches. Florida Marine Research Institute Technical Report. St. Petersburg, FL: Florida Marine Research Institute.

Wolf CM, Garland T, Jr., Griffith B. 1998. Predictors of avian and mammalian translocation success: reanalysis with phylogenetically independent contrasts. Biological Conservation 86:243–255.

Wolf CM, Griffith B, Reed C, Temple SA. 1996. Avian and mammalian translocations: update and reanalysis of 1987 survey data. Conservation Biology 10:1142–1154.

Woodroffe R. 2001. Strategies for carnivore conservation: lessons from contemporary extinctions. In: Carnivore conservation (Gittleman JL, Wayne RK, Macdonald DW, Funk S, eds.). Cambridge: Cambridge University Press.

Woodroffe R, Ginsberg JR. 1998. Edge effects and the extinction of populations inside protected areas. Science 280:2126–2128.

Woodroffe R, Ginsberg JR. 1999. Conserving the African wild dog, *Lycaon pictus*. II: Is there a rôle for reintroduction? Oryx 32:143–151.

Woodfoffe R, Ginsberg JR. 2000. Ranging behaviour and vulnerability to extinction in carnivores. In: Behaviour and conservation (Gosling LM, Sutherland WJ, eds.). Cambridge: Cambridge University Press; 125–140.

Woodroffe R, Macdonald DW, da Silva J. 1995. Dispersal and philopatry in the European badger, *Meles meles*. Journal of Zoology 237:227–239.

Wotschikowsky U. 1978. Rot- und Rehwild im Nationalpark Bayerischer Wald. Nationalpark Bayerischer Wald, Heft 7. Grafenau: Verlag Morsak.

Wright S. 1965. The interpretation of population structure by *F*-statistics with special regard to systems of mating. Evolution 19:395–420.

Wright S. 1969. The theory of gene frequencies. Evolution and the genetics of populations, vol. 2. Chicago: University of Chicago Press.

Wright S. 1978. Variability within and among natural populations. Evolution and the genetics of populations, vol. 4. Chicago: University of Chicago Press.

Wright S, Gompper ME, De Leon B. 1994. Are large predators keystone species in neotropical forests? The evidence from Barro Colorado Island. Oikos 71:279–294.

Yahner RH. 1988. Changes in wildlife communities near edges. Conservation Biology 2:333–339.

Yalden DW. 1993. The problems of reintroducing carnivores. In: Proceedings of the mammals as predators (Dunstone N, Gorman ML, eds.). New York: Oxford University Press; 289–306.

Yamazaki K. 1996. Social variation of lions in a male-depopulated area in Zambia. Journal of Wildlife Management 60:490–497.

Yates MG, Goss-Custard JD, Rispin WE. 1996. Towards predicting the effect of loss of intertidal feeding areas on overwintering shorebirds (Charadrii) and shelduck (*Tadorna tadorna*): refinements and tests of a model developed for the Wash, east England. Journal of Applied Ecology 33:944–954.

Zahavi A. 1975. Mate selection: a selection for a handicap. Journal of Theoretical Biology 53:205–214.

Zakrisson C. 2000. Do brown bear (*Ursus arctos*) females with cubs alter their movement pattern in order to avoid infanticidal males? Undergraduate thesis, Swedish University of Agricultural Sciences, Umeå.

Zanette L, Doyle P, Trémont SM. 2000. Food shortage in small fragments: evidence from an area-sensitive passerine. Ecology 81:1654–1666.

Zedrosser A, Gerstl N, Rauer G. 1999. Brown bears in Austria: 10 years of conservation and actions for the future. Umweltbundesamt and WWF Austria.

Zhang Z, Usher MB. 1991. Dispersal of wood mice and bank voles in an agricultural landscape. Acta Theriologica 36:239–245.

List of Contributors

Marco Apollonio, Department of Zoology and Biological Anthropology, University of Sassari, Via Muroni 25, 07100 Sassari, Italy

Peter Arcese, Centre for Applied Conservation Research and Department of Forest Sciences, 2424 Main Mall, University of British Columbia, Vancouver, British Columbia V6T 1Z4 Canada

Bruno Bassano, Centro Studi Fauna Alpina, Ente Parco Nazionale Gran Paradiso, Via della Rocca 47, I-10123, Torino, Italy

Joel Berger, Wildlife Conservation Society–North America; PO Box 340, Moose, Wyoming 83012 USA

Isabelle M. Côté, School of Biological Sciences, University of East Anglia, Norwich NR4 7TJ UK

André Desrochers, Centre de recherche en biologie forestière, Faculté de foresterie et de géomatique, Université Laval, Sainte-Foy, Québec, G1K 7P4 Canada

F. Steven Dobson, Department of Biological Sciences, Auburn University, Auburn, Alabama 36849 USA

Marco Festa-Bianchet, Département de biologie, Université de Sherbrooke, Sherbrooke, Québec J1K 2R1, Canada

Jean-Michel Gaillard, Unité Mixte de Recherche N°5558 "Biométrie et Biologie Evolutive," Université Claude Bernard Lyon 1, 43 Boulevard du 11 novembre 1918, 69622 Villeurbanne Cedex, France

Leonard Morris Gosling, Evolution and Behavior Research Group, School of Biological Sciences, University of Newcastle, Newcastle upon Tyne, NE2 4HH UK

John T. Hogg, Montana Conservation Science Institute, 5200 Upper Miller Creek Road, Missoula, Montana 59803 USA

Anne Loison, Unité Mixte de Recherche N°5558 "Biométrie et Biologie Evolutive," Université Claude Bernard Lyon 1, 43 Boulevard du 11 novembre 1918, 69622 Villeurbanne Cedex, France

Paolo Luschi, Department of Ethology, Ecology and Evolution, University of Pisa, Via A. Volta 6, I-56126 Pisa, Italy

Steve L. Monfort, Smithsonian National Zoological Park, Conservation & Research Center, 1500 Remount Road, Front Royal, Virginia 22630 USA

Andrea Mustoni, Adamello-Brenta Regional Park, Via Nazionale 12, 38086 Strembo (TN), Italy

Norman Owen-Smith, Centre for African Ecology, Department of Animal, Plant and Environmental Sciences, University of the Witwatersrand, Wits 2050, South Africa

Tom Roffe, Biological Resource Division, U.S. Geological Survey, Montana State University, Bozeman, Montana 59717 USA

Peter B. Stacey, Department of Biology, University of New Mexico, Albuquerque, New Mexico 87131 USA

Brian M. Steele, Department of Mathematical Sciences, University of Montana, Missoula, Montana 59812 USA

Jon E. Swenson, Department of Biology and Nature Conservation, Agricultural University of Norway, Box 5014, N-1432 Ås, Norway

J. Ward Testa, Division of Wildlife Conservation, Alaska Department of Fish and Game, 333 Raspberry Ridge Road, Anchorage, Alaska 99158 USA

Carole Toïgo, Unité Mixte de Recherche N°5558 "Biométrie et Biologie Evolutive," Université Claude Bernard Lyon 1, 43 Boulevard du 11 novembre 1918, 69622 Villeurbanne Cedex, France

Claus Wedekind, Institute of Cell, Animal and Population Biology, University of Edinburgh, West Mains Road, Edinburgh, EH9 3JT, Scotland

Rosie Woodroffe, Department of Wildlife, Fish and Conservation Biology, University of California, One Shields Avenue, Davis, California 95616-8751 USA

Bertram Zinner, Department of Discrete and Statistical Sciences, Auburn University, Auburn, Alabama 36849 USA

Index